Olive Horticulture: Research and Results

Olive Horticulture: Research and Results

Edited by **Thelma Bosso**

R CALLISTO
REFERENCE

New York

Published by Callisto Reference,
106 Park Avenue, Suite 200,
New York, NY 10016, USA
www.callistoreference.com

Olive Horticulture: Research and Results
Edited by Thelma Bosso

International Standard Book Number: 978-1-63239-490-3 (Hardback)

Contents

Preface

I am honored to present to you this unique book which encompasses the most up-to-date data in the field. I was extremely pleased to get this opportunity of editing the work of experts from across the globe. I have also written papers in this field and researched the various aspects revolving around the progress of the discipline. I have tried to unify my knowledge along with that of stalwarts from every corner of the world, to produce a text which not only benefits the readers but also facilitates the growth of the field.

This book presents various techniques and methods involved in olive horticulture worldwide. Olive (Olea europaea) is being greatly deemed as a crop of significant health and economic importance across the globe. The cultivation of olive is extremely essential in Italy. However, there still remains a lot of confusion about the genetic identity of cultivars. This book serves as a compilation of latest information regarding olive trees and olive oil industry. The aim of this book is to present information which is relevant for scientists, horticulturists, students, and readers wishing to attain knowledge regarding olive cultivation in order to enhance productivity and product quality. The book encompasses several topics including the cultivation process of olives, table olives and description of olive oil industry in Italy. A catalogue of variants of olives across Italy has been presented in this book. The information presented in this book has been contributed by eminent professionals engaged in the field of olive cultivation, olive oil production, table olives and associated fields. It includes all the aspects of olive fruit production, ranging from site selection, pest and disease control, recommended varieties, to primary and secondary processing. The book has been compiled in a manner to serve as an all-inclusive guide on olives.

Finally, I would like to thank all the contributing authors for their valuable time and contributions. This book would not have been possible without their efforts. I would also like to thank my friends and family for their constant support.

Editor

Olive Germplasm

Origin and History of the Olive

Catherine Marie Breton, Peter Warnock and André Jean Berville

Additional information is available at the end of the chapter

1. Introduction

To begin, the methodology followed to reconstruct the origin and history of the olive is presented. The genetic structure (density of alleles across the geographic distribution of individuals) based on allele frequencies of the present oleaster tree in the Mediterranean Basin computed with different methods of comparison with the genetic structure of cultivars grouped based on their geographic and genetic origins infers several possible scenarios for the transition from the oleaster to the olive. To screen among the scenarios requires solid dating in oleaster presence, diffusion and physical remains (from oleaster and cultivar trees) from different sites. Consequently, reconstructing the origin of the olive is based upon data from diverse disciplines and integrating them appears fruitful. Genetic data show that an event (such as bottleneck, migration, differentiation, adaptation) has occurred, but it cannot be dated. Thus it requires crossing genetic data with data gathered from different biological disciplines to make a strong case for this history. We examine successively:

1. The present distribution of the olive and its counterpart the wild olive,
2. Archaeological records of wood charcoals and artifact remains, ethno-botanical methods, pollen databases, and chemical methods for oil traces;
3. Molecular data obtained through 1995 to the present from *Olea europaea* and including relationships between varieties. We examined successively the evolution of methods to analyze data, the data sets examined through Bayesian methods, and the relationships between the oleaster and the olive in order to propose a wide scenario.

This article does not attempt to review all relevant literature on the history or background of the research, but rather focuses on the history of the olive tree and infers some shortcuts in the references of the work published. We apologize and readers can refer to recent general publications that fill the gap here [1, 2, 3, 4].

2. Dogma on the olive tree history

Domestication is based upon conscious behavior by humans over several centuries aimed at selecting among a species natural diversity those individuals that satisfied human requirements [5] such as in yield (seeds or other organs), composition (sugar, starch, fats, …), and to harvest and maintain the cultivation of said species (seed attachment to spike or capitulum, dormancy of seeds, …).

"However, human practices have had effects on the plant genome other than those intended through the conscious behavior of human domestication, effects which are highly documented for many crop species, especially cereals [6] barley, maize [7] Maize, [8] sunflower [9, 10], as examples."

For wheat and grains which were weeds in wheat fields, the main domestication center has been the Near-East in the Fertile-Crescent documented by [11, 12, 13]. Botanists have therefore inferred that the olive tree may have followed a similar history. The presence of Olea cuspidata in the Mountains of Iran suggested a relationship between this species and the olive (Figure 1). Initial molecular results have eliminated this hypothesis and definitively anchored the origin of the olive in the wild olive only [14, 15]. Relationships between the subspecies of *Olea europaea* are not addressed in this chapter.

Present populations of wild olive (called "oleaster") have questioned researchers on their origins. In the East of the Mitterrand they were considered natural. The famous botanist Pelletier has written, "the motherland of the wild olive tree is Anatolia", where numerous wild olive trees make up dense groves, and De Candolle opines that "olive was originated from Asia Minor and spread from Syria over to Greece via Anatolia" [16]. The Flora of the Mediterranean basin is split into eastern and western halves by a line between the Cyrenaica basin of Libya and the Adriatic sea [17, 18, 19, 20]. Effectively, Botanists believe that the oleaster was native to the eastern or oriental half, and once domesticated the olive was transported by humans in the western half, where it thrives as feral form. Thus, all oleaster trees in the western Mediterranean would be feral trees [21].

This supposition has been widely accepted on plenty of open commercial internet sites selling olive products. We believe that this assertion is false and will demonstrate that idea through the results given from various teams working on the olive (Italy, Spain, France, Portugal, Morocco, Tunisia, …).

Using the first molecular markers, namely isozymes, [22] has shown genetic differences between the wild and the crop in the west, but the results did not support an exclusive crop origin in the East, although in many internet sites, such as wikipedia, there is quotation of [23] Wikipedia to sustain the theory.

Readers may therefore encounter confusing theories dealing with the olive history. A number of commercial sites favor the olive's origin as belonging to the commercial site's country, but this has no scientific support. According to some commercial sites, the motherland of the olive is the island of Crete, according to others Southern Caucasia, Iran, the Atlas Mountains in North Africa, Lower Egypt, the Sudan or even Ethiopia. Further, The wild-olive is a tree of the maquis shrubland, itself in part the result of the long presence of mankind." [24].

In the west of the Mediterranean basin they were considered today, as a result of natural hybridization and the very ancient domestication and extensive cultivation of the olive throughout the Mediterranean Basin, wild-looking feral forms of olive, called "oleasters", constitute a complex of populations, potentially ranging from feral forms to the wild-olive. We believe that feral forms cannot be called "oleaster" due to their origin from the domesticated olive.

3. State of the art on the olive tree

The olive tree contributes in shaping landscapes and has deep importance in the agro-economy, including the industrial economies based upon its by-products. However, the wild olive thrives in most of the domesticated olive's cultivation area and the wild olive's contribution to landscapes is far from neglected even though its contribution to agro-economy is weak. One faces some confusion on their respective identification [25].

The olive tree is now used for oil and canned-fruit production, with minor use of the wood for handcrafts. The leaves are used in medicine as herb tea, due to mainly their high phenolic compound content as oleuropein and hydroxytyrosol, which are beneficial in nutrition and medicine (see chapter 15, this book). What was the first use of the olive tree that justified initial care to wild trees? This remains a question and we suggest some tracks. First, the use of the wood and oil as a fuel, since the wood burns green and further the oil produces little smoke, an great advantage in caves in comparison to using animal fats for light and to warm [26]. The spread of olive oil has been documented in the Bronze Age by the features and artifacts (stones, pottery) and later by the containers (aryballos and alabasters) of the perfume industry which used olive oil as a perfume base [27].

All European civilizations have tree symbols: Ash tree (Scandinavian), Sycomore (Egyptian), Plane tree (Sparta, Greece), Oak tree (for the Gauls, Druids, to harvest mistletoe), Pinus (Japanese), (For the Buddha, India) and for Adam and Eve, ... The olive tree is markedly present in all religions (Christian, Judaism, Islam) symbolizing peace, aging, longevity, rejuvenating, authority, ... and plenty of legends and stories are anchored in its history in Mediterranean cultures [28]. However, a primary feature of the olive is that olive oil may also be sacred and has many religious associations. Chrism (consecrated or anointing oil) is made of olive oil, usually includes balsam, and spices. There are many legends on the origins of the olive tree, and all started with the myth of a spontaneous (Athena) or a foreign origin, as Arbequina cv. (Museum Borges Blanca, Catalonia, Spain). Chrism is used for Blessed Sacrament, unction (anointing) oil for baptism, confirmations, Eucharist or communion, marriage, for doing penance, ordination of priests, and extreme unction or the last rites. If olive oil did exist during the Bronze Age, its exact purpose is not well known [27].

Figure 1. Europaea subsp. cuspidata x O. e. subsp. europaea grown in INRA nursery (Near Montpellier, France) See [33].© André Bervillé.

The area where the wild olive thrives is restricted in comparison to the area where the olive is presently grown. Indeed, one of the consequences of the 7,000 to 8,000 years of olive domestication was to spread the cultivated olive out of the biological area of the wild olive, whereas the wild olive did not spread (Figure 3). Indeed, the history of the domesticated olive is tightly linked to mankind and their trend to colonize all the environments, even harsh ones, in order to avoid conflicts with other populations.

It could appear simple to recognize an olive tree [29, 30, 31, 32]. However, out of orchards and in the ecological area of the wild olive, it is not so easy due to many tricks that could lead to confusion with a wild olive. Now, we have abandoned the idea to make the differences rational for all criteria retained (morphological, phenological, molecular) a continuous variation between the two forms is recorded. Consequently, we have to keep in mind that all a priori discrimination between the two forms is questionable and that their confusion does not bias the results exposed here.

Other chapters develop the taxonomy of *Olea* that naturally thrives over all continents except the Americas, and the species *Olea europaea* L. that spreads over Asia, Africa and Europa and is used for its fruit in the Mediterranean basin, but is used for wood charcoal in the mountains of India and Africa [33, 34].

From a scientific point of view the olive is an orphan species, that means there is no model plant surrounding the genus *Olea*. Although several thousand DNA sequences are deposed in databases, little is known from the *Olea* genome, which remains to be sequenced.

Figure 2. First plane: landscape of abandonned Medieval olive groves surrounded by stone walls returned to natural appearance (Near Montpeyroux South of France). © Catherine Breton.

4. The present distribution of the olive and its counter partner the wild olive

At present the oleaster is native in the following regions, and we can consider that since the last ice age the distribution has not changed, due to agricultural development the oleaster has disappeared in the agro-ecosystem, but it has not declined in the natural ecosystem (Figure 2). It is not an endangered species [30]. The wild olive tree thrives along the Mediterranean coasts. It is genuine in Spain, continental France and Corsica, continental Italy, Sardinia and Sicily islands, Greece and Turkey with Cyprus Island, and in all the east and south Mediterranean countries (Jordan, Lebanon, Syria, Israel, Egypt (Sinai) and Libya, it is present in plant formations. In Tunisia, Algeria and Morocco (Moulay-Idriss, Cascade d'Ouzou, Morocco; Ichkheul, Tunisia) once other tree species have been eliminated it may thrive as dense populations but is not a colonizing species.

Its current dispersal depends upon the wild olive trees that survived after the last ice age in refugee populations. However, its spread during the middle-Pleniglacial (Late Pleistocene) before the ice age is based upon wood charcoal and pollen sequences [17, 29, 35, 36] and it was present both in the Levant and in Spain. Based on this evidence, the oleaster is also genuine in the west. During the Holocene it is noticeable that it spread quickly and became abundant or dominant [35]. From a botanical point of view, there is no difference between the oleaster in the east and in the west.

Moreover, [35] analyzed botanical data to clearly identify the oleaster's associated with other thermophilous trees (*Ceratonia, Lentiscus, Phillyrea, Rhamnus*…) in the Mediterranean climate zone in comparison to those of the Atlantic formation (*Pinus, Betula*), which enabled them to define zones where the *Olea* has probably thrive.

However, the olive tree expanded widely outside the oleaster's limits and the famous French writer Georges [37] Duhamel has said " Là où l'olivier renonce, finit la Méditerranée "or 'There the olive has given up, the Mediterranean finishes', that means the olive tree is an excellent indicator of the Mediterranean climate. There is little knowledge on the spread of the olive, it was probably slow following the human colonization of harsh territories by populations seeking shelter to escape wars, and they were patient to adapt their cultivar set to the harsh environments. The spread of the olive follows the trade and settlement patterns of the Phoenicians from the Levant westward – to North Africa and Spain especially. Olive oil as both a trade good and utilitarian household item would have been a premier crop for any colonizer. The present diversity of the olive - probably around 2,000 cultivars - is a witness of this permanent fight between peoples and Nature [38]. The distribution of the olive tree around the Mediterranean basin goes in latitude northern and southern [39,40] and in elevation higher than the distribution of the oleaster tree (max 500m in Spain [31] (Figure 2).

The olive tree was introduced into the New World in South America by the Spanish (explorers and monks) at the beginning of the 1500's (Colombia, Peru, but later on the west coast of the USA). The common perception is that historic olive trees in California are dominated by the 'Mission' cultivar originally introduced by Spanish missionaries to the present day Caribbean and central Mexico in the early 1500's [41, 42]. Thomas Jefferson wrote to James Ronaldson on January 13, 1813, "it is now twenty-five years since I sent them (southern planters) two shipments of about 500 plants of the olive tree of Aix (Aix-en-Provence, France), the finest olive trees in the world." [43]. Olive seeds are believed to have been brought to California in 1769 to grow into trees hardy to 12 degrees Fahrenheit. Those olive trees were cultivated in the Franciscan Spanish monasteries. It was the Spanish who spread the olive to America. Catholic missionaries spread the olive to Mexico and later to California, as well as to South America. The late Earnest Mortensen of the Texas Agricultural Experiment Station brought olive trees to the Winter Garden area in the 1930's. It was introduced in South Africa after the Boer colonization and there it coexists with the subspecies *cuspidata*. In Australia the olive has been introduced by 1812 [21] and later cultivars were introduced in China, Japan, Argentina and Chili and in all countries with a

Mediterranean climate. When introduced as cuttings the cultivars were maintained, but when introduced as seeds unreferenced cultivars were obtained.

5. Archaeological records: wood charcoals, pollen sequences and artifact remains

The Oleaster

Due to oleaster wood being used as biofuel during the prehistoric age, abundant evidence exists to assess its presence from the pleniglacial to the Middle Holocene (for review see [35]). These authors stated that these kinds of remains constituted safe indicators for the presence of *Olea*, although endocarps and pollen grains may have accumulated due to human and wind transport them over undetermined distances.

The first record of olive wood is by [44] who found a fireplace dating to around 790,000 years ago containing (wild) olive wood charcoals. Wood-charcoal analyses carried out at prehistoric sites would reflect the local flora and therefore the frequency of *Olea* wood indicates its presence. Wood charcoals may be due to natural fires or from fireplaces in prehistoric sites depending on the sites and other remains in the site.

As reported by [35] the oldest site where *Olea europaea* thrived in Klissouva cave 1 (Southern Greece) which is dated to 61,140 – 55,230 Cal. Yr. BP. At Higueral de Valleja *Olea europaea* has been dated to 42,630 –41,390 Cal. Yr. BP. *Olea europaea* has been present on both sides of the Mediterranean Basin, but obviously, such sites are scarcely distributed and do not allow us to draw an accurate map for the presence of *Olea europaea*.

Pollen sequences that contain Oleaceae family pollen may include pollen from *Phil'yrea*, Jasmine (*Jasminum fruticans*), and Mediterranean ash tree (*Fraxinus angustifolia*) which thrives along rivers. The pollen is frequently transported long distances and accumulation sites (ponds, swamps, peat-lands) are often far from forests where the oleaster thrives. Thus some bias in pollen data may exist. However, pollen sequences are accurate for dating sites.

However, some of the oldest remains have been dated about to around one million years (wood charcoals from Israel [44] and leaf fragments from Tuff conglomerate [45]. Tuff does not give accurate aging of the site. These remains cannot attest to the presence of the actual oleaster, they belong to an *Olea europaea*, but the sub-species cannot be given. Moreover, such remains have been conserved due to exceptional favorable conditions and are too scarce to infer any model of distribution from them.

The Olive

When the oleaster first was tamed and received care marks the beginning of the domestication process. [29, 46] have shown that the wood charcoal kept traces of pruning practices because of specific vessel architecture and shape, as early as 7500 BP in the Portuguese Extremadura. Their results push back by 1,500 years the preceding estimation of olive domestication given in the eastern Mediterranean region.

Recently, Terral's team has revealed that wood charcoal could also reveals traces of watering in the Middle Ages [40]. If the reasons people were pruning the oleaster are unknown, the consequences of pruning probably appeared to these peoples by more regular blossoming over years and more fruits. Today olive cultivars display a wide variability in response to pruning methods that raises questions on the origin of the diversity [15, 47, 48, 49] at the Neolithic C site of Atlit-Yam on the Levantine coast (dated to 7100-6300 yrs. BP, uncalibrated C14) found underwater wells constructed of alternating layers of tree branches and stones, stone-installations, some lined with undressed stones and others dug into the clay sediment. Some of the crushing installations contained thousands of crushed olive-stones and waste resulting from the extraction of olive oil. So far this is the oldest known evidence for olive oil extraction.

Remains that enabled us to trace the olive tree are more and more numerous from the Mesolithic to the Historical periods. The most informative remains are olive endocarps (stones) that are frequently found in fireplaces (they are charred or carbonized). Terral's team has developed morphometric methods that appeared efficient in analyzing such remains [50]. The main features that result from their analyses are based on the fact that the morphometry of the endocarps has change during the domestication process. Using modern and ancient reference samples they screened and pointed out domesticated remains and unraveled some cultivar relationships. If many stones together in an archaeological site can reveal some transition phase between the wild to the domesticated olive (many broken stones together probably represent oil processing), numerous remains are single or a few stones and consequently these methods are limited on such samples. The accumulation of a few stones probably represents eating olives. However, secondary usage of olive pressing wastes may limit finding traces of olive oil production based on olive remains alone [51].

Pottery types absorb indications of the type of fat they have stored. Pottery types devoted to olive oil as containers for perfumes are aryballos and alabasters, which are widely present throughout the Mediterranean basin due to their diffusion by Greek and Roman cultures [52]. Ceramic chronologies are strict and factories are well recorded, providing a large corpus of data on exchange and trade during the historic periods. Documentation indicates that people used several plant oils (at least flax, saffron, safflower, castor oil and poppy), however, it is possible to differentiate plant oils from animal fats and to identify plant oils by the fatty acid composition obtained from pottery remains [53, 54, 55, 56].

Remains are concrete and their preservation is of importance for future diagnostic methods. The materials from sites studied by all the authors will probably tell more in the future. Furthermore, archaeology continues to uncover new sites and materials and this is likely to continue especially for the southern and eastern parts of the Mediterranean coasts.

In conclusion, the archaeological materials have enabled researchers from different disciplines to anchor the wild and cultivated olive in regions where they naturally thrived and colonized, respectively. Moreover, biologists and archaeologists have defined the basic elicit statistical differences between the wild and the cultivated olive for key historical periods. All this information lays the foundation required to set up genetic models derived from the present genetic diversity recorded in the oleaster and in olive cultivars.

6. Molecular data enabling genetic inference for subsp. europaea in Mediterranean basin

All molecular data based on several types of methods (isozymes, RAPD, RFLP, ISSR and SSR) have been obtained since 1995 from *Olea europaea* [see 1-4 for review]. Obviously, we neglected reports that established relationships between cultivars unless they are informative in reconstituting the olive's history. Molecular data should be based on samples of wild trees and of cultivars representative of the present genetic diversity. If molecular data are bias in the sampling the conclusion will be probably biased. There is no *a priori* rational criterion for sampling oleaster and olive cultivar trees due to the ignorance of their origin. Several hundreds of publications have reported data based on different types of molecular markers on various samples of wild trees and cultivars (depending on the country, the easy sampling of wild trees (from the local region or covering several regions) and with different methods to analyze data.

Evolution of methods to get and to analyze molecular data

Evolution of methods is permanent due to progresses in their development. We will not reiterate all methods of studying the history of the olive since the last ice age through domestication, but will try to enable non geneticists to follow our reasoning. The progress in developing molecular markers over the last twenty years has made some techniques lapsed, although they have released plenty of information [4, 57].

Whatever the techniques used to visualize the genetic diversity, the main feature is to aggregate data from the three DNA supports in the olive tree: the mitochondrial DNA (mt,), the Chloroplast DNA (cp, [58] and the nuclear DNA (nu-, [15, 59, 60, 61]. The information brought by the three compartments is not proportional to the length of the DNA, but by the mode of inheritance and by their mode of evolution. The mt-DNA is maternally inherited as the cp-DNA [62], and evolves by recombination and by mutation and deletion, respectively. These DNA pools are constituted from several copies of the same molecule (they are haploid) and define 'haplotype'. The nuclear DNA is made of two halves from each parent. If the two alleles at one locus are discernible they are said to be codominant, and if only one is discernible due to the other one being absent then the discernible allele is dominant over the hinted allele. If the same dominant allele is found in two different individuals they are said to be similar, whereas if the same two alleles are found in two different individuals they are said to be identical.

Population genetic methods based on similarities (established for dominant alleles) compares at each locus two heterogeneous groups, one homogenous (the double recessive) and one heterogeneous (the double dominant and the heterozygous). All methods used to structure the genetic diversity are based on the allelic frequencies that are firm with codominant markers but are estimated with dominant markers, i.e., Correspondence (FCA) and Hierarchic analyses (leading to dendrograms). Bayesian methods for nuclear DNA (nu-DNA) appeared in 2000 to analyze the data sets and they enable to constitute clusters (based on inferences of allele frequencies) and to check in each individual under examination the proportion of the genome that is coming from the different groups made by the software [63]. These clusters under some hypotheses may correspond to ancestor origins. However, Bayesian method can mix data from nu-, and cp- or mt-DNA. All the methods have contributed in eliciting the olive's origin and they have opened the way to use more adequate methods [64]. Obviously, old data sets could be treated with new methods to get new information.

By the year 2000, after several completed projects (European projects and country projects), the molecular diversity in the wild form appeared deeply structured, that means the geographic distribution of the molecular markers in the wild tree was not homogenous [1, 24, 65, 66]. The genetic structure (estimated by the Fst) was stronger with mt- and cp-DNA markers than with the nu-DNA. Moreover, the mt- and cp-DNA distribution in the eastern and the western halves of the Mediterranean Sea appeared strongly structured. Even if sampling problems for all the studies had biased their data, the trend from the whole data supports that clines for allele frequencies do exist in the wild olive diversity. The clines could be due to different causes and as for other tree species the spread of the wild olive at the end of the last ice age may explain its present distribution.

Data sets examined

All data on the olive and oleaster can be analyzed as i) botanical samples to look for a key to differentiate them ;ii) similarity records to try to differentiate the two forms from a statistical point of view, and iii) genetic relationships using Bayesian methods.

The botanical differentiation between the oleaster and the olive, although with numerous attempts, does not reveal key traits neither morphologic nor molecular [31, 47, 67].

Similarity records have been shown more effective in differentiating between the two forms. Clear cut separation has occurred though some trees remained not clustered, which may indicate that there are hybrid forms between trees from the two groups or that all trees did not share all the traits recorded either morphological or molecular. Low frequency markers should be eliminated in such analyses as they may weigh too much and may distort the results. These methods lead to references of phylogenetic relationships between different levels of taxa, even though, they are not accurate for distant taxa (species, genera), but they have shed light on variety relationships for the olive [47, 64, 68, 69].

[57] Breton has examined a wide sampling of about 1900 trees including wild olives (950 from 55 sites), old trees (50) with undetermined status, and cultivars (about 900, either abandoned, feral forms or from collections) sampled in most places around the Mediterranean basin (Figure 3). Using 16 nu-DNA loci (Single Sequence Repeat or SSR) and 3 cp-DNA loci (single base repeats), the comprehensive data set was examined using

structure software [63] by different a priori packages made of oleaster trees, cultivars, oleaster and feral trees. All clusters revealed by this study were checked by other aggregation methods (FCA, Dendrograms) to verify their consistency. However, these methods cannot ensure the biological existence of such groups. [70] have examined about 250 cultivars and two oleaster populations with AFLP markers (there are mostly dominant markers) from the central Mediterranean with similar methods. [71] has examined 171 wild trees with 8 SSR from the north-western Mediterranean [30] has studied about 32 cultivars and 70 oleaster trees from Tunisia, with morphological and molecular methods.

Figure 3. Anti Atlas Morocco elevation 1525masl. © Catherine Breton

7. Relationships between the oleaster and the olive

[70] Baldoni et al. concluded that most cultivars have been introduced into Central Italy regions from the outside and that Umbrian cultivars have originated by selection from local oleaster trees.[71] Belaj et al. concluded that the genetic structure (=density of alleles across the geographic distribution of individuals) is not strong enough to positively establish relationships between true oleaster trees and cultivated varieties. The impact of these studies has probably been limited due to the limited sampling of the wild forms. [30] Hannachi et al. (2009) has revealed that the cultivar sets can be split into those of local

origins and those introduced from the Near-East and western regions, making Tunisia in central Mediterranean a key-place for olive and oleaster diversity. [57] Breton concluded that the oleaster populations were structured in at least eleven ancestral populations, which colonized the Mediterranean basin after the last ice age, following mostly the sea-coasts. Based on coincidence of the sampling area and the clusters, some geographic zones for the refugee populations have been suggested: 1) Four in the East (Turkey, Cyprus (2) and Israel+ Lebanon), 2) four in the central Mediterranean (including North of Africa, and main islands Sicily, Sardinia, and Corsica) and 3) three in the western Mediterranean (Continental Spain (2) and South France). The zones were well defined in the East and wide in the West, probably due to limited sampling in Spain and continental Italy.

Crossing models based on historic methods and genetic based-hypotheses

[57] Breton applied the same methods to the cultivar set. Although other methods could not split the cultivars into different groups based on biological criteria [47], nine clusters were clearly defined. Furthermore, domestication centers have been revealed by crossing the oleaster and cultivar clusters. Nine domestication centers appeared. The main features from all these results are that the genetics suggests clusters and relationships that remain obscure without data to confirm them. Coincidences between the pre-domestication evidence in Portuguese Extremadura [29] and one refugee zone in central Spain [57] strengthen each other. Carrión et al. points out that the accurate records of all archaeological sites where the oleaster or the olive was found sustains some other sites in the west (Spain, see fig. 3 in [35]) and in the east (Cyprus).

Using the same methods (Wide sampling, Bayesian clustering, FCA and dendrograms) [64] have shown that by admixture analyses for some olive cultivars it is feasible to attribute different origins in the glacial refugees and furthermore the proportion of each origin is quantitatively computed by the Structure software. The application of this method to cultivars will enable us to have a clear view of the cultivars' origins. Data supports that in a western country most cultivars have been introduced from the eastern Mediterranean, but that some cultivars have their origin in local oleaster [72] Ozkaya. [30] Hannachi et al. used this method to reveal three olive origins in Tunisia from the north of Africa (Maghreb), the Near-East, and the west (Spain).

Archaeological remains could release more information by studying the DNA for other species. The method has been applied successfully to olive stones [73]. Many olive remains could be analyzed, but the method is still risky. Today, all these data converge to sustain that in each region the present olive cultivar diversity is either or both the result of ancient introductions from the Near-East and/or from other area (North of Africa, Cyprus, Turkey), local selection from oleaster trees, and from crosses between oleaster and ancient cultivar trees. More details on the history of the oleaster tree could be obtained at local levels and with reference to sampling the whole Mediterranean. Dialogue between researchers from the different fields will be required.

All these data converge to sustain that in each region the present olive cultivar diversity is either or both the result of ancient introductions from the Near-East and/or from other area (North of Africa, Cyprus, Turkey), local selection from oleaster trees and from crosses between oleaster and ancient cultivar trees. However, the self-incompatibility system in the wild olive and the olive is still not yet known, leaving the selection pressures that occurred along the domestication processes unknown, which are required to gather enough S-alleles in a region to enable fruit set. [74] Breton & Bervillé have recently deciphered S-allele pair wise combinations for a few varieties, and it appeared which varieties may combine efficiently, at least *in silico*, but it remains to experimentally check coincidence in blossoming and other compatibility levels, which may affect development of pulp and embryo. The model developed infers which genotypes may coexist to ensure correct fruit set, even though self-compatibility appears inherent to most varieties.

Figure 4. Abandonned olive trees along the Mediterraean coast (North of Catalogna, Spain)
© Catherine Breton

8. Conclusion

The origin of the olive tree displays singularities in comparison with other tree species. As well detailed by [35] the thermophilous requirements of the oleaster has constrained its diffusion. The domestication process has spread out the crop into harsh environments (in northern latitude, deserts, higher altitude) creating plenty of cultivars. About ten domestication centers may be at the origin of this diversity for adaptation to these environments. Recent findings in olive S-allele relationships have not been taken into account here to show the olive's history. The mode of reproduction of the species has probably played a major role enabling self-progenies and thus narrow local adaptation, thus explaining logically the huge diversification encountered in this species.

Author details

Catherine Marie Breton
Present address: CNRS ISE-M UMR 5554, Montpellier, France
Address: INRA , TGU AGAP, Equipe DAVEM, Montpellier, France

Peter Warnock
Missouri Valley College, USA

André Jean Bervillé
INRA, UMR DIAPC, Montpellier, France

Acknowledgement

This work was supported by the ANR project PATERMED (2011-2014) coordinated by Stéphane Anglès (UMR LADYSS) in the frame of the call SYSTERA.

9. References

[1] Breton, C., Tersac, M., Bervillé , A., 2006a. Genetic diversity and gene flow between the wild olive (oleaster, Olea europaea L.) and the olive: several Plio-Pleistocene refuge zones in the Mediterranean basin suggested by simple sequence repeats analysis. Journal of Biogeography 33, 1916–1928.

[2] Doveri S, Baldoni L. Olive. In: Kole C, editor. Genome mapping and molecular breeding in plants, fruits and nuts. Vol. IV. Berlin, Heidelberg: Springer-Verlag; 2007. p. 253-264.

[3] Berti L, Maury J, Advances in olive resources. Kerala, India: Transworld Research Network; 2009, pp 172.

[4]   Breton CM , Bervillé AJ The life history of the olive tree examined through molecular marker data In: Berti L, Maury J, editors. Advances in olive resources. Kerala, India: Transworld Research Network; 2009, pp 105-135

[5] Zeder M.A., Emswhiller E., Smith B.D., Bradley D.G. 2006. Documenting Domestication, the intersection of genetics and archaeology trends in genetics, 22: 139-155.

[6] Varshney RK, Paulo MJ, Grand S, van Eeuwijk FA, Keizer LCP, Guo P, Ceccarelli S,. Kilian A, Baum M, Graner A. Genome wide association analyses for drought tolerance related traits in barley (Hordeum vulgare L.). Field Crops Research 2012; 126 171–180

[7] Camus-Kulandaivelu L, Chevin L-M, Tollon-Cordet C, Charcosset A, Manicacci D, I. Tenaillon M. Patterns of Molecular Evolution Associated With Two Selective Sweeps the Tb1–Dwarf8 region in maize. Genetics 2008;180, 1107–1121.

[8] Wills-Burke[] Anonym fatty acid analysis from pottery http://www.texasbeyondhistory.net/varga/images/fattyAcid.html

[9] Dorian Q. Fuller, Yo-Ichiro Sato, Cristina Castillo, Ling Qin, Alison R. Weisskopf, Eleanor J. Kingwell-Banham, Jixiang Song, Sung-Mo Ahn, Jacob van Etten. Consilence of genetics and archaeobotany in the entangled history of rice. Journal: Archaeological and Anthropological Sciences , vol. 2, no. 2, pp. 115-131, 2010

[10] Lentz DL, DeLand Pohl M Alvarado JL, Tarighat S, Bye Robert. Sunflower (Helianthus annuus L.) as a pre-Columbian domesticate in Mexico. Proceedings of the National Academy of Sciences 2008 . 105(17) 6232-6237.

[11] Zohary D., Spiegel Roy P. Beginnings of fruit growing in the old world. Science 1976; 187 : 319-327.

[12] Zohary D, Hopf M Domestication of plants in the old world: the origin and spread of cultivated plants in West Asia, Europe, and the Nile Valley, 3rd edn. Oxford University Press, Oxford UK, 2000.

[13] Willcox G. & Ken-Ichi Tanno., 2006. How Fast Was Wild Wheat Domesticated? Science. 31. 311 no 5769, p 1886 .

[14] Besnard G., Baradat P., Bervillé A "Genetic relationships in the olive (Olea europaea L.) reflect multilocal selection of cultivars", Theoretical and Applied Genetics, 2001.102 (2001) 251-258.

[15] Contento A, Ceccarelli M, Gelati MT, Maggini F, Baldoni L, Cionini PG. Diversity of Olea genotypes and the origin of cultivated olives. Theor. Appl. Genet 2002 ; 104, 1229–1238.

[16] de Candolle A. Origine des plantes cultivées, 1882

[17] Blondel J & Aronson J 1995 BiodiversIty and ecosystem function in the Mediterrauean Basin Human and non-human determinants, in Mediterranean-Type Ecosystems The Function of Biodiversity Davis, GW and Richardson, DM, eds, pp 43-l 19, Springer-Verlag

[18] Rodrıguez-Ariza, M.O., Montes, E., 2005. On the origin and domestication of Olea europaea L. (olive) in Andaluci´a, Spain, based on the biogeographical distribution of its finds. Vegetation History and Archaeobotany 14, 551–561.

[19] Terral J-F, Newton C, Durand A, Bouby L, Ivorra S. Les origines de la culture et l'histoire de la domestication de l'olivier (Olea europaea L.) en Méditerranée nord-

occidentale au révélateur de l'archéobiologie. . In L'Olivier l'arbre des temps Eds C Breton & A. Bervillé, Quae, Versailles. 2012.

[20] Breton & Bervillé histoire de l'olivier. In L'Olivier l'arbre des temps Eds C Breton & A. Bervillé, Quae, Versailles. 2012.

[21] Breton C, Guerin J, Ducatillion C, Medail F, Kull CA, Berville´A. Taming the wild and 'wilding' the tame: tree breeding and dispersal in Australia and the Mediterranean. Plant Sci 175:197–205.

[22] Lumaret R., N. Ouazzani, H. Michaud, G. Vivie, "Allozyme variation of oleaster populations (wild olive tree) (Olea europaea L.) in the Mediterranean Basin", Heredity, 2004.

[23] wikipedia Wikipedia 2011 http://en.wikipedia.org/wiki/Olea_oleaster)

[24] Sesli M. & E.D. Yeğenoğlu Determination of the genetic relationships between wild olive (Olea europaea oleaster) varieties grown in the Aegean region Genetics and Molecular Research 9 (2): 884-890 (2010).

[25] Breton C, F Médail, C Pinatel, A Bervillé.2004c. In Crop ferality and Volunterism: A threat to food security in the transgenic Era. Ed J Gressel, Chapter 15 example 10: Olive - oleaster gene flow and risks of ferality in olive CRC Press, Boca Raton, USA.

[26] Breton C, Pinatel C, Terral J-F, Médial F, Bonhomme F, Bervillé A The Olive domestication in the Mediterranean basin. CR Biologies 2009 Doi : 10.1016/ j.crvi.2009.08.001.

[27] Riley F.R, Olive oil production on Bronze Age Crete: Nutritional properties, processing methods and storage life of Minoan olive oil, Oxford J. Archaeol. 21 (1) (2002) 64.

[28] Charlot C L'olivier dans l'histoire : chamanisme, religion, médecine et pharmacie. In L'Olivier l'arbre des temps Eds C Breton & A. Bervillé, Quae, Versailles. 2012.

[29] Figueiral I, JF Terral (2002) Late quaternary refugia of Mediterranean taxa in the portuguese estremadura: charcoal based paleovegetation and climatic reconstruction Quaternary Science Reviews, 21, 549-558.

[30] Hannachi H, Breton C, Msallem M, Ben El Hadj S, El Gazzah M, Genetic Relationships between Cultivated and Wild Olive Trees (Olea europaea L. var. europaea and var. sylvestris) Based on Nuclear and Chloroplast SSR Markers Natural Resources, 2010, 1, 95-103.

[31] Rubio, R., Balaguer, L., Manrique, E., Pe´ rez, M.E., Vargas, P., 2002. On the historical presence of the wild olive [Olea europaea L. var. sylvestris (Miller) Lehr. (Oleaceae)] in the Eurosiberian region of the Iberian Peninsula. Anales del Jardı´n Bota´ nico de Madrid 59 (2), 342–344.

[32] Breton C, Terral J-F, Newton C, Ivorra S, Bervillé. A Les apports décisifs de la morphométrie (éco-anatomie et morphométrie géométrique) et de la génétique (marqueurs moléculaires microsatellites) dans la reconstruction de l'histoire de la culture et de la domestication de l'olivier. Eleiva, oleum, olio Alle origini del patrimonio olivicolo toscano. San Quirico d'Orcia Palazzo Chigi Zondadari 8 Dicembre 2007 Giornata di studi, 2012.

[33] Hannachi, H., H. Sommerlatte, C. Breton, M. Msallem, M. El Gazzah, S. Ben El Hadj and A. Bervillé. 2009. Oleaster (var Sylvestris) and subsp. cuspidata are suitable genetic resources for improvement of the olive (Olea europaea subsp. europaea var europaea). Genetic Resources and Crop Evolution 56: 393-403.

[34] Mukonyi KW, Kyalo NS, Lusweti AM, Situma C, Kibet S. Framework and practical assessement of sustainable wild harvest of Olea europaea ssp Africana in Loldaiga ranch, Laikipia, Kenya. A preliminary report.
http://www.biotrade.co.ke/pdfs/Sustainable%20Wild%20harvest%20of%20Olea%20eur opea%20ssp%20Africana%20in%20Loldaiga%20Ranch.pdf

[35] [35 Carrión Y, M Ntinou, Badal E. Olea europaea L. in the North Mediterranean basin during th Pleni-Glacial and the Early-Middle Holocene. Quaternary Science reviews 2010. 20:952-968.

[36] Tzedakis P.C., Lawson I.T., Frogley M.R., Hewitt G.M., and Preece R.C., 2002. Buffered Tree Population Changes in a Quaternary Refugium: Evolutionary Implications. Science. 297, 2044-2047.

[37] Duhamel G. Le temps de la recherché. Ed Hartmann, Paris, 1947.

[38] Bartolini G., G. Prevost, C. Messeri, G. Carignani, U.G. Menini, Olive germplasm: Cultivars and world-wide collections, FAOSPGRSPPPD coordination, 1998.

[39] Camps-Fabrer, H. 1953. L'olivier. 1ère partie, In «L'olivier et l'huile dans l'Afrique romaine». pp: 1-93, Gouvernement général de l'Algérie. Direction de l'intérieur et des beaux arts. Service des Antiquités. Imp. Off., Alger.

[40] Camps-Fabrer, H. 1997. La culture de l'olivier en Afrique du Nord. Evolution et histoire In. « Encyclopédie Mondial de l'Olivier », C.O.I. eds., pp : 30-33, Madrid, Espagne.

[41] Soleri D, Koehmstedt A, Aradhya M. K, Polito V, Pinney K. Comparing the historic olive trees (Olea europaea L.) of Santa Cruz Island with contemporaneous trees in the Santa Barbara, CA area: a case study of diversity and structure in an introduced agricultural species conserved in situ Genet Resour Crop Evol DOI 10.1007/s10722-010-9537-9

[42] Taylor, K.C. The Holocene-Younger Dryas transition recorded at Summit, Greenland. Science 1997; 278 825-827.

[43] Mc Eachern GR, Stein LA. Growing Olives in Texas Gardens Extension Horticulturists,1997 http://aggie-horticulture.tamu.edu/extension/fruit/olive/olive.html

[44] Goren-Inbar's & Alperson (2004) Science Earliest Known Use of Fire Discovered by Israeli Scientists - 2004-04-29 Science

[45] Andlauer P. Musée AFIDOL Nyons.

[46] Terral, J.-F., Alonso, N., Buxo' i Capdevila, R., Chatti, N., Fabre, L., Fiorentino, G., Marinval, P., Pérez Jordà , G., Pradat, B., Rovira, N., Alibert, P., 2004. Historical biogeography of olive domestication (Olea europaea L.) as revealed by geometrical morphometry applied to biological and archaeological material. J. Biogeogr. 31, 63–77.

[47] Besnard G., Baradat P., Breton C., Khadari B., Bervillé A. 2001 Olive domestication from. structure of oleasters and cultivars using RAPDs and mitochondrial RFLP. Genet Sel Evol 33 (Suppl. 1): S251 – S268.

[48] Breton C, Besnard G, Berville AA Using multiple types of molecular markers to understand olive phylogeography. In: Zeder MA, Bradley DG, Emshwiller E, Smith BD (eds) Documenting domestication: new genetic and archeological paradigms. University of California Press, California, pp 143–152.2006a.

[49] Galili Ehud and Baruch Rosen, Israel Antiquities Authority Marine Archaeology In Israel, Recent Discoveries 2011
http://www.emu.edu.tr/underwater/Symposiums/symposiums1/abstracts/marineachol ogy.html

[50] Newton C., Terral J.-F., Ivorra S. 2005. The Egyptian olive (Olea europaea subsp. europaea) in the later first millennium BC: origins and history using the morphometric analysis of olive stones Antiquity 80, p. 405-414.

[51] Warnock P, Identification of Ancient Olive Oil Processing Methods Based on Olive Remains, BAR International Series 1635, 2007.]

[52] Brun Jean-Pierre "Une parfumerie romaine sur le forum de Paestum" http://www.centre-jean-berard.cnrs.fr/article/article_developpe.html.

[53] Evershed R P. *, Dudd S N., Copley M S. and Mutherjee A Praehistorica XXIX Identification of animal fats via compound specific δ 13 C values of individual fatty acids: assessments of results for reference fats and lipid extracts of archaeological pottery vessels Organic Geochemistry Unit.

[54] Anonymb http://www.olivetolive.com/Asp/Content.Asp-MS=1&Content=1&MN01=4&MN02=0&MN03=0&MN04=0&MN05=0&ID=39.htm

[55] Gregg M. W. 2010a new method for extraction, isolation and transesterification of free fatty acids from archaeological pottery Archaeometry, issue 5.

[56] see Rottlander, R, Lipid Analysis in the Identification of Vessel Contents – in Biers and McGovern, Organic Contents of Ancient Vessels: Materials Analysis and Archaeological Investigation, MASCA vol. 7, 1990).

[57] Breton C., Reconstruction de l'histoire de l'olivier (Olea europaea subsp. europaea) et de son processus de domestication en région méditerranéenne, étudiés sur des bases moléculaires, pp 210. Thèse Doctorat Biologie des populations et Écologie, Université Paul Cézanne, 2006, France. 2006 PhD thesis.

[58] Mariotti R, Cultrera NGM, Muñoz Díez C, Baldoni L, Rubini A. Identification of new polymorphic regions and differentiation of cultivated olives (Olea europaea L.) through plastome sequence comparison. BMC Plant Biol. 2010;10:211-

[59] Bronzini de Caraffa V, Giannettini J, Gambotti C, Maury J (2002) Genetic relationships between cultivated and wild olives of Corsica and Sardinia using RAPD markers. Euphytica 123, 263-271.

[60] Bronzini de Caraffa V., Maury J., Gambotti C., Breton C., Bervillé A., Giannettni J. Mitochondrial DNA variation and RAPD mark oleasters, olive and feral olive from Western and Eastern Mediterranean. Theor Appl Genet 104 1209-1216.

[61] Erre Patrizia , Chessa Innocenza , Muñoz-Diez Concepción, Belaj Angjelina, Rallo Luis and Trujillo Isabel 2009 Genetic diversity and relationships between wild and cultivated olives (Olea europaea L.) in Sardinia as assessed by SSR markers Genet Res Crop Evol 2009.

[62] Besnard, G., Khadari B., Villemur P., and A. Bervillé, 2000 A Cytoplasmic Male Sterility in olive cultivarsOlea europaea L:phenotypic, genetic and molecular approaches Theoretical and Applied Genetics 100: 1018-1024.

[63] Pritchard J.K., Stephens M., Donnelly P., 2000. Inference of population structure using multilocus genotype data. Genetics. 155, 945-959.

[64] Breton pinatel PLS 2008

[65] Besnard, G., Khadari, B., Baradat, P., Berville' , A., 2002. Olea europaea (Oleaceae) phylogeography based on choroplast DNA palymorphism. Theor. Appl. Genet. 104, 1353–1361.

[66] Besnard G., A.Bervillé, Multiple origins for Mediterranean olive (Olea europaea L. subsp europaea) based upon mitochondrial DNA polymorphisms, CR Acad Sci Paris série III 323 (2000) 173-181.

[67] Lumaret, R., Ouazzani, N., 2001. Ancient wild olives in Mediterranean forests. Nature 413, 700.

[68] Angiolillo A, Mencuccini M, Baldoni L. 1999. Olive genetic diversity assessed using amplified fragment length polymorphisms. Theoretical and Applied Genetics 4 98: 411–421.

[69] [69]Belaj A, Z Satovic, G Cipriani, L Baldoni, R Testolin, L Rallo, I Trujillo Comparative study of the discriminating capacity of RAPD, AFLP and SSR markers and of their effectiveness in establishing genetic relationships in olive. Theoretical And Applied Genetics 2001 107: 736-744,

[70] Baldoni L, Tosti N, Ricciolini C, Belaj A, Arcioni S, Pannelli G, Germana MA, Mulas M, Porceddu A. 2006. Genetic structure of wild and cultivated olives in the Central Mediterranean Basin. Annals of Botany 98:935–942.

[71] Belaj A, Muñoz-Diez C, Baldoni L, Porceddu A, Barranco D, Satovic Z. 2007. 12 Genetic Diversity and Population Structure of Wild Olives from the North-western 13 Mediterranean Assessed by SSR Markers. Annals of Botany 100:449-458

[72] Özkaya et al. (2009) Özkaya M T, Ergülen E, Nejat S Öz_Lbey Ü, Molecular Characterization of Some Selected Wild Olive (Olea oleaster L.) Ecotypes Grown in Turkey Tarim B_L_Mler_ Derg_S_ 2009, 15 (1) 14-19

[73] Elbaum R., Melamed-Bessudo C., Boaretto E., Galili E., Lev-Yadun S., Levy, A.A., Weiner S. 2006 Ancient olive DNA in stones: preservation, amplification and sequence analysis. J. Archaeol. Sci. 33: 77-88.

[74] Breton CM,& AJ Bervillé 2012 New hypothesis elucidates self-incompatibility in the olive tree regarding S-alleles dominance relationships as in the sporophytic model. Comptes Rendus Biologies, 335: 9, 563–572.

Chapter 2

Olive Tree Genomic

Rosario Muleo, Michele Morgante, Riccardo Velasco,
Andrea Cavallini, Gaetano Perrotta and Luciana Baldoni

Additional information is available at the end of the chapter

.

1. Introduction

The cultivation of olive trees dates back to ancient time. Mythology ascribes to the divine will the domestication of this species; the goddess Athena taught the people of the city of Athens, as a gift, the cultivation of the tree and the treatment of the drupe (Kakridis, 1986). Despite the economic, cultural and ecological importance of olive groves in the Mediterranean area, now extending to other regions, olive has been a poorly characterized species at genetic and genomic level among other fruit tree crops. Therefore, still remains unknown the inheritance of most genes controlling the agronomical performance and quality traits, even though in the last thirty years a wide molecular survey has been performed on the olive germplasm (Rugini et al., 2011). In the Mediterranean Basin, in fact, is conserved the majority of a large number of olive cultivars estimated in more than 1,200 (Bartolini et al., 2004).

Olea europaea subsp. *europaea* is present in two forms, namely wild (*Olea europaea* subsp. *europaea* var. *sylvestris*) and cultivated (*Olea europaea* subsp. *europaea* var. *europaea*); it is a diploid species ($2n = 2x = 46$), and the genome size range between 2.90 pg/2C and 3.07 pg/2C, with 1C = 1,400-1,500 Mbp (Loureiro et al., 2007). Crosses with other subspecies and with the wild plants are possible and may produce fertile offsprings, providing access to an enormous pool of genetic variability. Over the last two decades, new knowledge on olive genetics has been produced, with the development of nuclear and plastidial molecular markers and linkage maps.

The long generation time of the species has severely restricted breeding strategies to clonal or varietal selection and, in a very few cases, to inter-varietal crosses. Approaches of marker assisted selection could speed up the cross breeding programs but QTL markers are not yet available. The first linkage map of *Olea europaea* was constructed by de La Rosa and co-workers (2003), through the use of dominant PCR markers, such as RAPDs and AFLPs, and codominant marker as RFLPs and SSRs on a cross progeny between two highly heterozygous cultivars. Other maps have been constructed by the use of RAPDs, microsatellites and SCAR markers on a Frantoio x Kalamata progeny (Wu et al., 2004) and, more recently, a new maps has been derived through SSR, AFLP, ISSR, RAPD and SCAR marker, scored on a 140 F1 progeny from a Picholine Marocaine x Picholine du Languedoc cultivars cross (El Aabidine et al., 2010). In any case, no QTLs of agronomical interest for olive breeding have been detected.

The Italian project, OLEA, is an initiative, mainly supported by Italian Minister of Agricultural, Food and Forestry Policies, dedicated toward the development genomic resources of olive, and it aims to identify, isolate and determine the function of genes that are associated with both vegetative and reproductive phenotype. Therefore, the knowledge of the genetic structural basis is the first step to identify the relevant differences in the control of gene expression of the same sets of genes that exist among different genotypes. The development of new molecular tools through approaches of structural and functional genomics, together with those from proteomics, metabolomics, mapping and genotyping, will allow to advance in molecular breeding of olive, pull out under-exploited natural diversity that is present in the *Olea* complex and in olive germplasm, dissect the molecular mechanisms underlying traits related to high valued compounds and those involved in plant-environment interactions, establish a platform for a rapid and cost-effective transfer of knowledge and technologies.

2. Genome sequencing and assembly

The olive genome is being sequenced using a combination of Next Generation Sequencing (NGS) technologies and a combination of assembly approaches, using the cultivar Leccino as the genotype to be sequenced. The Whole Genome Shotgun approach to assembling the genome is being pursued using Illumina and 454 sequencing with a combination of long single reads, paired end reads and mate pairs until a coverage of at least 40 genome equivalents is reached. The assembly is being performed using Abyss and CLC assemblers. A BAC pooling approach is being used to sequence random pools of 384 BACs using Illumina paired end reads. A BAC coverage of approximately 3-4 genome equivalents is going to be sequenced, with each BAC clone sequenced on average at a 50X coverage. The advantages of the BAC approach are of two types: on one hand each BAC pool is much smaller in size than the total genome size, reducing the assembly complexity, on the other hand within each BAC pool we should not face the problem posed by sequence heterozygosity among maternal and paternal-derived genomes that strongly affects WGS approaches. The advantage of the WGS approach is the much more complete and homogeneous coverage of the entire genome. The two assemblies derived, the WGS and the pooled BAC assembly, will therefore be combined using a proprietary algorithm (GAM) to produce a consensus assembly. The consensus assembly will finally be anchored to the genetic map through the use of high throughput genotyping technologies.

As of today we have produced all of the data needed for the Whole Genome Shotgun component. We have produced approximately 90 Gbp of Illumina sequence data, corresponding to a nominal coverage of 60X of the olive genome. The Illumina sequences were obtained from two paired-end libraries with 500-600 bp inserts that were sequenced on the Illumina Genome Analyser IIx producing 150 bp reads for a total coverage of 43X (65 Gbp) and from one paired-end library with 1000 bp inserts that was sequenced on the Illumina HiSeq2000 system producing 100 bp reads for the remaining 17X coverage (25 Gbp). Finally two mate-pair libraries with 3 Kbp inserts were constructed and sequenced on the HiSeq2000 to produce 100 bp reads and reach a coverage of 4 genome equivalents (6 Gbp).

We have produced approximately 18 Gbp of Roche-454 sequence data, corresponding to 12X coverage approximately. 12 Gbp were obtained as long single reads of which approximately one third were 400 bp long reads (FLX TITANIUM technology) and two thirds were 700 bp long reads (FLX XL PLUS technology). Additionally 6.2 Gbp of sequence data were obtained as paired end reads from 3 libraries with 3 Kbp inserts (3.8 Gbp) and 10 libraries with 8 Kbp inserts (4.4 Gbp).

The 454 single reds and the Illumina paired-end reads are being used in a traditional WGS assembly. The Illumina mate-pair and the 454 paired end sequenced, i.e. all those sequences that have been obtained from inserts of larger size, will be utilised in order to scaffold into larger assemblies the contigs obtained from the assembly of the reads from the shorter inserts and try to overcome the assembly problems posed by the occurrence of repetitive elements. Since many of the transposable elements in plant genomes are larger than 3 Kbp the larger inserts are going to be of crucial importance.

We have performed a number of assemblies to test different strategies and to obtain a first rough draft of the olive genome. We tested assemblies both using the Illumina data only, as well as using Illumina and 454 data. All data sets have been initially filtered for low quality sequences and for chloroplast DNA contamination and then subject to assembly using the CLCBio assembler. When only the Illumina data were used (53X coverage after filtering), we produced an assembly of total size of 1.1 Gbp and N50 size of 1.7 Kbp. The scaffolding using the mate pair and paired end information on the same assembly using the SSPACE tool increased the N50 size to 2.3 Kbp. The addition of an initial set of 454 data (3.5 genome equivalents after filtering, single reads only) increased the total assembly size to 1.5 Gbp and the N50 size of contigs and scaffolds to 2.8 and 3.7 Kbp, respectively. We expect that the addition of the remaining 454 sequenced from the large insert libraries (3 and 8 Kbp inserts) should greatly improve the assembly by increasing considerably the N50 size of the scaffolds. However, due to the problems posed by the high levels of sequence heterozygosity present in the olive genome of cultivar Leccino, we consider the sequencing of the pools of BACs a necessary component of our strategy in order to obtain a satisfactory assembly. The problems here are represented by the difficulties in obtaining BAC libraries with large insert sizes (>100 Kbp) from cultivar Leccino. Should this not prove feasible we will anyhow resort to using a fosmid library (40 Kbp inserts).

3. Analysis of the repetitive component of the genome

3.1. Assembly of olive repetitive sequences

Some of the biggest technical challenges in sequencing eukaryotic genomes are caused by repetitive DNA (Alkan et al., 2011): that is, sequences that are similar or identical to sequences elsewhere in the genome.

The first step in characterizing and sequencing large genomes has to be a genome survey, from which important information about common repeat sequences can be obtained. NGS data are particularly suitable to identify sequences present in many copies per genome, by assembling reads according to their sequence.

The olive genome is largely uncharacterized, despite the growing importance of this tree as oil crop. Concerning repeated sequences, the most characterized are tandem repeats belonging to 4 families, isolated from genomic libraries and, in some instances, localized by cytological hybridization on olive chromosomes (Katsiotis et al., 1998; Minelli et al., 2000; Lorite et al., 2001; Contento et al., 2002). Also putative retrotransposon fragments have been isolated and sequenced (Stergiou et al., 2002; Natali et al., 2007), but a comprehensive picture of RE landscape in the olive genome is still lacking.

We have performed a deep analysis of the repetitive component of olive genome, using NGS techniques (454 and Illumina). We have used around 25 million Illumina paired-end reads of 75 nt, corresponding to 1.8 billion nt and a 1.3 x coverage, and around 8 million 454 single reads, with mean read length of 407 nt, corresponding to a total of 3.3 billion nt and a 2.3 x coverage.

This large amount of sequencing data cannot be sufficient for whole genome assembly, but it enables representative sampling of elements present in a genome in multiple copies. Moreover, the proportion of individual sequences in the reads reflects their genomic abundance, thus providing a simple and reliable means for quantification of repetitive elements (Macas et al., 2007).

In our experiments, we performed de novo repeat identification and reconstruction by direct assembly of the reads. Due to the relatively low genome coverage of the sequencing, most of the contigs that are obtained do not represent specific genomic loci; instead, they are probably composed of reads derived from multiple copies of repetitive elements, thus representing consensus sequences of genomic repeats (Novak et al., 2010). Even though the exact form of this consensus does not necessarily occur in the genome, this representation of repetitive elements has been shown to be sufficiently accurate to enable amplification of the full length repetitive elements using PCR (Swaminathan et al., 2007).

We assembled Illumina and 454 sequence reads by overlapping DNA sequence fragments using CLC-BIO and CAP3 as aligners. In spite of recent progresses, a major challenge remains when reads map to multiple locations, i.e. with multi-reads. The occurrence of multi-reads is strongly dependent on the read length: they are most common in the Illumina sequence packages, and less common in 454 sequence packages, in which sequence length is rapidly growing to lengths similar to those achieved by classical Sanger sequencing, though at higher costs than Illumina.

The sequencing coverage affects heavily the possibility to recover repeated sequences. Obviously, the larger is the coverage, the higher is the possibility that multi-reads are not resolved and discarded. For example, it has been demonstrated, in pea, that a very low coverage (0.008 ×) of the genome allows to obtain repetitive sequences present with at least

1000 copies (Macas et al., 2007). Hence, we decided to proceed to the final assembly of Illumina reads after having splitted the sequence read datasets into subpackages of different genome coverages.

In a first assembly, we assembled the complete pool of Illumina reads using CLC-BIO and subsequently CAP3 assembler. In other experiments, the pool of Illumina reads was splitted into 8, 16, 32, 250, or 500 subpackages and assembled separately (indicated as split 1, 2, 3, 4, and 5, respectively); for each splitting, the resulting contigs were assembled on their turn using CAP3 assembler obtaining 210,063 supercontigs.

All supercontigs were then mapped with all Illumina 75 nt long reads (Table 1). It can be observed that major splittings allow to recover the most redundant supercontigs, that are not found in the lower splittings, because of their too large coverage and, hence, the occurrence of multi-reads. Due to the different redundancy observed in the different subpackages, we decided to use all supercontigs in the final assembly.

Split	Nr. of sub-packages	Subpackage coverage	Nr. of assembled supercontigs	Mean length	Mean nr. of mapped reads	Average coverage	N_{50}
0	0	1.309 x	44336	235.6	19.61	6.07	243
1	4	0.327 x	78983	200.5	19.96	6.01	201
2	8	0.163 x	50698	204.2	31.72	9.56	204
3	32	0.041 x	22749	252.3	68.58	14.35	265
4	244	0.005 x	14748	240.6	218.57	74.42	258
5	489	0.003 x	11819	223.6	212.77	67.99	239

Table 1. Characteristics of supercontig sets obtained by CLC Bio Workbench and CAP3 assembly after different splitting of Illumina reads.

Concerning 454 sequence reads, we did not proceed to such a subdivision, estimating that the superior length of reads compared to that of Illumina ones allowed to recover also highly repeated sequences. In fact, in longer sequences, the occurrence of multi-reads is naturally reduced.

All Illumina- and 454-derived supercontigs and contigs longer than 80 nt were masked against an in-house made database of chloroplast and mitochondrial sequences using RepeatMasker, and organellar sequences were removed. Then, a final assembly was performed, using CAP3, among all datasets, i.e. six Illumina datasets (split 0-5) and one 454 dataset. The resulting whole genome dataset included 238,914 supercontigs, with mean length of 667.9 nt and $N_{50} = 1.331$.

3.2. Estimation of copy number of assembled sequences

Assuming that Illumina sequence reads in our experiments are sampled without bias for particular sequence types, mapping the whole genomic dataset with Illumina sequence reads provides a method of estimating the copy number of any genomic sequence in the dataset (Swaminathan et al., 2007).

Data in the literature and slot blot experiments previously performed in our lab (Giordani, personal communications) allowed estimation of the copy number per haploid genome of 16 sequences. The 16 sequences with known redundancy were inserted in the whole genomic database and used as reference for the estimation of copy number by mapping on them a pool of around 270 million Illumina 75 nt reads (coverage 14.4 x). We adopted a classification commonly used in biochemical experiments (Britten & Kohne, 1968) and defined supercontigs as highly repeated (HR, redundancy > 10.000 copies per genome, 3,619 supercontigs), medium repeated (MR, redundancy ranging between 100 and 10,000 copies per genome, 67,045 supercontigs) and "unique" (U, redundancy < 100 copies per genome, 168,250 supercontigs).

3.3. Olive genome composition

HR and MR supercontig datasets were annotated to produce the OLEAREP 1.0 database. The annotation pipeline is reported in Figure 1.

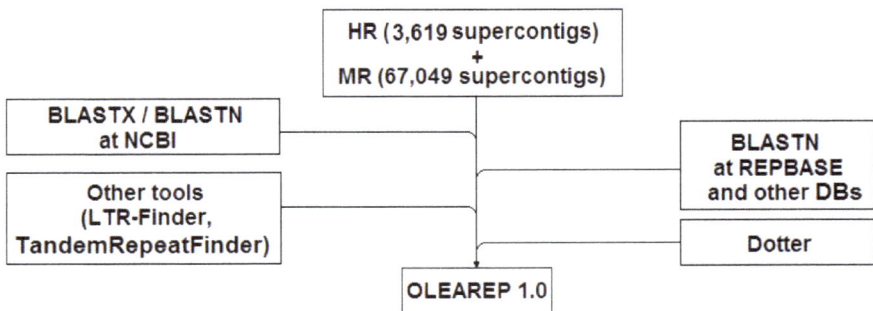

Figure 1. The annotation pipeline for the production of OLEAREP database.

The distribution of sequence type in the HR and MR datasets and in the whole OLEAREP 1.0 database is reported in Table 2.

The average coverage of each HR and MR sequence was used to estimate the redundancy of the various types of repeat classes. Concerning the whole olive genome, around 50% appears to be made of highly repeated sequences. Of these, around 2/3 are tandem repeats belonging to five major families and other minor families (Figure 2). Such extreme redundancy of tandem repeats appears a peculiar feature of olive genome, not found in the other plant species whose genome has been sequenced. On the contrary, medium repeated component is mainly composed of LTR-retrotransposons, while tandem repeats are much less represented in this genome portion.

Sequence type		Nr. of sequences (%)			
		HR		MR	
DNA transposons		31	(0.86)	2,183	(3.26)
Retrotransposons	LTR-*Copia*	134	(3.70)	7,569	(11.29)
	LTR-*Gypsy*	258	(7.13)	8,066	(12.03)
	Non-LTR	29	(0.80)	949	(1.42)
Tandem repeats		1,535	(42.42)	6,718	(10.02)
rDNA		29	(0.80)	555	(0.83)
Putative genes		46	(1.27)	2,729	(4.07)
Unknown repeats		317	(8.76)	1,795	(2.68)
No hits found		1,240	(34.26)	36,481	(54.41)
Total		3,619		67,045	

Table 2. Functional percentage distribution of the supercontigs in OLEAREP 1.0.

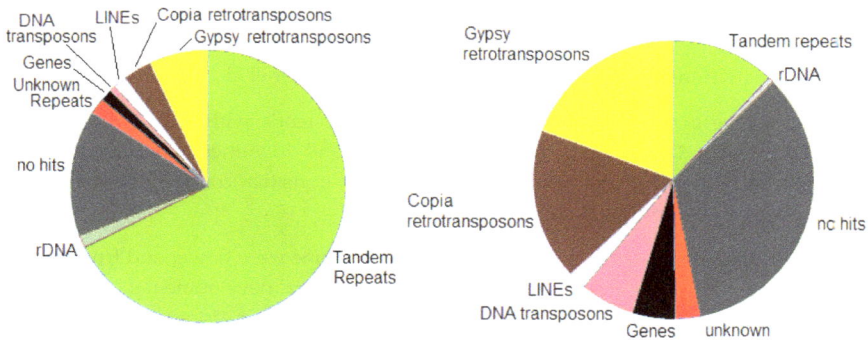

Figure 2. HR (left) and MR (right) fraction composition.

4. Olive chloroplast genome

The chloroplast genome of the olive has an organisation and gene order that is conserved among numerous Angiosperm species and do not contain any of the inversions, gene duplications, insertions, inverted repeat expansions and gene/intron losses that have been found in the chloroplast genomes of the genera *Jasminum* and *Menodora*, from the same family as *Olea* (Mariotti et al., 2010). 40 polymorphisms have been identified in the plastome sequence, poorly able to differentiate among olive cultivars.

5. miRNA

A first inventory of sRNAs in olive has been obtained from juvenile and adult shoots, revealing that the 24-nt class dominates the sRNA transcriptome and atypically accumulates to levels never seen in other plant species, suggesting an active role of heterochromatin silencing in the maintenance and integrity of its large genome (Donaire et al., 2011). A total of 18 known miRNA families were identified in the libraries.

6. Analysis of transcriptome

Despite its global importance, genomic sequence resources available for olive are still scarce, though an increasing number of expressed gene functions are being described in the last few years through limited NGS approaches. Recently, many EST sequences from large scale transcriptomic analyses of different organs, such as fruits and leaves have been released (Alagna et al., 2009; Galla et al., 2009; Ozgenturk et al., 2009).

While these studies have highlighted the utility of cDNA sequencing for candidate gene discovery and gene function, a comprehensive description of genes expressed in *Olea europaea* remains unavailable.

Over the past several years, the NGS technology has emerged as a cutting edge approach for high-throughput sequence determination and this has dramatically improved the efficiency and speed of gene discovery (Ansorge, 2009). Furthermore, NGS has also significantly accelerated and improved the sensitivity of gene-expression profiling and, is expected to boost collaborative and comparative genomics studies (Strickler et al., 2012).

In this study, we generated over one million sequence reads with 454 FLX technology (Roche Diagnostics Corporation, Basel, Switzerland) and identified a number of gene functions potentially involved in the expression of major traits that control productivity and quality of olive and oil production.

The starting materials used to explore the olive transcriptome were flower and fruit samples from five different genotypes. Flower tissues at different developmental stages were sampled from Leccino, Dolce Agogia and Frantoio varieties. Two 454 sequencing libraries were obtained from retro-transcribed pooled RNA samples, extracted from flower buds at all stages of development until anthesis of Leccino and Dolce Agogia genotypes, respectively. Furthermore, pooled flower samples of Leccino and Frantoio genotypes, collected after anthesis, were used for synthesis of cDNAs and for the subsequent preparation of two additional 454 sequencing libraries.

In order to gain information on genes expressed in the drupe, with particular regard to those involved in response to pathogen infections, another set of four 454 sequencing libraries was obtained from fruit samples (at about 17 weeks after flowering) of Ortice and Ruveia genotypes collected before and after the infection caused by the olive fruit fly (*Bactrocera oleae*).

The eight 454 cDNA libraries, four from flower and four from fruit tissues, were sequenced in two separate runs by using 454 GS FLX Titanium Sequencer (Roche Diagnostics Corporation, Basel, Switzerland); each library was loaded on ¼ sector of a picotiter sequencing plate.

We identified a total of more than 1 million sequence reads with an average length cf 356 bp, corresponding to a little less of half billion bases, about 60% of them are from fruit and 40% from flower samples (Table 3).

Sample	Reads	Total Bases	Length Average (bp)
Flower_1	113,134	42,568,083	376.55
Flower_2	146,576	54,807,276	374.23
Flower_3	67,797	21,939,320	323.92
Flower_4	137,824	48,459,361	351.82
Flower_total	465,331	167,774,040	360.55
Fruit_1	173,118	63,611,956	367.45
Fruit_2	197,782	73,464,937	371.44
Fruit_3	146,765	52,792,933	359.71
Fruit_4	177,846	61,220,898	344.23
Fruit_total	695,611	251,090,724	361.02
Olea_total	1,160,942	418,864,764	356.63

Table 3. Raw sequencing data

Assembling of adaptor-trimmed 454 sequence data was performed using GSAssembler Software (Roche Diagnostics Corporation, Basel, Switzerland). To build a compilation of gene structures and functions expressed in *Olea*, we first assembled row data from all the eight libraries together (Table 4).

Samples in Assembly	Reads in Assembly	%	Contigs		Singletons	
			Number	Length Average (bp)	Number	Length Average (bp)
Flower + Fruit samples	964.266	83,05	25.342	892	112.717	323

Table 4. Total assembling

More than 83% of raw sequences were included in the assembly with 112,717 remaining as singletons. This produced a set of 25,342 contigs with an average length of 892 bp (Table 4). As expected, when sequences from flower and fruit samples are assembled separately, the number of EST sequences assembled in contigs are significantly lower; however the average length of contigs and singletons remains similar (Table 5).

Samples in Assembly	Reads in Assembly	%	Contigs		Singletons	
			Number	Length Average (bp)	Number	Length Average (bp)
Flower	338.853	72,82	14.599	804	91.999	345
Fruit	570.878	82,01	15.058	884	72.662	333

Table 5. Flower and Fruit Assembling

To assess the representativeness and the overall quality of the assembling, three randomly chosen gene sequences, among those already characterized in *Olea*, were used as a reference to map contig and singleton sequences produced by the assembling (Figure 3).

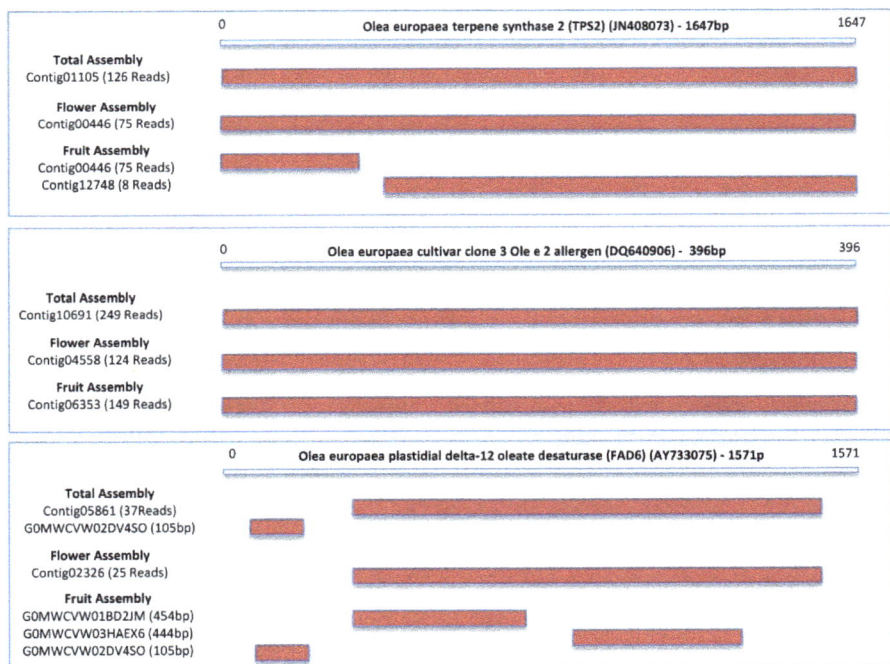

Figure 3. Overview of assembling procedure

The fact that two out of three selected genes are 100% covered by the total assembly with a single contig composed by a great number of EST's, indicates the coverage of the assembly is sufficient to characterize the full coding sequence of high-medium expressed transcripts. Only FAD 6 shows partial coverage, especially in the fruit assembly where only three matching singleton sequences were found (Figure 3). This is most probably due to the sharp decrease of FAD 6 transcript abundance in fruits sampled at late developing stages, from 15 to 20 weeks after flowering (Matteucci et al., 2011).

To predict gene functions, we used a BlastX-based annotation (E-value ≤ 1-e^{-5}) of unigenes comparing them to NCBI non-redundant (nr) database (http://www.ncbi.nlm.nih.gov/). About 52% of the unigenes match to known functional genes; while the remaining 48% has no function assigned (Figure 4).

The majority of the BlastX annotated unigenes matches most to *Vitis vinifera*, *Populus trichocarpa* and *Ricinus communis* counterpart sequences, in decreasing order (Figure 4).

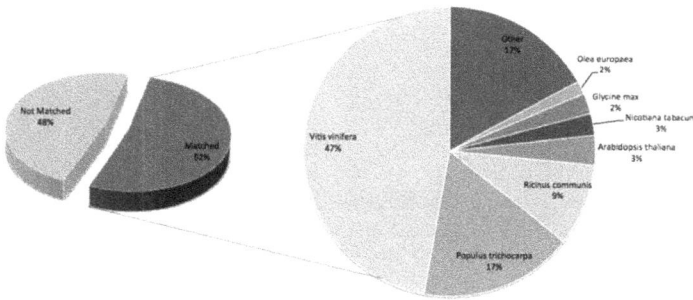

Figure 4. Overall profile of unigenes based on homology with GenBank sequences

We also mapped the GI identifiers (http://www.ncbi.nlm.nih.gov/) of the best BlastX hits to UniprotKB protein database (http://www.uniprot.org/) in order to extract Gene Ontology (GO, http://www.geneontology.org/). Approximately one-fourth of the unigene set was assigned to GO terms. This allowed us to group unigenes in 14 sub-categories of biological processes, 9 sub-categories of cellular components and 11 sub-categories of molecular functions (Figure 5).

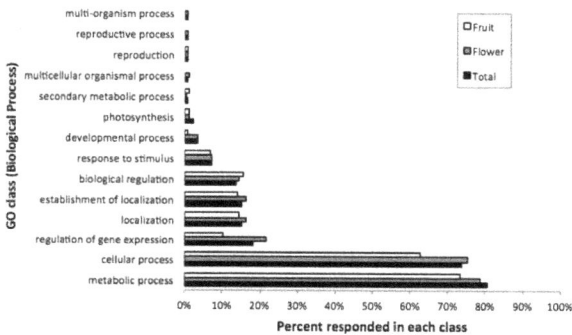

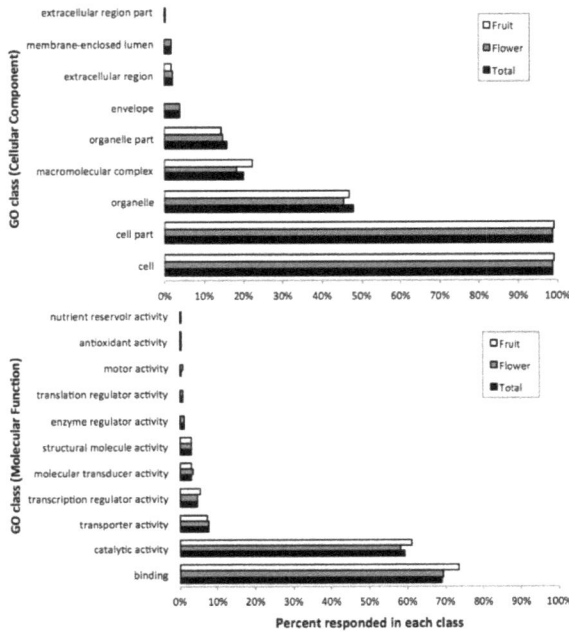

Figure 5. GO terms distribution in the cellular components, biological processes and molecular functions vocabularies.

Metabolic process sub-category, consisting of more than 11,000 genes, is dominant in biological process. While, binding and cell part subcategories, consisting of about 13,0000 and 8,000 genes, are dominant in molecular function and cellular component, respectively. We also noticed an appreciable number of genes included in cellular process, catalytic activity and organelle sub-categories (Figure 5). However, at this level of detail no dramatic differences are evident between flower and fruit tissue transcriptomes.

Next generation RNA sequence from additional organs/tissues/genotypes of *Olea europaea*, as well as, full comparative data analysis between and within the sequenced samples are currently in progress. These studies will certainly provide valuable information about gene functions that trigger key metabolic pathways for the expression of desired traits.

7. Conclusion

Genomic sequences for olive will enable researchers to explore the breadth of genetic diversity present within the species and within the breeding germplasm using high throughput methods of resequencing based on NGS technologies. This will give access to all types of variations, namely Single Nucleotide Polymorphisms (SNPs), small insertion/deletions and structural variants (large insertion/deletions). It will allow a better assessment of the relationships among the different accessions, of the geographical patterns of distribution of genetic variation and of the genetic consequences of olive trees domestication. It will finally form the basis for the development of novel molecular marker assays.

They will also allow the analysis of global gene expression and specific gene expression of olive tissues in diverse developmental stages and conditions. The identification and characterization of expression of important genes involved in agronomic and productive traits affecting fruit production and quality, biotic and abiotic stress resistance, important development characters (e.g., juvenility, self-incompatibility, ovary abortion, chill response), it may offer a significant amount of tools and open new opportunities for improvement either through molecular breeding and/or genetic engineering.

Many researches will be focused on gene network activities, using olive microarray and/or qPCR to address the expression patterns of genes, during plant and fruit development and ripening of drupe fruits, fatty acid metabolism as well as phenylpropanoid metabolism. Moreover, new generation of molecular markers will be developed, helpful to localize genes involved in both monogenic and polygenic agronomic traits, to construct genetic fine-maps; these markers will be also used for marker-assisted selection (MAS) to obtain elite genotypes by allowing the analysis of cross progenies at earlier stages.

Author details

Rosario Muleo
University of Tuscia, Dept. DAFNE, Viterbo, Italy

Michele Morgante
IGA-Institute of Applied Genomics, Udine, Italy

Riccardo Velasco
Edmund Mach Foundation, IASMA, San Michele all'Adige, Italy

Andrea Cavallini
Dept. Crop Species Biology, Pisa, Italy

Gaetano Perrotta
ENEA, Trisaia, Rotondella (MT), Italy

Luciana Baldoni
CNR- Institute of Plant Genetics, Perugia, Italy

Acknowledgement

This research was partially supported by Progetto Strategico MIPAF "OLEA - *Genomica e Miglioramento genetico dell'olivo*", D.M. 27011/7643/10, and by the Province of Trento and Edmund Mach Foundation. We thank the Roche Diagnostic Spa, Applied Science to support the OLEA Italian Project.

8. References

Alagna, F.; D'Agostino, N.; Torchia, L.; Servili, M.; Rao, R.; Pietrella, M.; Giuliano, G.; Chiusano, M.L.; Baldoni, L.; Perrotta, G. (2009). Comparative 454 Pyrosequencing of Transcripts from Two Olive Genotypes During Fruit Development. *BMC Genomics*, Vol. 10, pp. 1-15, ISSN 1471-2164

Alkan, C.; Coe, B.P. & Eichler, E.E. (2011). Genome Structural Variation Discovery and Genotyping. *Nature Reviews Genetic*, Vol. 12, No. 5, pp. 363-376, ISSN 1471-0056

Ansorge, W.J. (2009). Next-generation DNA sequencing techniques. *New Biotechnology*, Vol. 25, No. 4, pp. 195-203, ISSN 1871-6784

Bartolini, G.; Prevost, G. & Messeri, C. (1994). Olive tree germplasm: descriptor lists of cultivated varieties in the world. *Acta Horticulturae*, Vol. 365, pp. 116-118, ISSN 0567-7572

Britten, R.J. & Kohne, D.E. (1968). Repeated sequences in DNA. *Science*, Vol. 161, pp. 529-540, ISSN 0036-8075

Contento, A.; Ceccarelli, M.; Gelati, M.T.; Maggini, F.; Baldoni, L.; Cionini, P.G. (2002). Diversity of *Olea* genotypes and the origin of cultivated olives. *Theoretical and Applied Genetics*, Vol. 104, No., 8, pp. 1229-38. ISSN 1432-2242

De La Rosa, R.; Angiolillo, A.; Guerrero, C.; Pellegrini, M.; Rallo, L.; Besnard, G.; Bervillé, A.; Martin, A. & Baldoni, L. (2003). A first linkage map of olive (*Olea europaea* L.) cultivars using RAPD, AFLP, RFLP and SSR markers. *Theoretical and Applied Genetics* Vol. 106, No. 7, pp. 1273-1282. ISSN 1432-2242

Donaire, L.; Pedrola, L.; de la Rosa, R.; Llave, C. (2011). High-throughput sequencing of RNA silencing-associated small RNAs in olive (*Olea europaea* L.), *PLoS One* Vol. 6, No. 11, pp. e27916. ISSN 1832-6203

El Aabidine, A.Z.; Charafi, J.; Grout, C.; Doligez, A.; Santoni, S.; Moukhli, A.; Jay-Allemand, C.; El Modafar, C.; Khadari, B. (2010). Construction of a genetic linkage map for the olive based on AFLP and SSR markers, Crop Science Vol. 50, No. 6, pp. 2291-2302. ISSN 1435-0645

Galla, G.; Barcaccia, G.; Ramina, A.; Collani, S.; Alagna, F.; Baldoni, L.; Cultrera, N.G.M.; Martinelli, F.; Sebastiani, L.; Tonutti, P. (2009). Computational annotation of genes differentially expressed along olive fruit development. *BMC Plant Biology* Vol. 9, pp. 128. ISSN 1471-2229

Kakridis, I.Th. (1986). *Greek Mythology*, Tome II. Athens: Editons Ekdotiki of Athens, Greece.

Katsiotis, A.; Hagidimitriou, M.; Douka, A. & Hatzopoulos, P. (1998). Genomic organization, sequence interrelationship, and physical localization using in situ hybridization of two tandemly repeated DNA sequences in the genus *Olea*, *Genome* Vol. 41, No.4, pp. 527-534. ISSN 1480-3321

Lorite, P.; Garcia, M.F.; Carrillo, J. A.; Palomeque, T. (2001). A new repetitive DNA sequence family in the olive (*Olea europaea* L.), *Hereditas* Vol. 134, No.1, pp. 73-78. ISSN 1601-5223

Loureiro, J.; Rodriguez, E.; Costa, A. & Santos, C. (2007). Nuclear DNA content estimations in wild olive (Olea europaea L. ssp. europaea var. sylvestris Brot.) and Portuguese cultivars of O. europaea using flow cytometry, *Genetic Resources and Crop Evolution*. Vol. 54, No.1, pp. 21–25. ISSN 1573-5109

Macas, J.; Neumann, P. & Navratilova, A. (2007). Repetitive DNA in the pea (*Pisum sativum* L.) genome: comprehensive characterization using 454 sequencing and comparison to soybean and *Medicago truncatula*, *BMC Genomics*, Vol. 8, pp. 427. ISSN 1471-2164

Mariotti, R.; Cultrera, N.G.M.; Muñoz Díez, C.; Baldoni, L.; Rubin,i A. (2010). Identification of new polymorphic regions and differentiation of cultivated olives (*Olea europaea* L.) through plastome sequence comparison, *BMC Plant Biology*, Vol. 10, pp. 211. ISSN 1471-2229

Matteucci, M.; D'Angeli, S.; Errico, S.; Lamanna, R.; Perrotta, G.; Altamura MM. (2011). Cold affects the transcription of fatty acid desaturases and oil quality in the fruit of *Olea europaea* L. genotypes with different cold hardiness. *Journal of Experimental Botany*. Vol. 62, No. 10, pp. 3403-20. ISSN 1460-2431

Minelli, S.; Maggini, F.; Gelati, M.T.; Angiolillo, A.; Cionini, P.G. (2000). The chromosome complement of Olea europaea L.: characterization by differential staining of the chromatin and in-situ hybridization of highly repeated DNA sequences, *Chromosome Research*, Vol. 8, No.7, pp. 615-619. ISSN 1573-6849

Natali, L.; Giordani, T.; Buti, M. & Cavallini, A. (2007). Isolation of Ty1-*Copia* putative LTR sequences and their use as a tool to analyse genetic diversity in *Olea europaea*. *Molecular Breeding*, Vol. 19, No.3, pp. 255-65, ISSN 1572-9788

Novak, P.; Neumann, P. & Macas, J. (2010). Graph-based clustering and characterization of repetitive sequences in next-generation sequencing data. *BMC Bioinformatics*, Vol. 11, pp. 378, ISSN 1471-2105

Ozgenturk, N.O.; Oruc, F.; Sezerman, U.; Kucukural, A.; Korkut, S.V.; Toksoz, F., Un, C. (2010). Generation and analysis of Expressed Sequence Tags from *Olea europaea* L., *Comparative and Functional Genomics*, Article ID 757512, 9 pages doi:10.1155/2010/757512, ISSN 1532-6268

Rugini, E.; De Pace, C.; Gutiérrez-Pesce P.; Muleo R. (2011). *Olea*. In: *Wild Crop Relatives: Genomic and Breeding Resources, Temperate Fruits*, Chittaranjan Kole (Ed.), 79-114, Springer-Verlag, ISBN 978-3-642-16057-8, BERLIN-HEIDELBERG, HEIDELBERG, DORDRECHT, LONDON

Stergiou, G.; Katsiotis, A.; Hagidimitriou, M. & Loukas, M. (2002). Genomic and chromosomal organization of Ty1-*Copia*-like sequences in *Olea europaea* and evolutionary relationships of *Olea* retroelements. *Theoretical and Applied Genetics*, Vol. 104, No.6-7, pp. 926-33, ISSN 0040-5752

Strickler, S.R.; Bombarely, A. & Mueller, L.A. (2012). Designing a transcriptome next generation sequencing project for a nonmodel plant species. *American Journal of Botany*, Vol. 99, No.2, pp. 257-266, ISSN 1537-2197.

Swaminathan, K.; Varala, K. & Hudson, M.E. (2007). Global repeat discovery and estimation of genomic copy number in a large, complex genome using a high-throughput 454 sequence survey. *BMC Genomics*, Vol. 8, pp. 132, ISSN

Wu, S.B.; Collins, G.; Sedgley, M. (2004). A molecular linkage map of olive (*Olea europaea* L.) based on RAPD, microsatellite, and SCAR markers. *Genome*, Vol. 47, No.1, pp. 26-35, ISSN 1480-3321

Chapter 3

Botanical Description

Adriana Chiappetta and Innocenzo Muzzalupo

Additional information is available at the end of the chapter

1. Introduction

The olive (*Olea europaea* L.) is an emblematic species that represents one of the most important fruit trees in the Mediterranean basin (Loumou & Giourga, 2003). The Mediterranean form, *Olea europaea*, subspecies *europaea*, which includes wild (*Olea europaea* subsp. *europaea* var. *sylvestris*) and cultivated olives (*Olea europaea* subsp. *europaea* var. *europaea*), is a diploid species (2n = 2x = 46) (Kumar *et al.*, 2011).

The origin of the olive tree has been lost over time, coinciding and mingling with the expansion of Mediterranean civilizations which for centuries governed the destiny of mankind and left their imprint on Western culture. From the eastern of the Mediterranean basin, olive trees spread west throughout the Mediterranean area and into Greece, Italy, Spain, Portugal, and France. In 1560, the Spanish Conquistadors carried olive cuttings and seeds to Peru. From there or independently, olive trees were found in Mexico at Jesuit missions. The Franciscan padres carried olives and other fruits from San Blas, Mexico, into California. Sent by Jose de Galvez, Father Junipero Serra established the San Diego de Alcala Mission in 1769. Though oil production began there in the next decade, the first mention of oil was written in the records of the San Diego de Alcala Mission in 1803 as described by Father Lasuen (Winifred, 1967).

Currently, a renewed emphasis of the health benefits of monosaturated olive oil has lead to a resurgence of olive oil production. The olive tree has been widely used for shade around homes and as a street tree in cities. Its distribution is only limited by cold weather in winter, indeed temperatures below 10 °C are lethal (Denney *et al.*, 1993). Most olive-growing areas lie between latitudes 30° and 45° north and south of the equator, although in Australia some of the recently established commercial olive orchards are nearer to the equator than to the 30° latitude and are producing a good yield; this may be because of their altitude or for other geographic reasons.

Olive (*Olea europaea* L.) is the main cultivated species belonging to the monophyletic Oleaceae family that includes 30 genera and 600 species (Cronquist, 1981), within the clade of *Asterids*, in which the majority of nuclear and organellar genomic sequences are unknown. The *Olea* genus comprises 30 species and has spread to Europe, Asia, Oceania and Africa (Bracci *et al.*, 2011).

The olive is a member of the *Oleaceae*, the family that contains the genera *Fraxinus* (ash), *Forsythia* (golden bell), *Forestiera* (*F. neomexicana*, the California "wild-olive"), *Ligustrum* (privet), and *Syringa* (lilac) as well as *Olea* (olive). Its primary genetic resources are taxonomically classified in the *Olea europaea* complex in which six subspecies are recognized (Green, 2002) (scheme 1).

Kingdom:	*Plantae*
Phylum:	*Magnoliophyta*
Class:	*Rosopsida*
Order:	Lamiales
Family:	*Oleaceae*
Sub-family:	*Oleideae*

Genus: *Olea*

    Sub-genera: *Paniculatae*
                    Tetrapilus
                    Olea

              Sections:*Ligustroides*

                  Olea
                  Sub-species: *cuspidata*
                                laperrinei
                                maroccana
                                cerasiformis
                                guanchica
                                europaea
                                varieties: *sylvestris* (wild olive)
                                europaea (cultivated olive)

Scheme 1. Taxonomic scheme of *Olea europaea* L.

The *Olea europaea* subsp. *europaea* include the wild form, also named *sylvestris*, and the cultivated form, called *Olea europaea* subsp. *europaea* var. *europaea*. The olive tree is self-incompatible. Out - crossing is mediated by the wind that transports pollen over long distances, with cytoplasmic male - sterile cultivars being pollinated efficiently by surrounding cultivars or even by sylvestris (Besnard *et al.*, 2000). It is assumed that cultivars originated from the wild Mediterranean olive and have been disseminated all around the Mediterranean countries following human displacement. It is also presumed that crosses between wild and cultivated forms could have led to new cultivars around Mediterranean countries (Besnard *et al.*, 2001).

Nowadays, there are more than 2000 cultivars in the Mediterranean basin that displays huge diversity based on fruit morphology and pit size and morphology and several modern cultivars display small pits such as the *sylvestris*, making the distinction criteria doubtful (Bartolini *et al.*, 1998; 2002; Ganino *et al.*, 2006).

Until recent years, cultivar identification was based only on morphological and agronomic traits. However, recognition of olive cultivars based on phenotypic characters appeared to be problematic, especially in the early stages of tree development. Traditionally diversity within and between olive tree cultivars was determined by assessing differences in the olive tree, namely leaf shape and color, and olive fruits morphology. These measures have the advantage of being readily available, do not require sophisticated equipment and are the most direct measure of phenotype, thus they are accessible for immediate use, an important attribute. However, these morphological and phenological markers have the disadvantage of the small number of polymorphism detected and of being environmentally dependent (Mohan *et al.*, 1997; Tanksley & Orton, 1983). Besides that, some of the phenological characteristics are only accessible for a limited period (e.g., olive fruits) or when the olive tree achieves a mature stage, which may delay correct identification. Due to the high genetic diversity level observed in olive germoplasm and the presence of homonym and synonym cases, efficient and rapid discriminatory methods are urgently required.

2. Description of the *Olea europaea* sub-species

2.1. *Olea cuspidate*

Olea europaea subsp. cuspidata is a native from South Africa, from which it spread through the Middle East, Pakistan, India to China. Subtropical dry forests of *Olea europaea subsp. cuspidata* are found in the Himalayan subtropical broadleaf forests ecoregion.

In the nineteenth century it was introduced to the Australian territory for economic purposes (Spennemann & Allen, 2000; Bass *et al*, 2006). Since 1960, cuspidata naturalized populations have been found in the Hawaii Archipelago (Starr *et al.*, 2003). The *Olea europaea* subsp. *cuspidata* includes much - branched evergreen trees, which vary their size between 2 to 15 m in height. The leaves have an opposite, decussate arrangement, and are entire. Their size is between 3 to 7 cm in length and 0.8 to 2.5 cm in width. The leave apex is acute with a small hook or point, and the base is attenuate to cuneate. Leaf margins are entire and recurved, the upper surface is grey-green and glossy, and the lower surface has a dense covering of silvery, golden or brown scales. Venation is obvious on the upper surface and obscure on the lower surface; the petiole is up to 10 mm long.

Fruit are borne in panicles or racemes 50 to 60 mm long. The calyx is four-lobed and is about 1 mm long. The corolla is greenish-white or cream; the tube is 1 to 2 mm long; lobes are about 3 mm long and reflexed at the anthesis. The two stamens are fused near the top of the corolla tube, with bilobed stigma.

The fruit is a drupe whose shape varies from globose to ellipsoid, it is 6 mm in diameter and 15 to 25 mm long. The drupe is fleshy, glaucous to a dull shine when ripe, and purple-black. The tree usually flowers in spring. The wood is much - prized and durable and it is used for fine furniture and turnery.

Figure 1. Phenotypic aspect of *Olea europaea* subsp. *cuspidata* trees

2.2. *Olea laperrinei*

The *Olea europaea* subsp. *laperrinei* is restricted to the massifs of central - southern Sahara and eastern Sahel (Wickens, 1976; Quézel, 1978; Maley, 1980; Médail *et al.*, 2001; Green, 2002). The *Olea europaea* subsp. *laperrinei* are present at high altitudes, from 1800 to 2800 m, on volcanic or eruptive rocks, generally in cliffs and canyon banks . This taxon is adapted to very dry conditions and in Hoggar, a highland region in southern Algeria, along the Tropic of Cancer, it persists in areas reaching a mean rainfall of about 20 – 100 mm *per* year (Quézel, 1965).

The *Olea europaea* subsp. *laperrinei* trees reach a height of 1.5 - 3 m and their trunk is mainly monocaulous. Leaves have a lanceolate - linear to linear aspect, 2.8 - 4 to 5 - 7 cm long and 0.3 – 0.5 to 1 - 1.5 cm wide. They are ashy-green above and whitish - silvery beneath in colour and their apex is clearly mucronate. The median vein is flat or canaliculated and the petiole is 0.2 - 0.4 cm in length. Flowers are 4 - 6mm in diameter, white, with bracteoles present and well developed. Fruit are borne in panicles. Their shape is ovoid - globose, they are 4 - 5 mm in diameter and 5 - 8 mm long. The pulp is purplish in colour (Medail *et al.*, 2001).

2.3. *Olea maroccana*

The *Olea europaea* subsp. *maroccana* is located in the South - west of Morocco, in the western part of the High Atlas. Its area of distribution is mainly on the southern slope of the Ida – ou - Tanane massif and in the western Anti Atlas (Maire, 1933; Jahandiez & Maire, 1934). The existence and the taxonomic position of this tree have long remained uncertain, but the combination of several morphological traits is unique.

The *Olea europaea* subsp. *maroccana* tree is arborescent or shrubby in appearance and evergreen. The trunk is 4 - 5 m high and generally pluricaulous. Branches and ramifications are erect, internodes of terminal ramifications are elongate, reaching 20 - 60 mm. The bark is smooth, grey-brown. Leaves are lanceolate or lanceolate-acute aspect; they are 3 - 4 to 7 - 8 cm long and 0.4 – 0.6 to 1.2 - 1.3 cm wide. They are slightly contracted into an acuminate reddish-brown apex and progressively contracted into a medium petiole 0.2 - 0.7 cm long. Lateral veins are not visible or scarcely visible, the median vein is partly canaliculate on the lower side. Leaf-blades have a revolute margin; glossy dark-green and very sparsely scattered with tectorous and star-like hairs above; whitish and densely covered by flattened tectorous hairs beneath.

Flowers are rather large, 4 – 6 mm in diameter, white - yellowish; inferior flowers are pedunculate and 2 - 4 mm long with 2 bracteoles ovoid - elongate of 1 - 2 mm, inserted either on the pedicel or beneath the calyx. The calyx is urceolate, erect, cylindrical - oval, 1 mm long. Fruit are borne in panicles or axillary and terminal racemes, elongate and flexuous; lateral ramets are 50 - 70 mm long, ramified; secondary ramets are 10 - 30 mm long; terminal ramets are reflected, and 60 - 120 mm long. Inflorescence bracts are lanceolate - obtuse, 3 - 4 mm long. The fruit is a globose - ovoid drupe; it is 5 - 7 mm in diameter and 9 -

11 mm long, obtuse at the apex, green then red - brown when unripe, becoming purplish - black. Drupe pulp is carmine and aqueous, sweet or slightly bitter taste (Medail *et al.*, 2001).

Figure 2. Morphological aspect of *Olea europaea* subsp. *maroccana* fruits

2.4. *Olea cerasiformis*

The *Olea europaea* subsp. *cerasiformis* tree is arborescent or shrubby, and evergreen. The trunk is 1 - 2.5 m high, generally pluricaulous, and is grey or whitish in colour. Leaves are oblanceolate to linear in shape, rarely suborbicular; they are 3 - 4 to 8 - 10 cm long and 0.4 – 0.6 to 1.0 - 1.4 cm wide. The leaf apex is acuminate and the colour is greyish-green above and paler beneath. They have a petiole 0.2 - 0.7 cm long. The main vein protrudes on the abassial surface.

Flowers are 4 mm in diameter, white; bracteoles are generally present and well developed. Fruit are borne in panicles. The fruit is an ellipsoid drupe; it is 9 - 12 mm in diameter and 12 - 22 mm long. Drupes are green then purplish - black; pulp with a bitter taste (Medail *et al.*, 2001).

2.5. *Olea guanchica*

Traditionally wild olive populations present in the Canary Islands are ascribed to the species *Olea europaea* subsp. *cerasiformis*. A recent genetic study concluded that populations of Madeira and the Canary Islands were genetically separate enough as to be separated into distinct subspecies, therefore the Canarian wild pass was renamed *Olea europaea* subsp. *guanchica*.

This subspecies is present throughout the islands forming part of transition forests or thermophiles. In Gran Canaria it is very abundant and it can be easily found around the north, forming clumps, but especially in the north-east. In the south of the island it is much more local and rare.

The *Olea europaea* subsp. *guanchica* is a small tree evergreen that can reach 6 m in height or more often it has a shrub appearance. Generally, the trunk is pluricaulous, grey or whitish. Leaves are bright green, oblanceolate to narrowly elliptic, 2 - 3 to 7 - 8 cm long and 0.4 – 0.6 to 1.1 - 2.1 cm wide. The apex is mucronate to cuspidate and the colour is greyish-brown above and paler beneath. Leaves have a petiole 0.2 - 0.5 cm long. The main vein partly protrudes on the abassial surface.

Flowers are 4 - 5 mm in diameter, white; bracteoles not well individualized or missing. Fruit are borne in panicles. The fruit is an ovoid - globose drupe; it is 9 - 12 mm in diameter and 12 - 22 mm in length; their colour is green then purplish - black; pulp with a bitter taste (Medail *et al.*, 2001).

Figure 3. Morphological aspect of *Olea europaea* subsp. *guanchica* leaves and trunk

2.5.1. *Olea europaea subsp. europaea var. sylvestris (wild olive)*

Olea europaea subsp. europaea var. *sylvestris* is a typical tree of the Mediterranean regions and it contributed to the Mediterranean forest. In fact, it is considered by many authors as a forest tree. With the olive being the most longeve plant crop species, numerous ultra-millennial still living *sylvestris* trees are present over all the European Mediterranean countries (Baldoni *et al.*, 2006).

However, forest fires and extensive urbanization that characterize the Mediterranean coast have endangered the *sylvestris* variety determining a decline of this genetic resource.

Olea europaea subsp. europaea var. *sylvestris* has not been recorded and evaluated and thus it is not used as a genetic resource although it spread to areas without olives and it seems well adapted to hard environments such as drought, cold, salt, poor soils, etc.

The *Olea europaea* subsp. *europaea* var. *sylvestris* tree is arborescent or shrubby. The plant is long-lived, despite the difficulty in determining the real age, in many cases it can exceed 1000 years old. The trunk is often twisted and cable, and it can reach a considerable size, up to 15 m in the monumental specimens (Baldoni *et al.*, 2000). The branches are numerous, they have thorns in young plants, and can have erect, intermediate or pendulous aspect. The bark is gray – ash - colored, more or less smooth in young trees, becoming rough in the adult ones. The leaves are opposite, leathery, with smooth margin. The lamina is elliptic - lanceolate in shape, the adaxial surphace is green and glabrous, the abaxial one has small silver shield-shaped scales. Flowers are white, pedunculated, very numerous and grouped in axillary racemes (inflorescence). The calix normally has four leaves ovoid, while the corolla, white, is formed by four petals of 2 - 4 mm. There are two stamens per flower, the stigma is bifid and the ovary has four niches. The fruit is an ovoid-globose drupe; it is 5 - 7 mm in diameter and 10 - 15 mm in length. Pulp is initially green then black - brown at maturity. The endocarp is hard and woody with a single seed, rarely two (Camarda *et al.*, 1983).

Figure 4. *Olea europaea* subsp. *europaea* var. *sylvestris* trees: one of the oldest *sylvestris* genotypes located in Sardinia

Figure 5. *Olea europaea* subsp. *europaea* var. *sylvestris*: higher magnification of the trunk of the plant represented in fig. 4

Figure 6. *Olea europaea* subsp. *europaea* var. *sylvestris* fruits

2.5.2. *Olea europaea* subsp. *europaea* var. *europaea* (cultivated olive)

Commercial olive fruits are products of *Olea europaea* subsp. *europaea* var. *europaea*, and only this sub-species of the *Olea* genus produces edible fruits.

The cultivated olive tree can reach heights ranging from just a few meters to 20 m. The wood resists decay, and when the top of the tree is killed by mechanical damage or environmental extremes, new growth arises from the root system. Whether propagated by seed or cuttings, the root system is generally is shallow, spreading to 0.9 - 1.2 m even in deep soils. The above - ground portion of the olive tree is recognizable by the dense assembly of limbs, short internodes, and compact nature of the foliage. Light does not readily penetrate to the interior of an olive tree unless the tree is well managed and pruned to open light channels toward the foliage. If unpruned, olives develop multiple branches with cascading limbs. The branches are able to carry large populations of fruit on terminal twigs, which are pendulous and flexible - swaying.

Figure 7. *Olea europaea* subsp. *europaea* var. *europaea*: morphological aspect of a cultivated olive tree (A), leaves (B); inflorescence (C); fruits (D); endocarp (E)

Olive leaves are thick, leathery, and oppositely arranged. Each leaf grows over a 2-year period. Leaves have stomata on their abassial surfaces only. Stomata are nestled in peltate trichomes that restrict water loss and make the olive relatively resistant to drought. Some multicellular hairs are present on leaf surfaces. Olive leaves usually abscise in the spring when they are 2 or 3 years old; however, as with other evergreens, leaves older than 3 years are often present.

Flower buds are borne in the axil of each leaf. Usually the bud is formed on the current season's growth and begins visible growth the next season. Buds may remain dormant for more than a year and then begin growth, forming viable inflorescences with flowers a season later than expected. When each leaf axil maintains a developing inflorescence, there are hundreds of flowers per twig. Each inflorescence contains 15 - 30 flowers, depending on the cultivar.

Olives are polygamo - monoecious. The flowers are born axially along the shoot, arranged in panicles. Perfect flowers, those with both pistillate and staminate parts, normally consist of a small calyx, 4 petals, 2 stamens with a filament supporting a large pollen-bearing anther, and a plum green pistil with a short thick style and a large stigma. Perfect flowers are borne apically in an inflorescence, and within the typical triple-flower inflorescence the middle flower is generally perfect. Imperfect flowers are staminate, with the pistil either lacking or rudimentary. The flowers are borne on the inflorescence and are small, yellow-white, and inconspicuous.

The perfect flower is evidenced by its large pistil, which nearly fills the space within the floral tube. The pistil is green when immature and deep green when open at full bloom. Staminate flower pistils are tiny, barely rising above the floral tube base. The style is small and brown, greenish white, or white, and the stigma is large and plumose in a functioning pistil.

Floral initiation occurs by November (Pinney & Polito 1990), after which, the flower parts form in March. The inductive phase of flowering in the olive may occur as early as July (about 6 weeks after full bloom), but initiation is not easily seen until 8 months later in February. Complex microscopic and histochemical techniques reveal evidence of floral initiation by November, but the process of developing all the flower parts starts in March. Some olive cultivars, such as those grown in Crete, southern Greece, Egypt, Israel, and Tunisia, bloom and fruit heavily with very little winter chilling; whereas those originating in Italy, Spain, and California require substantial chilling for good fruiting.

At full bloom, flowers are delicately poised for pollination, when some 500,000 flowers are present in a mature tree; a commercial crop of 7 metric tons/ha or more can be achieved when 1 or 2% of these flowers remain as developing fruit. By 14 days after full bloom, most of the flowers destined to abscise have done so. By that time, about 494,000 flowers have abscised from a tree that started with 500,000 flowers (Rosati *et al.*, 2010).

Cultivars vary, but most abscission occurs soon after full bloom and final fruit set nearly always occurs within 6 weeks of full bloom. Further fruit abscission can result from pest infestation and environmental extremes. When trees have an inflorescence at nearly every leaf axil a commercial crop occurs with 1 to 2% fruit set; with a small population of inflorescence, a commercial crop may require 10% fruit set.

Shot berries (parthenocarpic fruits) occur randomly and for reasons which have not been clearly understood. When shot berries occur, they may be seen in clusters on each inflorescence. Here, the inter-fruit competition for raw materials differs from that of normal olive fruits. Shot berries mature much earlier than normal fruit and may be more prevalent when conditions favor a second large crop in succession.

The olive fruit is a drupe, botanically similar to almond, apricot, cherry, nectarine, peach, and plum fruits. The olive fruit consists of a carpel, and the wall of the ovary has both fleshy and dry portions. The endocarp (pit) enlarges to full size and hardens by 6 weeks after full bloom. At that time, the endosperm begins to solidify and embryo development takes place, leading to embryo maturity by September. The mesocarp (flesh) and exocarp (skin) continue their gradual growth. The fruits begin changing from the green color to yellow-white (straw) and accumulate anthocyanin from the distal or base end. Fruit shape and size and pit size and surface morphology vary greatly among cultivars (see the elaiographic cards attached to chapter "Description of varieties")

The mature seed is covered with a thin coat that covers the starch-filled endosperm. The latter surrounds the tapering, flat leaf like cotyledons, short radicle (root), and plumule (stem). Seed size and absolute shape vary greatly with cultivar.

The seed undergoes most of its development starting in July and ending in about September. The fruit is horticulturally mature in October or November (in Italy) and if harvested and stratified at that time, it will achieve the maximum of germination. However, seeds are physiologically mature in January or February when its germination is greatly reduced (Lagarda *et al.*, 1983a).

Olea europaea L. subsp. *europaea* var. *europaea* is a species of great economic importance in the whole Mediterranean basin.

In fact, the genetic patrimony of the Mediterranean basin olive trees is very rich and is characterised by an abundance of varieties. Based on estimates by the FAO Plant Production and Protection Division Olive Germplasm (FAO, 2010), the world's olive germplasm contains more than 2.629 different varieties, with many local varieties and ecotypes contributing to this richness. It is likely that the number of cultivars is underestimated because of inadequate information about minor local cultivars that are widespread in different olive-growing areas.

Current scientific knowledge offers the possibility of introducing new assessment systems, based not only on the varietal character phenology, usually adopted, but also on genetic traits.

For seed production, the fruits should be harvested when ripe, but before they turn black. This period extends from late September to mid-November, depending on the cultivar (Largarda *et al.*, 1983a,b). Pits are removed from the flesh of the fruit with macerators. Pits can be stored in a dry place for years or planted directly, but germination is slow and uneven. Pre-germination treatments are designed to overcome both seed coat (mechanical) and embryo dormancy. Mechanical or chemical scarification is used to treat mechanical dormancy. During the scarification phase, the endocarp can be cracked mechanically or clipped at the radicle end, with care taken not to damage the embryo. Clipping just the cotyledonary end of the endocarp does not improve germination. Good germination results can be obtained using a seed cracking device before subsequent handling procedures (Martin *et al.*, 1986). Pits may be soaked in concentrated sulfuric acid to soften the endocarp. Soaking time depends on the thickness of the endocarp; typical soaking times for Manzanillo are between 24 and 30 hours. The acid bath is followed by 1 to 2 hours of rinsing in water (Crisosto & Sutter, 1985).

The pits can be planted directly after the endocarp treatments at a depth about 2 to 3 times their diameter. Seeds planted outdoors in December do not germinate until the following spring. Pits can also be planted in pots or seedbeds in a greenhouse maintained at a range of temperature between 21 - 24 °C. Germination takes up to 3 months (Hartmann, 1949).

Germination is quicker and more uniform when treatments to overcome internal dormancy are carried out in addition to scarification. The most successful of these treatments on a commercial scale is stratification. Pits are scarified as described above and then soaked in water at room temperature for 24 hours. The pits are mixed with moist sand or vermiculite and then placed in the dark in a controlled environment. The temperature is kept at 15 °C for 30 days. Stratification is thought to reduce abscisic acid, an inhibitor of germination, within the embryo or seed-coat. After stratification, pits can be planted outdoors if the weather is suitable; severe weather can cause losses. Pits can be planted in a greenhouse at 21 to 27 °C. Bottom heat is necessary. Germination should occur within 1 month. Transplanting seedlings from the greenhouse to the nursery should include steps to harden the seedlings, such as partial shade provided by a lath house. Adequate irrigation and fertilization are recommended to ensure continued rapid growth.

Virtually all olive trees are produced from rooted cuttings. Seed handling difficulties, low germination percentage, and slow initial seedling growth rate make seedling production impractical.

3. Conclusion

Olea europaea L. represents one of the most important trees in the Mediterranean basin and the oldest cultivated plant. Among cultivated plants, the olive is the sixth most important oil crop in the world, presently spreading from the Mediterranean region of origin to new production areas, due to the beneficial nutritional properties of olive oil and to its high economic value.

The Mediterranean basin is the traditional area of olive cultivation and has 95% of the olive orchards of the world. From the Mediterranean basin, olive cultivation is presently expanding into areas of Australia, South and North America (Argentina, Chile, United States), South Africa and even in exotic place, like Hawaii. Given its wide range of distribution, it is becoming increasingly urgent to identify plants into different ranges of distribution in the world to avoid cases of homonymy, synonymy and mislabeling so that a reliable classification of all varieties can be achieved without unnecessary confusion.

In this context, along with morphological characteristics the acquisition of additional information on biochemical markers is essential. This aspect represents a fundamental and indispensable step to preserve the main olive varieties and also to safeguard minor genotypes, in order to avoid a loss of genetic diversity.

Recent research has focused on using morphological markers associated with molecular ones to characterize and identify olive varieties (Ercisli *et al.*, 2009; Muzzalupo *et al.*, 2009). The identification of varieties by using molecular markers is a crucial aim of modern horticulture, because such a technique would greatly facilitate breeding programmes and germplasm collection management.

Author details

Adriana Chiappetta*
University of Calabria (UNICAL), Dep. of Ecology,Arcavacata di Rende (CS), Italy

Innocenzo Muzzalupo
Agricultural Research Council - Olive Growing and Oil Industry Research Centre, Rende (CS), Italy

4. Acknowledgement

The authors thank the CERTOLIO project and University of Calabria funds for financial support.

5. References

Baldoni, L.; Pellegrini, M.; Mencuccini, M.; Mulas, M. & Angiolillo, A. (2000). Genetic relationships among cultivated and wild olives revealed by AFLP markers. *Acta Horticulturae* Vol. 521, pp. 275-284, ISSN 0567-7572

Baldoni, L.; Tosti, N.; Ricciolini, C.; Belaj, A.; Arcioni, S.; Pannelli, G.; Germanà, MA.; Mulas, M. & Porceddu A. (2006). Genetic structure of wild and cultivated olives in the central Mediterranean basin. *Annals of Botany,* Vol. 98, No 5, (Novembre 2006), pp. 935-942, ISSN 0305-7364

Bartolini, G.; Petruccelli, R. & Tindall, HD.; (2002) In: Tindal HD.; Menini UG: (eds) Classification, origin, diffusion and history of the olive FAO Rome, Italy ISBN 978-92-5-106534-1

Bartolini, G.; Prevost, G.; Messeri, C.; Carignani, G. & Menini, UG. (1998). Olive germplasm: Cultivars and world-wide collections, FAO Rome, Italy ISBN 978-92-5-106534-1

Bass, DA.; Crossman, ND.; Latrie, SL. & Lethdridge, MR. (2006). The importance of population growth, seed dispersal and habitat suitability in determining plant invasiveness. *Euphytica*, Vol. 148: pp. 97-109, ISSN 0014-2336.

Bracci, T.; Busconi, M.; Fogher, C. & Sebastiani, L. (2011). Molecular studies in olive (*Olea europaea* L.): overview on DNA markers applications and recent advances in genome analysis, *Plant Cell Report* Vol. 30, pp. 449-462, ISSN 0721-7714

Camarda, I. & Valsecchi, F. (1983). Alberi e arbusti spontanei della Sardegna. (ed.) Gallizzi, Sassari: pp. 403-412

Crisosto, C. & Suffer, EG. (1985). Improving "Manzanillo" olive seed germination. *Horticoltural Science* Vol. 20, pp. 100-102, ISSN 0862-867X

Cronquist, A. (1981). An Integrated System of Classification of Flowering Plants. Columbia University Press, New York, ISBN 0-231-03880-1

Denney, JO.; Martin, GC.; Kammereck, R.; Ketchie, DO.; Connell, JH.; Krueger, WH.; Osgood, JW.; Sibbeft, GS. & Nour, GA. (1993). Freeze damage and coldhardiness in

* Corresponding Author

olive: findings from the 1990 freeze. *California. Agriculture* Vol., 47, pp. 1-12, ISSN 0008-0845

Ercisli, S.; Barut, E. & Ipek A. (2009). Molecular characterization of olive cultivars using amplified fragment length polymorphism markers. *Genetics and Molecular Research* Vol. 8, pp. 414-419, ISSN 16765680

FAO 2010. *The Second Report on the State of the World's Plant Genetic Resources for Food and Agriculture*. Rome, Italy, ISBN 978-92-5-106534-1

Ganino, T.; Bartolini, G. & Fabbri A. (2006). The classification of olive germoplasm. *Journal of Horticoltural Science and Biotechnology*, Vol. 81, pp.319-334, ISSN 1462-0316

Green, PS. (2002). A revision of *Olea* L. (*Oleaceae*). *Kew Bulletin* Vol. 57, pp. 91–140, ISBN 00755974

Hartmann, H.T. (1949). Growth of the olive fruit. *Proceedings of the National Academy of Sciences* Vol. 54, pp. 86-94, ISSN 1091-6490

Jahandiez, E. & René M. (1934). Catalogue des plantes du Maroc 3. *Imprimerie Minerva Algerie*, pp. 559 - 914

Kumar, S.; Kahlon, T. & Chaudhary, S. (2011). A rapid screening for adulterants in olive oil using DNA barcodes, *Food Chemistry* Vol.127, pp. 1335–1341, ISSN 0308-8146

Lagarda, A.; Martin GC. & Kester, DE. (1983a). Influence of environment, seed tissue and seed maturity on 'Manzanillo' olive seed germination. *Horticoltural Science* Vol.18: pp. 868–869, ISSN 0862-867X

Largarda, A.; Martin, GC. & Polito, VS. (1983b). Anatomical and morphological development of 'Manzanillo' olive seed in relation to germination. *Journal of the American Society of Horticultural Science* Vol. 108, pp. 741–743, ISSN 0003-1062

Loumou, A. & Giourga, C. (2003). Olive groves: "The life and identity of the Mediterranean". *Agriculture and Human Values* Vol. 20, pp. 87–95, eISSN 1572-8366

Maire, R. (1933). Etudes sur la flore et la végétation du Sahara central. *Mémoires de la Société d'Histoire Naturelle de l'Afrique du Nord* No. 3, Mission du Hoggar II, pp. 166–168

Maley, J. (1980). Les changements climatiques de la fin du Tertiaire en Afrique: leur conséquence sur l'apparition du Sahara et de sa végétation. In: *The Sahara and the Nile: Quaternary environments anil prehistoric occupation in nmthern Africa*. M.A.J. Williams, H. Faure (Eds.), 63–86, Rotterdam, Balkema, ISBN 2-86267-011-1

Martin, GC.; Kuniyuki, AH.; Mehlschau, JJ. & Whisler, J. (1986). Semi-automated pit cracking machine for rapid seed removal. *Horticoltural Science* Vol. 21, pp. 535-536, ISSN 0862-867X

Medail, F.; Quezel, P.; Besnard, G. & Khadar B. (2001). Systematics, ecology and phylogeographic significance of *Olea europaea* L. ssp. *maroccana* (Greuter & Burdet) P. Vargas *et al.*, a relictual olive tree in south-west Morocco. *Botanical Journal of the Linnean Society* Vol. 137, pp. 249-266, ISSN 0024-4074

Muzzalupo, I.; Stefanizzi, F.; Salimonti, A.; Falabella R. & Perri, E. (2009). Microsatellite markers for identification of a group of Italian olive accessions. *Scientia Agricola*, Vol.66, pp. 685-690, ISSN 0103-9016

Pinney, J. & Polito, VS. (1990). Flower initiation in "Manzanillo" olive. *Acta Horticulturae* Vol. 286, pp.203-205, ISSN 0567-7572

Quézel , P. (1965). La végétation du Sahara. Du Tchad à la Mauritanie. Stuttgart: Gustav Fischer Verlag, ISBN 3-437-30258-2

Quézel, P. (1978). Analysis of the flora of Mediterranean and Sahara Africa. *Annals of the Missouri Botanical Garden* Vol. 65, pp. 479–534, ISSN 0026-6493

Rosati, A.; Zipanćič, M.; Caporali, S. & Paoletti, A. (2010). Fruit set is inversely related to flower and fruit weight in olive (*Olea europaea* L.), *Scientia Horticulturae* Vol. 126, pp. 200–204, ISSN 0304-4238

Spenneman, DHR. & Allen, LR. (2000) Feral olives (Olea europaea) as future woody weeds in Australia: a review. *Australian Journal* of *Experimental Agriculture* ,Vol. 40, pp. 889-901, ISSN 0816-1089

Starr, F.; Starr, K. & Loope, L. (2003). Plant of Hawaii. http://www.hear.org/starr/hiplants/reports

Wickens, GE. (1976). The flora of Jebel Marra (Sudan) and its geographical affinities. *Kew Bulletin Additional Series* Vol. 5, pp. 1–368, ISSN 0075-5982

Winifred W. (1967). In: Fray Junipero Serra and the California Conquest, New York: Charles Scribner's Sons, pp. 154-155

Chapter 4

Cultivation Techniques

Caterina Briccoli Bati, Elena Santilli, Ilaria Guagliardi and Pietro Toscano

Additional information is available at the end of the chapter

1. Introduction

In an agricultural context, olive growing is emerging as a dynamic and interesting topic. The majority of Italian olive growing still requires organic renewal interventions such as farm restructuring and tree planting, varietal conversion, mechanization, technical assistance for the implementation of technological innovations, better organization and contractual weight in product marketing. Therefore, the main objective in this area can be obtained with a more economical management of olive orchards in order to achieve a high production per unit area at lower costs and while respecting the environment. Therefore, current olive growing must be based on two pillars such as the reduction of management costs and the use of cultivation techniques with a low environmental impact. More and more attention is being paid, by the EU, and also by Italy, to environmental sustainability, biodiversity and compliance requirements in agriculture. Indeed, in recent years the EU has issued a set of regulations aimed at environmental protection and enhancement of rural areas by improving the competitiveness of the agricultural sector in order to obtain high-quality products aimed at enhancing the peculiarities of the different territories of origin (PGI, PDO) and protecting agricultural and natural resources. Quality must always be considered a key resource for agriculture, which will enable farms to survive and compete in both Italian and International markets. The cultivation techniques used in olive orchards are directed to preserve and improve the physical and chemical characteristics of the soil (soil preparation and tillage, irrigation, fertilization) and to enhance plant production (training, pruning, fruiting, production and pesticide treatments). The knowledge of olive morphology and biology is a prerequisite for the rationalization of cultivation techniques to improve the quantity and quality of production.

Although the olive tree can be considered a hardy plant and is cultivated in marginal areas, it requires specific cultivation techniques coordinated and integrated with each other in order to exalt their productive potentialities. This is why the wise use of tools such as pruning, irrigation, nutrition and soil management plays an important role in achieving a greater vegetative and reproductive plant balance, and cost containment should be the main goal that guides management decisions.

2. Soil

Soil is defined as the top layer of the earth's crust. It is formed by mineral particles, organic matter, water, air and living organisms. It is in fact an extremely complex, variable and living medium, and represents a non-renewable resource which performs many vital functions: food and other biomass production, storage, filtration and transformation of many substances including water, carbon, nitrogen. As long as 100 years ago, Wollny (1898) described the positive effect of soil structure on root growth, water availability, gas transport in soils as well as the positive effects of soil structure on soil strength. He mentioned that the mechanisms involved in the interaction between soil structure and plant growth and yield needed to be investigated. Since then, the positive effects of a favorable soil structure and the negative effects of, for example, soil compaction on crop growth and/or yield have been repeatedly described (e.g. Blank, 1932-1939; Dexter, 1988; Hakansson et al., 1988; Kay, 1990). The anthropogenic activities such as tillage, mineral fertilization, waste disposal and industrial pollution, affect both chemical and physical natural soil properties (Kabata-Pendias & Mukherjee, 2007).

Recent improvements and new methods in analytical chemistry and increasing areas of environmental investigation have substantially added to our knowledge of agricultural soil science. For example, the soil characteristics of an olive plantation are especially important in terms of vulnerability to erosion and, to a lesser extent, to leaching of potentially contaminating elements contained in fertilisers and pesticides. The root system of the olive is concentrated in the top 50-70 cm of soil although it may send out roots to a depth of more than one meter to satisfy its water needs. Therefore, the soil must have an optimal texture, structure and composition to a depth of at least one meter. The management of a cropping system requires periodic evaluation that includes systematic testing, with the aim of determining the nutritional status of soils in order to assess the existence of any nutritional deficiency, excess or imbalance and form a basis for planning the nutrient supply as well as other practices (tillage, amendment, correction). The following is a brief description of the main chemical and physical soil properties.

2.1. Soil texture, porosity and density

The textural class is the first parameter that defines soil properties, and is determined by the relative percentage of the three major soil compounds: sand, silt and clay, defined by the respective particle diameter size.

Clayey grounds are characterized by particles of a diameter of less than 0.002 mm, constituted by flinty minerals with different capacities to inflate in the presence of water and to contract in dry conditions, forming cracks which are typical of vertisols. Clayey soils show a low water permeability and high plasticity, which can induce stagnation phenomena and root asphyxia in wet conditions; while in the dry state it has notable tenacity and cohesion.

Silty soils have elementary particles of greater dimensions than the clayey ones and, unlike these, they have greater difficulty in reaching a glomerular structure. The lower structural stability of silty soils causes a low macroporosity and a great bulk density, that determine conditions of low aeration, low permeability and water stagnation in the profile. This type of soil is subject to loosening conditions with greater facility in comparison to clayey soils. However, they have a greater tendency to pulverization in the dry state, and the formation of mud in the wet state.

Sandy soils have particles of a diameter between 2 and 0.02 mm; they are characterized by high permeability of rainwater, and fast mineralization of organic matter.

The Olive tree responds best to soil textures with balanced proportions of sand, silt and clay. Soils that are primarily sandy do not have good nutrient or water-holding capacities, but they do provide good aeration and olives do well, especially when water is available and the crop is properly fertilized to satisfy its mineral requirements. The soil should not contain too much clay to avoid limiting air circulation and to prevent soil management problems. The soil particles should aggregate in granules or crumbs to make the soil porous; this is ensured by sufficient quantities of organic matter and rational soil management to prevent compacting and erosion. Soil is composed of solid particles (mineral and organic matter) of different sizes, usually bound together into aggregates by organic matter, mineral oxides, and charged clay particles. The number and size of pores vary considerably among soils exhibiting different organic matter content, texture and structure and cultivation techniques have a great effect on bulk density and porosity: any management practice that increases organic matter will increase the granular structure of the soil, increase the pore space, and decrease the bulk density (Gisotti, 1988; Giordano, 1999; Hao et al., 2008).

2.2. Organic Matter and nitrogen ratio

The Organic Matter (OM) is a complex mixture of organic compounds deriving from metabolic wastes and decaying residuals of plants, animals and microorganisms, at different stages of decomposition. The OM percentage directly influences the structure and chemical-physic properties of soil in terms of water infiltration and retention, element absorption, particle aggregations, Cation Exchange Capacity (Al^{3+}, Fe^{3+}, Ca^{2+}, Mg^{2+}, NH_4^+), buffering power, over the nutrient source for the plant. The quantity and nature of OM is highly dependent upon farming practices and climatic conditions and is found as both chemically stable humus (or passive OM) and partially decomposed plants, microbes and animal residues (or active OM).

Measures to increase the organic content are a very important part of good soil management in Mediterranean regions, especially in order to reduce vulnerability to erosion (European Soil Bureau, 1999). Practical measures are based on the incorporation of organic matter such as farm-yard manure, cover crops, pruning and processing residues, and soil tillage.

Plant availability of organic N is dependent on OM breakdown, which is difficult to estimate. The ratio of total organic carbon and total nitrogen (C/N) is the traditional guide to the nature of the organic matter present in the soil.

The basic premise behind this ratio is that organic carbon is the primary source of energy for soil microbes, but these also require nitrogen to multiply and utilise this energy. The microbes utilise soil carbon via respiration, with the consequent loss of carbon dioxide from the soil. As the active fraction of the OM is thus degraded, the C/N ratio drops until a steady state (the passive fraction) is finally attained. Interpreting this ratio is complicated, as it also depends on the nature of the OM. The passive fraction of the OM can have a C/N ratio that is 'medium'. Consequently, medium C/N ratio soils can have a wide variation in mineralisable N status, and this is a limitation when considering the C/N ratio in isolation.

2.3. Cation Exchange Capacity (CEC), pH, electrical and hydraulic conductivity, water content

Plant nutrients usually exist as ions which carry an electrostatic charge. This electrostatic charge is a result of atomic substitution in the lattices of soil minerals and because of hydrolysis reactions on the broken edges of the lattices and the surface of oxides, hydroxides, hydrous oxides and organic matter (Hendershot et al., 2008a). These charges attract counterions (exchangeable ions) and form the exchange complex. Ions can be bound to the soil in varying degrees. At one extreme, they may be an integral part of the soil, strongly bound to silica and essentially unavailable to growing plants. At the other extreme, they may be fully soluble and not interact with the soil to any significant extent. Exchangeable ions are between these two extremes , and are weakly bound to soil particles. The bonds between soil particles and exchangeable ions are not permanent, and are continually broken reformed, as the ions move within the water surrounding soil particles. The bonding of these ions largely prevents their loss by leaching, but is not so strong that plants cannot extract them from the soil. In fact, plant roots absorb exchangeable ions by 'swapping' them for hydrogen cations (H^+).

The cation exchange capacity is often estimated by summing the major exchangeable cations (K, Ca, Mg, and Na) using units of cmol kg^{-1}, even if the common expression for CEC is in terms of milliequivalents per 100 grams (meq/100g) of soil. The CEC of soil can range from less than 5 to 35 meq/100g for agricultural type soils, and is related to clay and organic matter content.

CEC is important for maintaining adequate quantities of plant available calcium, magnesium, sodium and potassium in soils. For many crops the magnesium level should ideally be twice as much as that of potassium. When magnesium is lower than potassium, suppression of magnesium uptake can occur. Sodium is only of secondary importance in the soil test as its uptake by plants is largely dependent on the plant species involved and the potassium status of the soil, rather than the level of sodium extractable from the soil.

The Total Base Saturation is related to CEC, which represents the proportion of the soil's total capacity for cations that is actually occupied by these nutrients. It is calculated by summing together the levels of calcium, magnesium, potassium and sodium found in the soil and expressing this sum as a percentage of the CEC value.

Soil pH is one of the most common and important measurements in standard soil analyses (Hendershot et al., 2008b). The pH value expresses degree of acidity or alkalinity of the soil. It is important because it influences the chemical and physiological processes in the soil, and the availability of nutrients. Availability changes differently with pH levels: aluminium, copper, iron, manganese and zinc increase when the pH decreases; unlike magnesium that decreases when the pH decreases (Belsito et al., 1988; AA.VV., 1989; Jones, 2003).

Electrical conductivity (EC) is the ability of a material to conduct an electrical current and is commonly expressed in units of microSiemens per meter (μS cm^{-1}). It is used to estimate the level of soluble salts. The measurement of EC in the soil water extracted from the field-water content is theoretically the best measure of salinity as it indicates the actual salinity level experienced by the plant root (Miller & Curtin, 2008). However, this measurement has not been widely used because it varies as soil-water content changes over time and so it is not a single-valued parameter. A soil is considered saline if the EC of the saturation extract exceeds 4000 μS cm^{-1} at 25°C. The soil EC varies depending on the amount of moisture held by soil particles. Consequently, the EC correlates strongly to soil particle size and texture and affects crop productivity.

Soil water analyses can be organized into two main groups: analysis of storage properties and analysis of hydraulic properties. The water content of soil is part of the analysis of storage properties which refer to the soil's ability to absorb and hold water. Instead hydraulic conductivity is a hydraulic property which refers to the soil's ability to transmit or conduct water . It is more difficult for plants to absorb nutrient elements at low soil moisture levels, so nutrient element contents will be lower.

3. Soil management

In order for plants to live, two key functions can be attributed to soil: habitability and nutrition. The function of habitability mainly depends on the physico-chemical characteristics of the soil. The function of nutrition depends on the factors that make nutrients bio-available to the plants, described above, determining the fertility of the soil as productive attitude. Biological soil functions depend on the micro-organic pattern, responsible for processes on the organic matter such as: mineralization, humification, nitrification, nitrogen fixation, symbioses, and parasitism. The agricultural management systems of soils, such as crop rotation, nutrient application, plant species, kind of tillage, and use of pesticides may have a strong impact on the composition of the soil microbial community. Maintenance of sustainable soil fertility depends greatly on the ability to harness the benefits of rhizosphere microorganisms such as arbuscular mycorrhizal fungi (AMF), which form a symbiotic association with the roots of most plant families. Olive plants are known to form arbuscular mycorrhiza (Roldán-Fajardo & Barea, 1985; Briccoli Bati et al., 1992; Calvente et al., 2004), the most common mycorrhizal type involved in normal cropping systems, being considered as a key component in environmentally friendly agro-biotechnologies (Jeffries & Barea, 2001).

Mycorrhizae act as biofertilizers, bioregulators, and biocontrol agents (Lovato et al., 1996; Von, 1997). Arbuscular mycorrhizal fungi allow the plant to absorb greater quantities of water and nutrients, particularly those less mobile in the soil such as phosphorus. In addition to phosphorus, other elements such as nitrogen, zinc, calcium and sulfur are involved in the mechanisms of mobilization and uptake by mycorrhizal fungi.

Mycorrhizal symbiosis also acts as a bio-regulator, able to influence some physiological processes, growth regulators and the development of the plant, to modify the morphology of the roots, the roots/foliage ratio and sometimes branching foliage and flowering. These soil fungi improve the agronomic fertility of the soil through the formation and stabilization of particle aggregates, in particular in land lacking structure.

3.1. Tillage

Tillage consists of some mechanical operations, performed with different tools, which modify soil structure, according to the management needs, that can be summarized as: increase of the soil mass (active layer); increase in soil permeability, runoff and erosive phenomena; accumulation of water reserves; reduction of evaporation due to interruption of superficial capillarity; destruction or containment of weeds; burial of fertilizers, corrective, amendants, and crops residual.

Usually, for olive groves a deeper autumnal tillage is carried out, to increase the water reserve and to bury the phospho-potassic fertilizers; while during the spring-summer period, some harrowing is performed to reduce evaporation and to eliminate weeds. Soil tillage was classified according to the epoch and the type of performance, distinguishing into preparatory practices, performed before the plantation, to constitute suitable conditions to sustain the crop after the implant, and subsequent practices, performed during crop culture.

Soil tillage can be performed with different tools classified into three main groups: mouldboards, rippers and scramblers. Mouldboards cut and upset the soil; rippers cut the profile producing clods but without modifying soil stratigraphy; and scramblers break up and remix the worked layer.

The choice of the best tillage technique must be performed in order to: reduce costs, in terms of working times and fuel needs; increase the timeliness of intervention; maintain a suitable productive level of crops and soil fertility; contain erosive phenomena.

The soil water content strongly influences the choice of the epoch and the type of intervention performance. Dependent on this, there are different physical soil states: i) cohesive, when the soil is dry; in this state the soil does not stick to utensils, it has maximum tenacity, it is crushproof and resistant to breakup; ii) plastic, characterized by a progressive warping and stickiness, that increases with an increase in damp; iii) liquid, when the soil behaves like a suspension. Tillage is difficult and harmful to the soil when the soil is sticky; over this limit, the passing of machines provokes undoing of the structural aggregates. With dry soils the work needs higher powers, and forms compact clods of varying dimensions, according to the type of performance: plowing causes large clods, while with rotary hoeing a notable pulverization and formation of small clods occurs.

A soil is considered as loosening when it has an optimal damp for the execution of tillage, approximately corresponding to half of their field capacity, with a more or less ampler range depending on the soil type and intervention (Bonciarelli, 1981; Giardini, 1986).

Plowing is the most known and commonly used form of soil tillage in agricultural practice and is performed with three type of tools: the ploughshare plow and the disk plow that work by traction; and the rollover plow, that acts by the tractor power take-off (PTO).

The plow operates by cutting and overthrowing a slice of soil, with an angle varying with the type of plough, the operating velocity and the operation goal: the complete overthrow of slices is necessary in green manuring and in weed control; while vertical slices improve airing and rainwater infiltration. Using a cylindrical bending breast a greater crumbling action is achieved; while a helical breast favours the slice overthrow with smaller production of thin soil. The speed of ploughing acts on both the slice overthrow and crumbling: a fast ploughing enhances the inversion of layers and the pulverization of clods. The ploughshare action can cause compaction of the deep soil, called tillage pan. Such a drawback can be enhanced using tractors working "within furrow", and in conditions of high damp. The tillage pan hinders the vertical movement of rainwater and the gaseous exchanges in the soil and the growth of the roots.

The drawbacks of the ploughshare plow are: excessive clod level, that requires other refining tillage, with further passages of machinery that stamp on the soil and degrade its structure; formation of tillage pan; high requirement of traction power. Such negative aspects can be mitigated using the disk plow, in which the ploughshare and the breast are replaced by a spherical cap, free rotating on an axle angled in respect of the operating direction. The disk limits attrition and needs of traction power. During rotation the cap lifts a slice of soil that is then crumbled and remixed. In comparison to the classical plow it better prepares the bed for seeding; it is proper for light ploughing in loose or medium textured soil rich of skeleton and in groves.

There are other tillage techniques that can be carried out, using different types of machinery which can be complementary or alternatives to ploughing. Among these the most common are:

- Ripping technique is characterized by the vertical breakup of the soil without inversion of the layers. The tools used in this type of tillage are constituted by a varying number of anchors (from 1 to 5) that practice a different action on the soil depending on their shape and interaxle. Some machineries are endowed with vibrating tools that enhance shattering of the soil, and can be joined to rolls or harrows to finish up and to level the surface in a single pass.
- Rotary Hoeing consists in the shattering and remixing of the soil performed by machinery moved with a tractor PTO., with tools that work on a horizontal axle (millers) or vertical axle (rototiller). Their drawback is the excessive shattering of soil, that worsens its structure, with compaction, formation of superficial crust and tillage pan, fast tool usury; while requiring high working power. They are unsuitable in heavy and/or skeleton soils.

- Weeding is a tillage practice complementary to ploughing, carried out to reduce the clod of the soil, and to bring up weed roots. This machinery is formed by bent rigid or elastic rippers and with different types of feet.
- Harrowing is performed to refine clods, eliminate weeds, bury fertilizers and break the superficial crust. The tools for this operation are of different shapes and dimensions, according to soil types, soil conditions and the needs of soil refining.

All tillage up to 15 cm of depth is included in the minimum tillage group, , with the aim of energy saving, preserving soil structure and timeliness of work. Among the different operative options, various types of machinery are available that can perform tillage, fertilization, seeding and chemical weeding of the soil in a single pass (Toscano, 1998).

3.2. Tillage: Soil characteristics and erosion

The usable kind of tillage depends on the soil texture. In clayey soils, the minimum tillage can have positive effects on the containment of erosive phenomena, due to the residual crop on the soil surface, and on compaction, in order to reduce passing of the machinery.

In silty soils all the tilling techniques that do not involve the inversion of layers favour soil structuring, and the presence of residual crop, and avoids the destructuring caused by the beating action of rain water.

In sandy soils, the choice of tillage techniques should exclude deep intervention, while all the minimum tillage techniques generally guarantee best results.

The different handling of the soil can determine a different availability of the nutrient elements, as well as a different biological activity. The techniques that do not involve the inversion of layers, allow maintaining or increasing the organic matter in the soil. With regards to the availability of the principal nutrient elements, the effects produced by the different tilling techniques vary according to the different movement of each element in the soil.

Nitrogen results mostly available for plants in a worked soil, due to the high aeration of the mass and for the velocity whereby the residual and organic fertilizers are degraded into mineral elements which can be assimilated.

For the phosphorus, a low mobile element, there are strong differences in its stratification along the profile according to the type of tillage. With a plow this element has the tendency to distribute itself in a more homogeneous way in the soil in comparison to how it is distributed in soil that has not been worked, or only cracked soil, in which phosphorus remains in the most superficial layers.

The common good supply of natural Italian agricultural land, particularly clayey ones, leads to a substantial independence of availability of potassium from the plowing technique adopted.

All tilling techniques that improve soil permeability, and allow maintenance of a vegetable coverage, are very useful in the control of erosive phenomena (Stein et al., 1986; Rasiah &

Kay, 1995; Raglione et al., 2000; Toscano, 2000; Toscano et al., 2004a). Erosion consists in the removal of the most fertile soil layers by wind and/or rainwater action. The eroded amount is proportional to the intensity of the rainfall, to the slope and type of the soil.

3.3. Soil grassing

To protect the soil profile, structure and edaphic biocenosis, it is useful to apply less expensive cultivation techniques which have a lower environmental impact than traditional tillage. The minimum tillage or controlled soil grassing, generally determine a great activity of the soil biota, due to the greater presence of organic matter and the low trouble of soil.

It is possible to implement soil grassing, which can be either natural or artificial, and to partially or wholly cover the orchard surface. The benefits of controlled grassing in olive orchards are: improvement of the soil structure, increase of soil organic matter and water absorption, reduction of runoff and erosion, improvement of carrying capacity and reduction of compactness, enhancement of microbial activity and nutritional balance; simplified management at lower operating costs. The possible competition of turf for water and nutrients with regards to the olive plants (Pastor, 1989), can be prevented with proper grass management, such as cutting, or additional fertilization (Toscano et al., 2004b).

The simpler type of soil grassing is "permanent", whereby the soil is constantly covered by spontaneous vegetation that is periodically mowed or shredded.

Alternatively, in dry summer conditions, temporary grassing can be adopted, eliminating grass, when competition for water competition begins, with superficial tillage or using contact herbicides; the coverage naturally reestablishes itself with the resumption of the rainy season, retaining its beneficial effects up to the following spring.

For artificial grassing the choice of the essences is very important, these must have fast growth following planting and to be resistant to pounding and to mulching. However, the artificial grassing presents some economic and managerial limits such as the difficult choice of the essences and the seeding costs.

The machinery for grass management consists in the rotary mower, and the shredder. They are of great working capacity and have low power needs, compared to the tools that operate on the soil; the shredder has the advantage that it grinds the mown grass, thus reducing degradation times, and it can also be used for pruning residues. Both these machines can be equipped with intercept rotary mower, which allows cutting of the grass along the row, avoiding damage to the tree trunks. Long-time experiences of controlled grassing in different non-irrigated olive orchard soils have confirmed the effectiveness of this technique in improving soil properties, in the drainage system, in the control of soil erosion and on olive tree productivity (Briccoli Bati et al., 2002; Toscano et al., 1999, 2006). On the contrary other tests, comparing different application methods of grassing, have evidenced better agronomic and productive results with green manure in summer, to avoid the increased competition for soil water occurred by permanent grassing in this environment (Toscano, 2009). Therefore, the choice of the best soil management system must be made according to the specific soils and crop characteristics. In olive groves the replacement of tillage with other techniques is possible, according to water availability, in order to obtain the best maintenance of soil fertility, a reduction of the erosion in slopes, a timeliness of intervention, a reduction of the costs and, therefore, the attainment of greater incomes.

4. Nutrition

The olive tree is still often considered a rustic plant, having little nutritional requirements and capable to survive even in rough environments, with minimal care and management. The olive plant grows in most soil types as long as they are well drained. These plants, could also vegetate in the absence of fertilization, but require suitable nutrition to express their productive and qualitative potentialities.

In the traditional olive-grove plant nutrition is mainly based on systematic and massive inputs of chemical nutrients distributed to the soil, not always correctly and often unnecessarily (if not harmful) for plants and the environment (groundwater pollution). In many cases olive tree fertilization is often empirically approached and farmers apply much more fertiliser than the crop really needs (Tombesi et al., 1996).

The compilation of adequate fertilization programs, in terms of type, doses, epochs and disposal of the nourishing elements, are not of simple generalization and depend on local environmental and climatic factors, as well as on the effectiveness of the fertilizer composition and its application method.

It is indispensable to carry out a prior analysis of soil chemistry and of the nutrient contents of the plant by plant tissue analysis (usually leaves are used). These analyses will give significant data on the status of both soil and plant, indicating the most useful typology and doses of nutrients to apply in the fertilization plans.

Leaf analysis, is a reliable method for assessing the nutritional status of the crop (Bouat, 1968; Freeman et al., 1994). The content of the major nutrients in the leaves differs not only according to the cultivars, the soil and climatic conditions of the cultivation area, the time of sampling for the analysis but also in relation to the pruning and irrigation applied to olive orchard (Briccoli Bati et al., 1995). Some research on a regional scale has defined certain relationships between the time of leaf sampling, the foliar nutrient content and the quality of production (Failla et al., 1997; Soyergin et al., 2000). In fact, it was found that the leaf diagnostic at flowering is conclusive for the less mobile elements (Ca, Mg, Fe and Zn) while during the winter rest foliar analysis better shows the nutritional potential of the soil for nutrients with increased mobility as N, P and K (Failla et al., 1997). The level of global nutrition, generalizing, expressed as percentage amount on the leaf dry matter for nitrogen, phosphorus and potassium, results as 3,5% divided, respectively, in 2,1 - 0,35 - 1,05 with a physiological relationship of 6:1:3.

In profitable olive growing, the nutritional needs also vary in relation to the phenological phases, to the climate, to the cultivar, to the trees' productive potentiality, and to the olive orchard management, i.e. presence of soil grassing and irrigation. For these reasons, fertilization planning cannot be approached as a standard procedure and many authors report different evaluations about the nourishment needs of olive trees (Natali, 1993; Petruccioli & Parlati, 1983).

Fertilisation systems include: chemical fertilisers (NPK applied beneath the tree canopy projection, usually in the form of combined fertilisers), organic fertilisation (green and animal manures, leaves, compost, manufactured organic fertilisers), and fertilisation through watering systems and through foliage.

During the first three years of the olive plantation, when vegetative activity prevails on fructification, it is important to stimulate, with fertilization, rapid canopy and root growth of the tree to predispose the plants quickly to flowering and fruiting (Palese et al., 1997). In this phase, Nitrogen is the essential element, while phospho-potassic fertilizers at this time are less important, provided that during the preparatory work of the soil for planting, such fertilizers were distributed over the entire surface and buried with deep tillage.

When the plant completed the first phase of growth (5th - 6th year) and during the entire life of the orchard, the scope of fertilization is to induce and support the yield and, simultaneously, also to ensure the renewal of fruiting shoots and roots.

In order to calculate the amount of nutrient supply to plants it is helpful to adopt the returning criteria of nutrients removed with fruit harvesting, with pruned wood and abscised leaves: for 100 kg of drupes produced the olive tree needs around 900 g nitrogen, 200 g of phosphorus and 1.000 g of potassium. In fertilization planning, such doses must be triplicate, due to the losses leaching, volatilization, fixation, etc.

Traditionally nitrogen is supplied annually and divided in at least two doses. Most of the quantity to be given (2/3) at the end of winter before flower bud differentiation and before the growth of new lateral shoots, and the second during the flowering period (from the pre-flowering stage till fruit set). Usually the recommended nitrogen application ranges between 500-1500g for bearing tree, according to canopy volume.

Throughout the life of the olive-grove, phosphorus and potassium supply must be repeated every 5-6 years, with the doses defined by the results of soil and leaf analysis. These fertilizers are usually supplied in autumn, and burying with shallow tillage, on alternate inter-rows to limit damage to the roots, with doses of 200-400 units of potassium and 100-200 of phosphorus per hectare integrated with suitable doses of organic matter (manure, green manure or compost).

During the annual cycle, nitrogen absorption is more intense from the flowering up to the pit hardening, while the contents of N and P decrease in the leaves up to the pit hardening and at the same time they increase in the drupes. Subsequently, both in the leaves and fruits nitrogen and phosphorus decrease after veraison. Instead, potassium, constantly decreases in the leaves, while increasing in the fruits.

Biennial or triennial interventions for phosphorus and potassium, in the poor soils are useful, applied after harvesting in concomitance with deep tillage for rainwater storage, in old or dry-condition raised olive orchard; or at the end of winter, with lighter tillage in a young and intensive olive grove.

4.1. Nutrients typology and effects

The use of the appropriate fertilizer at the right time increases the efficiency and reduces the cost of fertilization, with a positive impact on the produced olive fruit and oil content.

Fertilization can be distinguished into organic and mineral. The first one has the purpose to improve the physical characteristics of the soil, such as the structure, the porosity, the permeability, the tackiness, the consistency, the water retention, and the pH. The second one is destined to feed the plants.

Nitrogen is fundamental in plant growth, it participates in the formation of amino acids and in the formation of proteins, therefore, it is crucial in the growing processes when the plant is young. In adult trees nitrogen supports the formation of shoots, a necessary condition to ensure constant productivity and positively influence flower formation, fruit setting and fruit development, especially during the early stages, up to the pit hardening. Nitrogen fertilization consistently increases the olive yield but only when leaf N is below the sufficiency threshold (Hartmann,1958). It is usually applied to the soil using urea, ammonium sulphate, or ammonium nitrate. Nitrogen can also be supplied with either organic materials such as feathers or blood meal, compost, or a leguminous cover crop. Its deficiency is manifested by decreased growth activity, leaf yellowing, high ovary abortion, low yield and alternate bearing (Cimato et al., 1990; 1995).

Phosphorus also has a role in growth, being essential for cell division and the development of the meristematic tissues, fruit set, fruit growth and maturation, and in lignification of the shoots. Even if absorbed by the olive tree in a relatively high quantity, the effects of phosphatic fertilization are nevertheless manifested with extreme slowness. The most used phosphate fertilizers are soluble phosphates and in particular superphosphates with 35-45% of phosphorus pentoxide, which is the form absorbed by the plant. A content of 50-100 ppm of phosphorus pentoxide in the soil detected by soil analysis, can be considered to be satisfactory. The symptoms of deficiency of this element, which is very rare, occur with a reddish or purplish coloration of the green parts of the plant, and it causes metabolic problems for growth and fructification, with delays in drupe maturation.

Potassium promotes the accumulation of carbon hydrates such as starch, an energetic reserve for metabolic processes. This element, regulates the water consumption of the plant through an increase in water retention in the tissues and it also controls transpiration. Potassium is an enzymatic activator, it increases the plants resistance to thermal extremes and to some fungal diseases, and it promotes oil accumulation in the fruits. This element is, usually, applied to the soil during winter in order to gradually reach the rooting zone with the rain. Regular potassium fertilization is necessary in order to maximize both yield and quality, especially in heavy yield years.

After nitrogen, phosphorus and potassium, other very important nourishing elements are magnesium and calcium. Magnesium is an essential component of chlorophyll and generally it is not considered in fertilization plans because it is already contained in many fertilizers. Occasionally, magnesium deficiency can be revealed in orchards growing on sandy, neutral soil. Fertilization based on magnesium sulfate corrects this deficiency.

Calcium is vital to olive plant growth, because it is an essential constituent of cell walls and contributes to the mechanical resistance of tissues, it also acts as an activator of some enzymes. Deficiencies of calcium due to soil acidity, can be corrected with an adequate lime supply as calcium carbonate.

Sulphur is present in plant amino-acids such as cystine, cysteine and methionine and is located in the soil in the organic matter. Fertilizers containing sulphur as ammonium or potassium sulphate, etc. are distributed against possible deficiencies of this element.

The most important microelements are iron, copper, zinc, manganese, molybdenum and especially boron, all developing a specific and exclusive role as enzymatic activators in the biochemical processes of the plants. These elements, present in small amounts in olive tissues, have a very narrow range between a sufficiency and toxicity level. Leaf tissue analyses provide excellent information in order to directly diagnose the toxicity or the lack of these microelements.

Above all it is very important to know the boron content of the leaves because it plays a major role in pollen growth, fruit set and plant productivity. Visible symptoms of boron deficiency are manifested with leaves with apical chlorosis, followed by necrosis and leaf drop. In the cases of a slight boron deficiency, the fertility of the flowers is reduced due to increased ovary abortion (Perica et al., 2001). Boron deficiency is nevertheless removable with extreme rapidity and effectiveness through leaf treatments during the pre-flowering stages. Foliar applications have had statistically significant effects on the yield and leaf B contents, therefore, the most economic dose was found to be 0.4% foliar application of sodium tetraborate.

The organic matter in soil plays a central role in controlling the availability of N, P and K and it can also act as a chelate, making certain micronutrients more available for the roots in the form of complexes.

4.2. Fertilization techniques

Plant nutrition is physiologically dependent on the absorption of nourishing elements through the roots; it is therefore necessary to ensure that in the active soil layer there is a suitable endowment of available nourishing elements for the plants. Normally fertilizers are spread on the soil. Nitrogenous fertilizers, nitric, ureic or ammoniacal, are used annually according to requirements and the time of intervention, the first one being easily soluble with a fast effect, while ureic and ammoniacal products have a longer acting time and greater persistence. The nitrogen amount usually provided is of 1kg N equal to, approximately, 5kg of ammonium sulfate, 3kg of ammonium nitrate, 4kg of calcium nitrate or 2kg of urea. The principal provisions of phosphorus and potassium, due to their low mobility in the soil, are provided before the plantation of the orchard establishment, further applications are made every 4 to 5 years in the autumn on the ground.

Fertilizers can also be supplied by the foliage, and in olive trees this characteristic can be effectively exploited in order to satisfy the needs of the plants in situations of particular demands (lacks of microelements), or as integration of soil fertilization in the different phenological phases. This technique is considered to be a valid support to increase the nutrient levels and the crop yield, reducing competition among metabolic sinks (shoots, inflorescences and fruits) and increasing the absorption of nutrients through the roots (Cimato et al., 1990, 1991, 1994, 1995; Toscano et al., 2000; Toscano, 2008). It provide nutrients quickly, uses low amounts of fertilizer, can be combined with pesticide applications, is well suited to rain-fed olive trees or when ground fertilizations would be useless due to a lack of soil humidity. The advantages of this technique are manifold: timely intervention, nutrients are given at the moment of greatest necessity, and is effective in a short time, and allows an integral use of the administered element. If only a foliage solution is applied, several applications take place (Fernández-Escobar, 1999; Ben Mimoun et al., 2004). Some results demonstrate, however, that foliar fertilization cannot entirely replace nutrition through the roots, even though it permits a reduction of the fertilizer required to be applied to the soil (Fontanazza, 1988; Toscano et al, 2000).

Many authors have studied the efficiency of olive foliar nourishment and for specific nutrients good results have been achieved using urea solution (Cimato et al, 1991; 1994).

Potassium is easily absorbed and distributed through leaf tissues (California Fertilizer Association, 1998) and foliar application is helpful to satisfy plant requirement having a high efficiency (Inglese et al., 2002). Phosphorous is given during summer fruit growth, in order to be readily absorbed and translocated to the fruits for quality purposes, therefore its application is better in the form of a foliar fertilizer.

In addition, recent studies have assessed the effectiveness of some commercial products on different olive cultivars, which behave differently. A helpful example of this efficiency is shown using NutriVant (NV) foliar fertilizers in addition to soil fertilization and showed that better results are obtained on the 'Carolea', than on 'Nocellara del Belice' cultivar. This difference is more clear in the "off year" orchards, during which an increase in the vegetative parameters and yield entity, in comparison to the control tests, was recorded. Conversely, in the "on year", the NV test had good results on both observed cultivars (Toscano et al., 2002b; Toscano, 2005; Toscano & Godino, 2010).

In irrigated orchards it is possible to supply nutrients to the plant by watering systems (fertigation) (Toscano et al., 2002a). The advantages of such practice consist in the easiness of application and in the efficiency of fertilizers, being able to reduce the needs of fertilizers by up to 30% in comparison to soil distribution. Fertigation implies a sensitive reduction of the management costs both in terms of purchase, transport and distribution of fertilizers , enhancing their efficacy in order to grant a better nutritional level to the trees, to maximize yield, oil production and profitability.

5. Irrigation

The efficient use of water resources in agriculture is extremely important in order to improve the economical and environmental sustainability of agricultural activity. Mediterranean regions of Italy are characterized by a high evaporative demand of the atmosphere, water scarcity and increasing negative consequences of climate change. In Italy the rainfall can vary annually from less than 400 to over 800 millimeters, and the lack of precipitation that is often manifested during the summer, involves the use of irrigation during dry periods to ensure the constant productivity of olive orchards.

In traditional olive cultivation areas, characterized by water scarcity, rainfall and underground water resources are the only supplies for the olive tree water requirements. Rainfed olive groves, therefore, are characterized by low plantation density which allows the exploitation of an adequate soil volume by the root system, minimizing competition for water among plants.

The olive, a sclerophyllous evergreen tree, is able to tolerate the low availability of water in the soil by means of morphological and physiological adaptations acquired in response to coping with drought stress (Connor & Fereres, 2005; Bacelar et al., 2007). Under semi-arid conditions olive trees were able to restrict water loss by modulating stomatal closure at different levels of soil moisture and evaporative demand and show a non-balanced allocation of dry matter among the different plant organs, resulting in a reduction of the vegetative growth and a significant decline in the productive performance (low yield and alternate bearing behaviour) in favour of development of the root system. Indeed, olive tree roots can extend and go deep into the soil to exploit a wider soil volume (Fernández et al., 1991; Dichio et al., 2002). Olive plants maintain a high rate of photosynthesis during long drought stress periods. The high efficiency of the olive is also due to its ability to continue to absorb carbon dioxide and to produce carbohydrates in water deficit conditions that determine the complete stomatal closure and threaten the survival of other species (Xiloyannis & Dichio, 2006). A higher photosynthetic rate under drought is a decisive factor for better drought tolerance in olive cultivars (Bacelar et al., 2007). Generally, when water is not restricting growth, plants invest a considerable fraction of photoassimilates in the expansion of photosynthetic tissues, maximising light interception and, as a consequence, growth (Dale, 1988). The capacity to withstand severe and prolonged drought periods, however, is negatively associated with olive tree growth and productivity, owing to the decrease of assimilates under water deficit conditions. Reductions in photosynthetic performance under water stress have also been observed by several authors (Inglese et al., 1999; Patumi et al., 1999; Tognetti et al., 2005; Bacelar et al., 2006; Lavee et al., 2007; Ben Ahmed et al., 2009).

Much research shows the productivity benefits of irrigation. Irrigation is highly effective in increasing yield and yield components such as fruit size, fruit number and oil content, moreover, irrigation affects the pulp-to-pit ratio, phenology and time of fruit maturation (Agabbio, 1978; Goldhamer et al., 1994; Michelakis et al., 1994; Inglese et al., 1996; Gòmez-Rico et al., 2006, 2007; Dag et al., 2008; Servili et al., 2007).

A proper soil water availability enhances vegetative growth, such as shoot length, allowing the olive trees to produce a higher number of buds able to provide the opportune basis for the next year's production (Patumi et al., 2002; D'Andria et al., 2004; Gucci et al., 2007;Ben-Gal et al., 2008;). Stress levels and water requirements are highly dependent on fruit load and best irrigation management must account for biannual bearing effects. Although biennial bearing is basically genetically determined, the degree to which it occurs is greatly affected by environmental conditions, especially the weather and cultivation practices (Pandolfi et al., 2000). Alternate fruit bearing occurs under both extensive and intensive growing conditions (Pannelli et al., 1996; Lavee, 2006). With irrigation, olive production can increase up to five times that of olive groves in dry arid climates, in the Italian climate on average a double production must be expected (Bini et al., 1997). Obviously the scale of production will depend on soil conditions, average rainfall, evapo-transpiration and temperatures, cultivars, planting distances and other cultural practices (Nuzzo et al., 1997). Proper management of irrigation, especially during the summer drought, keeps leaves in activities promoting fruit growth and accumulation of reserves in the various plant organs (Xiloyannis & Palese, 2001), in any case, table olives cannot be cultivated without irrigation.

5.1. Olive tree water needs

The unitary water consumption of the olive tree, namely the quantity of water that must be transpired in order to synthesize a gram of dry substance or commercial product, have been estimated to be 1 liter of water by 1 m^2 of leaf, daily transpired in August. Such indexes, with opportune calculations, can be useful to help establish watering volumes. Best indications are drawn by the compilation of water budgets that, from the comparison among the entity of the rains and the losses of damp from the soil by evapo-transpiration, allow the determining of the water deficit or excess in the different periods of the year.

The criteria to be adopted in watering planning must be based on respect of the water requirement of the crop, and on the knowledge of the critical phases of the vegetative cycle of the plants, over that of the quantity of available water for irrigation, for the evaluation of the economic convenience of the intervention. As for all the other production factors, the economic principle of marginal productivity is in force also for water.

Olive tree water requirements are variable and depend upon factors such as soil type, climate, planting density, age of trees, cultural management (e.g. fertilizing, pruning) and the method of irrigation.

In the olive tree there are nevertheless some critical periods during the annual cycle, during which the plant mostly needs water. The first one extends from bud differentiation up to flowering and therefore to the fruit set: in these phases a water deficit can create trouble in flower development with a smaller number of flowers for inflorescence, increasing ovary abortion, and a lower fruit set. Generally during this period in Italy there are no deficiencies in soil water. Subsequently, at the second phase of fruit growth, corresponding to the pit hardening period, olive trees are most resistant to water deficit (Goldhamer et al., 1994; Moriana et al., 2007), on the contrary the third phase, when olive oil is accumulated, the olive tree again seems to be sensitive to water stress (Lavee & Wodner, 1991)

The inolition process starts around the pit hardening phase and reaches a maximum before ripening. The effects of irrigation on oil content are nevertheless quite controversial depending on different experimental conditions

Some authors did not find any difference in oil content between irrigated and non-irrigated trees (Michelakis et al., 1994; d'Andria et al., 2004), while Inglese et al. (1999) reported a lower oil content in the fruits of trees grown under high soil water deficit conditions. The literature suggests that the fruit and oil yield response to irrigation is highly cultivar specific (Lavee et al., 2007). Despite the increasing use of irrigation in olive groves, there is still a poor understanding of the effect of irrigation deficit on the qualitative parameters of olive oil.

Increasing irrigation leads to fruits with a greater water content (lower oil percentage), and irrigation has been found to decrease the polyphenol content (Patumi et al., 1999; Gómez-Rico et al., 2006; Ben-Gal et al., 2008; Dag et al., 2008), which then changes the oil bitterness and spicy tastes.

Several studies, which focused on the effect of irrigation on olive oil composition, report that irrigation increases free fatty acids in oil (Dag et al., 2008), can affect the fatty acid composition (Ranalli et al., 1997; Aparicio & Luna 2002; Servili et al., 2007) and the accumulation of secondary metabolites, that are fundamental in improving the organoleptic characteristics of the oil, is increased (Pannelli et al., 1996; Inglese et al., 1996).

For the calculation of the water needs in an olive-grove, some formulas are used that consider climatic environmental data, such as the rains and the potential evapo-transpiration (ETP), adopting different coefficients in relation to the spacing of trees, the age and shape of the plants, and season. The water deficit, will be given by the difference between the water used by the crop and the water availability in the soil: such a deficit will be therefore compensated for with irrigation to optimize the productive potentialities of the plants.

The calculated seasonal watering volumes, will be nevertheless reduced considering the threshold of convenience, in relation to the efficiency of the irrigation system, the cost and availability of water and the value of the product. For the intensive olive growing in South Italy, it increased from 1.500 up to 3.000 m3 hectare^{-1} per year (Agabbio, 1978).

An evaluation of the water needs, such as the water consumed by the crop (evaporation and transpiration), can rationalize the irrigation technique. The evaporation potential (ET$_0$) must be determined through the compilation of the soil hydrological balance and the search for an empirical correlation between the potential evapo-transpiration and one of the climatic factors.

To satisfy the needs of an intensive olive-grove the results of different watering trials pointed out that for the olive tree it is enough to supply 30-50% of the evaporated water.

The beginning of the irrigation season should take place when the soil is still wet (60-70% of available water) to ensure the maintenance of adequate reserves even in deeper layers and at points not covered by providers in order, however, to maintain roots present in those areas.

Irrigation can be realized in different ways and the choice of the optimal method should be made according to each single olive-grove typology and environments. Sprinkling methods, with giant irrigators or wings, have the advantage of adapting to any soil condition, the facility of moving and transfer, and the timeliness of intervention, but generally with high costs and low efficiency of water. On the contrary localized irrigation, that allows water distribution evenly in sloping land, is a technique which offers the possibility to intervene in certain biologically critical phases for the plants (flowering, fruit setting, pit hardening, etc.), allowing a significant reduction, of about 25-30%, in the consumption of water. Furthermore it allows a more uniform distribution of water over time, with shorter shifts and increases the efficiency of irrigation up to 90%, avoiding losses due evapo-transpiration, runoff, etc.

With sprinkling the volumes are calculated for the whole surface; with the localized irrigation on the wet surface equal to 10% with drip irrigation and to 25% with microjets. Microjets enable irrigation of a rather large surface to meet the different needs of the olive tree during its development, but also creates constraints for tillage and weeds removal by mechanical means and increase water loss by evapo-transpiration.

In the center-northern olive-grove environments of Italy, natural water availability is often such to allow satisfactory production, even without resorting to irrigation. In the South of Italy, generally the annual average rainfall is rather low, with rains concentrated in the autumn-winter season, that does not coincide with the needs of the plant, therefore it is necessary to apply irrigation during summer.

The irrigation intervals depend, more than from the evaporative demand of the environment, on the type of soil and therefore from the quantity of water that it can retain.

In the case of localized irrigation shifts are on average 2-4 days with about 30 m^3 ha^{-1} of water, the turns will progressively be shorter passing from slimy-sandy to sandy soil.

The most critical phases in which water stress should be avoided are at floraison, at fruit set, at fruit growth and at inolition. An effective watering season could start, according to watering water availability, at the end of flowering (May-June) and continuing until late September.

In recent years many studies have tried to apply deficit irrigation strategies to olive trees. These are based on the observation of Chalmers et al. (1981), who reported for peach trees that the maintenance of a slight plant water deficit can improve the partitioning of carbohydrate to reproductive structures, such as fruit, thus controlling excessive vegetative growth. The asynchronous growth of olive fruits and shoots reduces competition for assimilates at critical stages, providing a sound basis for the application of irrigation deficit.

The controlled irrigation deficit is a water management method that does not completely satisfy the tree's water requirements during the growing season. It causes a temporary and regulated water deficit in a specific phenological stage. When it is applied in the pit-hardening period, the olive oil yield is not affected while the water use efficiency (WUE) is improved. On the contrary when the controlled irrigation deficit is applied from fruit set to harvest, the oil yield decreases but the WUE and certain olive oil quality parameters improve. As the productive tree responses are not affected by moderate levels of water stress, irrigation deficit strategies are recommended in arid and semi-arid areas to save the scarce conventional water resources (Angelakis et al., 1999; Massoud et al., 2003).

Finally it is interesting to note that the olive is quite resistant to salinity. This plant tolerates brackish water (up to a salty residue of 4 g/liter), and therefore can allow the realization of irrigated olive-groves, valorizing waters which are not usable for other crops (Basta et al., 2002; Perica et al., 2008).

6. Pruning

Pruning is a very expensive practice in olive grove management, reaching up to 40% of total cultivation costs, but it is also essential for olive grove profitability. It is finalized to modify the natural shape and structure of the trees, to reduce to the least one skeletal structures, to balance the vegetative and productive activity, and to maximize fructification. To reach the best results, pruning must be rationally managed, and based on the harvesting system. A modern approach to this practice allows to form and maintain the tree structure at a relatively low cost, reducing and simplifying pruning operations without negatively affecting yield, oil quality, or orchard sustainability.

Strategies of "minimum pruning" can be developed at a farm level independently of the type and size of the orchard. Managing the canopy according to the criteria of "minimum pruning" is suitable both for traditional olive groves and modern, high-density orchards.

The growing habit of the cultivar, the natural tendency for high vegetative activity, the type of buds and branches, and alternate bearing are all important biological features of the olive tree that it is important to consider for pruning.

In practice, pruning is distinguished into a formation and a production pruning: formation pruning has the purpose to give the selected form to the olive tree; while the production pruning is finalized to preserve the form and the size of the canopy, to eliminate inefficient or unproductive structures, to facilitate the functional positioning of fruiting shoots to enhance harvest efficiency, to maintain the trees' vegetative and productive balance. Olive trees bear fruits on the previous year's shoots, so to have fruit every year an adequate vegetative growth must be achieved. Annual interventions should be faster, smaller and easier cuts using small tools such as shears and saws.

The execution of pruning should avoid the accumulation of too much wood caused by an excess of primary branches and an excessive overlapping of secondary branches. To stimulate olive production pruning must be reduced to strictly necessary interventions, leaving the most possible greater number of leaves. Periodically some return cuts made on the branches return the plants to their assigned volume to maintain the volume and the shape of olive trees.

Extraordinary kinds of pruning are practiced when it is necessary to restructure the canopy in another form held to be more convenient. Pruning old trees requires drastic cuts to rejuvenate or to restore the health of the plants so as to stimulate their growth and renew fruit-bearing shoots and branches.

Pruning also contributes to reducing the occurrence of pest and disease. Dense canopies encourage the presence of parasites due to high relative humidity, whilst well-aerated canopies considerably decrease the attack of pest and disease such as the 'olive knot' which appears on branches and is otherwise very difficult to control.

6.1. Training system

Around the Mediterranean basin, a traditional area of olive tree cultivation, there are many different training systems for olive trees, but now in modern orchards the most common shapes are:

-   The "vase" with several different variants is by far the most popular shape and the most practical for hand or semi-mechanical harvesting. Usually the vase has a single trunk varying in height from 50 cm to 120 cm, branching into some primary branches, in adult plants generally three, equally spaced so as to intercept as much sunlight as possible. These branches, tilted about 45-50 degrees, support the scaffolding of the tree.

The only pruning required is in the centre of the canopy to allow enough sunlight to come through and removing cross branches leaving the greatest possible number of leaves on the plants, because productivity and the oil quantity in the drupes is dependent on them.

-   The "single-trunk" is constituted by a central axis of the tree that raises a series of primary branches chosen amongst those that grow vigorously inserted in a spiral, alternated by 50-60 cm between them. To achieve this form few thinning cuts are made during the training phase. When the plants are in production the primary branches are periodically pruned by cutting or heads back, or eliminating them in order to renew the fruit-bearing surface. This shape is good for the intensive olive growing systems where the canopy cone reduces distances between the plants and for the mechanical harvest with shakers, but it is difficult to manage.
-   The "bush" shape is the result of natural basal sprouting of the olive tree with numerous branches and those that arise from the bottom. This system is suitable for intensive cultivation models to be collected by hand or with tool facilitators it requires very little pruning during the training phase, but it is unsuitable for mechanical harvests with shakers due to lateral branches shooting from the proximal part of the trunk.
-   The hedgerow is a training system in which trees grow freely, so that the canopy forms a productive wall along the row, usually managed with mechanical pruners to maintain the trees with an available volume.

Lately the training system suitable for olive orchards with over 1000 trees per hectare, is a single axis, obtained by thinning out the side branches in the apical part and by also removing those located below 0.5 m from soil during the first year of planting the to allow the passage of the machine. Once full production has been reached, plants are pruned so as to contain vegetation within 2.2 m in height and prevent the development of little branches of a diameter greater than 3 cm. In these groves mechanical pruning alternating with manual interventions is used to contain the development of foliage.

The training system is the result of the growing habit of the tree and pruning practices and it must be chosen before the planting as a function of the harvesting method and mainly of the area climate. Indeed, experimental tests of comparison between the "vase" and the "single-trunk" conducted over several years in the experimental field of CRA-OIL, located in Mirto Crosia (CS-Italy), showed the extreme difficulty in maintaining the last shape due to the climatic characteristics which are strongly favorable to olive plant development.

6.2. Pruning scopes and effects

A first objective of pruning is to provide a shape and structure to the tree which guarantees proper illumination of the canopy to enhance photosynthesis, good circulation of air, avoidance of pest disease, and a better disposition of fruiting shoots to facilitate and maximize the harvest.

At plantation, the first cuts are executed to plan the scaffold and the principal branches are chosen according to the selected shape. In the following years, pruning will be limited to the elimination of unfit shoots, favoring correct skeletal development. After the third-fourth year, with the beginning of the yield, annual pruning will have to balance volumetric growth with the vegetative-productive equilibrium of the trees.

Pruning intensity increases with the age of the tree. Pruning is light on young trees to allow the shaping and to grow and build energy reserves.

As general guidelines, for adult trees, given the light need of this species to perform photosynthesis, it is necessary to reduce the density of the foliage, allow sunlight to penetrate into every part of the olive canopy and promote air circulation. All the suckers around base of the trunk and branches that have already produced should be removed. It is also important to keep the upper parts of the tree open to allow the lower parts to remain productive.

In adult olive-groves, in full production, annual pruning allows better regulating of the balance between vegetative and reproductive activities and so contributes to reduce alternate bearing. This phenomenon is more dramatic the wider the pruning shift. The shift of pruning cycles should be chosen based on factors such as the growth of branches, the fruit-laden, the training system, soil fertility and climate and structural aspects of the farm. The execution of pruning every 2 or more years allows a reduction of the cost of such practice but favors the occurrence of alternate bearing.

Tests for several years showed that it is necessary to maintain a large leafy area of the plant, and pruning of medium intensity in shifts of two - three years, depending on the cultivar, gives the best results (Tombesi et al., 2002; Tombesi et al., 2007). These pruning cycles, compared with annual pruning, allow an increase in the production efficiency of the plant, and also a net saving of human work, since the execution times are not very different, and the number of cuts per plant is almost similar.

According to whether the mechanical harvest are adopted, with shakers or mechanization, in the first instance it is necessary to build the canopy towards erect forms, with shoots which are relatively short and rigid, to favor the transmission of vibrations up to the drupes in the whole volume of the canopy, while in the second case, the pruning will have the aim of bringing up the external wall of the canopy longer and pending fruiting shoots, to facilitate harvesting with pneumatic combs(De Simone & Tombesi, 2006; Tombesi et al., 2008).

Pruning must be carried out at the end of winter, before the restart of vegetative growth. It must be avoided after the harvest, because it reduces the cold resistance of the plants and does not allow wounds to heal, favoring diseases from fungi or other parasites. Traditionally in late summer a second pruning is performed on adult plants to eliminate suckers inside the plants, with special reference to the vase shape, where the formation of shoots within the canopy becomes a rule.

In profitable olive orchards pruning mechanization is essential to reduce management costs and regain timeliness in working, even though it penalizes the productive efficiency of trees. To balance the economic needs of management there are several ways to carry out mechanical pruning that should not be performed simultaneously on the entire plant. To ensure a good production of fruiting branches the canopy that remains after pruning should not excessively reduced. Mechanical pruning can be applied alternately in different years and/or rows by carrying out topping and hedging, reserving annual hand pruning to eliminate shoots and suckers.

Several experiments have been performed in different types of olive orchards in order to study the feasibility of mechanical pruning (Giametta & Zimbalatti, 1997; Ferguson et al., 1999; Ferguson et al., 2002; Peça et al., 2002; Tombesi et al, 2008; Dias et al., 2008; Farinelli et al., 2009). In our experience, the technical-economical convenience of pruning mechanization, also considering equipment integrated with pneumatic combs for olive harvesting, was evaluated. The results showed the good effectiveness of the pruning equipment in terms of cutting quality and working capacity (Pascuzzi et al., 2007; Toscano 2010a). Other trials have been carried out to assess the performances and the effects of mechanical pruning, that was performed both with toothed disks and scissor blades pruning machines. Both resulted in great efficacy and work productivity; nevertheless, the first ones are more efficient for woody vegetation, up to 10 cm diameter, while the second ones work better on thin branches, even though it can cut woody branches up to 5 cm diameter (unpublished data).

7. Harvesting

In olive grove management, the harvest is the other most expensive practice, together with pruning. Harvesting systems, can be considered rational only when they can reconcile the operation costs with the necessity to pick up the maximum yield in respect of the product quality. In the past, the availability of low cost manpower with a manual harvest allowed the satisfaction of these two demands, but the low availability and increase of the cost of labor, have made such operations excessively onerous, and applicable only to table olives.

Manual harvest can be improved using hand-held pneumatic combs to detach the olives from the plant that assuage the work, and give maximum flexibility in terms of harvesting time, and increase the operators productivity, but it is time-consuming and costly.

Mechanical harvesting is executed with shaker, also equipped with a reverse umbrella as an olive interceptor, that has considerable economic advantages compared with traditional manual picking procedures(Tombesi & Tombesi, 2007). In this way a great reduction in labour costs, harvesting timeliness and good performance, is achieved. Nevertheless it is difficult to apply in the majority of traditional olive groves due to the presence of malformed, voluminous plants, or those of an unsuitable cultivar.

In the new intensive olive groves, with trees optimized for cultivar and structure, mechanical harvest is instead applicable with positive results, usually in a step, getting up to 80-90% of yield (Hartmann & Reed, 1975; Ferguson et al., 1999; Giametta & Pipitone, 2004; Toscano & Casacchia, 2006). In these orchards with well pruned trees it is possible to harvest up to 50 trees/hour with a suitable shaker and collecting system (Lavee, 2010).

Using shakers, the more efficient harvesting yard is constituted by 5 or 7 operators, of which one operates the shaker, and the others the nets and the moving of olives, reaching a productivity up to 0,4 tons/man by hour (Briccoli et al., 2006; Tombesi, 2006; Toscano, 2007).

In super intensive olive orchard, or intensive olive orchard with trees structured in a productive wall, continuous harvesters , derived from grape pickers that work on both walls of a row (Bellomo et al., 2003; Arrivo et al., 2006) can be used. Wall pickers, that work on a single side of tree walls, also reach a working productivity up to 1 hectare by hour, with yield percentages similar to the shakers (Toscano, 2010b).

7.1. Ripening physiology of drupes

The olive tree fruits (Olea europaea L.) are oval or round drupes, of variable weight according to the cultivar, the yield, the nutritional and health state of the tree. The drupe is constituted by the external exocarp (peel), by the fleshy mesocarp (pulp) and by the internal stone (pit). Fruit development and ripening are a combination of biochemical and physiological changes that occurred during maturity of fruit. The development process is characterized by changes in size, weight, composition, color, flavor and physical proprieties of the fruit (Connor & Fereres, 2005) and is critical for final yield and oil quality. Oil accumulation, which occurs from pit hardening to harvesting may be early or late depending on the variety, generally it starts in the northern hemisphere from the month of August and continues up to November-December, subsequently the increase of oil content in the drupes is apparent being due to water reduction.

The maturation of olives also varies depending on the crop load, environmental conditions, which are subject to strong annual variations, soil moisture, and cultural practices. At harvest, within a tree, not all the fruit are at the same ripening stage, in fact this parameter also depends on the position of the fruits.

For the olives to be used for the oil extraction, the harvest must have been carried out at the beginning of the ripening phase of fruits, at veraison, when the pulp changes its color from green to purplish red. This stadium corresponds to the maximum oil yield per hectare, since, even if subsequently a slight increase in the oil content of the drupes is had, the loss for natural fruit drop undoes the advantage. In many instances the oil quality also decreases. An early harvest allows the production a good oil, rich in antioxidants and aromatic flavors, that confers resistance to oxidation, and a "fruited" taste. Instead, oils obtained from olives harvested at an advanced ripening stage are less intense, less bitter with a lower percentage of mono-unsaturated and saturated fatty acids and a higher percentage of poly-unsaturated fatty acids, that penalize its stability.

With regards to the harvesting method, oils with excellent quality can be produced with both manual and mechanical harvesting, as long as the drupes are intact and healthy.

For table olives harvesting is carried out manually from the plant, to avoid damage to the fruits that would consequently depreciate their market value. In some cases harvesting in the olive-grove is done in different steps as a consequence of the ripening scale, or performed at the same time and the sub-size fruits are sent to the crusher for oil extraction. To facilitate the harvest it is necessary that the plants are of a contained dimension, and with suitable forms that assemble the fruits on the outside of the canopy and on lean shoots.

In table cultivars for green fruit processing the harvest is done when the peel color changes to light green, that corresponds to the beginning of pulp softening and the maximum content of sugars, fundamental during tanning, for the fermentation process that follows sweetening, with lye or brine. For the olives destined for tanning to black, the harvest must be effected when the pulp is also colored, based on the physiological maturation of the fruits.

The most important parameters to determine the stage of maturation of the drupes are the fruit separation force and the development of natural fruit drop. Before the natural fall of olives there is an attenuation of the force with which they are attached to branches and shoots. These physiological changes do not occur simultaneously on all the drupes of the same tree but occur with a certain scaling. Therefore, the decrease of the attachment strength of the fruit and the drop of the first fruits are the most important indices of the final stage of ripening. These indices are easily determined and able to predict with sufficient reliability the time to start harvesting.

The optimal time of harvest can be, further, defined as one in which there is a high amount of fruit on the plant capable of being detached by the machines in considerable percentages and with a high content in good quality oil when ripening the pulp becomes less consistent (Farinelli et al., 2006).

8. Conclusions

The information contained in this paper highlights that it is possible to achieve some improvements in olive tree productivity (in terms of quantity and quality) and a reduction in costs, spreading more rational agronomic practices. Increasing the olive groves income can be achieved through updated cultivation techniques. All these must be coordinated and integrated with each other to obtain a rapid formation of the tree production structures that allow the maximum expression of their productive potential and provide a high level of mechanization. Soil management, plant structure, fertilizer, irrigation, pruning and mechanical harvesting must be chosen according to variety and environmental features. The paper provides useful indications on the introducing of the cover crop to better soil management in order to control erosion and maintain soil fertility.

Moreover, olive trees respond very strongly to irrigation and take advantage of very low volume of water also with regulated irrigation deficit. With regards to olive pruning, this cultural practice must be managed rationally based on the harvesting system and both these techniques (pruning and harvesting) must be done mechanically to reduce the running costs for better crop competitiveness.

Author details

Caterina Briccoli Bati, Elena Santilli and Pietro Toscano
Agricultural Research Council - Olive Growing and Oil Industry Research Centre, Rende (CS), Italy

Ilaria Guagliardi
National Research Council - Institute for Agricultural and Forest Systems in the Mediterranean, (ISAFOM), Rende (CS), Italy

Acknowledgement

Financial support for this study was provided by the Italian Ministry of Agriculture, Food and Forestry Policy through the project GERMOLI "Salvaguardia e valorizzazione del GERMoplasma OLIvicolo delle collezioni del CRA-OLI".

9. References

AA.VV. (1989). Paolo Sequi (a cura di), Chimica del suolo,. Pàtron (ed.),Bologna, Italy.

Agabbio, M. (1978). Influenza dell'intervento irriguo sul ciclo produttivo dell'olivo. Nota II: Influenza del regime idrico sulla biologia fiorale e sui caratteri morfo - qualitativi dei frutti. Studi Sassaresi. Sez. III. Vol. XXV: 266-272.

Angelakis, A.N.; Marecos do Monte, M.H.F.; Bontoux, L. & Asano, T.(1999). The status of wastewater reuse practice in the Mediterranean basin: need for guidelines. *Water Res.* 10: 2201–2217.

Aparicio, R. & Luna, G. (2002). Characterisation of monovarietal virgin olive oils. *Eur. J. Lipid Sci. Tech.* 104: 614–627.

Arrivo, A.; Bellomo, F. & D'Antonio, P. (2006). Raccolta meccanica dell'oliveto superintensivo. *Inf. Agrario*, 1, pp. 68-71.

Bacelar, E.A.; Santos, D.L.; Moutinho-Pereira, J.M.; Gonçalves, B.C.; Ferreira, H.F. & Correia, C.M. (2006). Immediate responses and adaptative strategies of three olive cultivars under contrasting water availability regimes: changes on structure and chemical composition of foliage and oxidative damage. *Plant Sci* 170: 596–605.

Bacelar, E.A.; Moutinho-Pereira, J.M., Gonçalves, B.C.; Ferreira, H.F. & Correia, C.M. (2007). Changes in growth, gas exchange, xylem hydraulic properties and water use efficiency of three olive cultivars under contrasting water availability regimes. *Environ. Exp. Bot.* 60: 183–192.

Basta, P.; Toscano, P.; Ciliberti, A. & Turco D. (2002). Osservazioni pluriennali sulle risposte vegeto-produttive di cinque cultivar di olivo irrigate con acqua salmastra a differenti concentrazioni. *Conv. Internaz. di Olivicoltura, Centenario I.S.Ol.* Spoleto, (PG) pp. 235-240.

Belsito A. et al., (1988). Chimica agraria. Zanichelli (Ed.) Bologna, Italy.

Bellomo, F.; D'Antonio, P. & D'Emilio, F. (2003). Spagna, quando l'olivicoltura è superintensiva. *Olivo e Olio*, 11: 8-13.

Ben Ahmed, C; Ben Rouinab, B.; Sensoyc, S.; Boukhrisa, M. & Ben Abdallah, F. (2009). Changes in gas exchange, proline accumulation and antioxidative enzyme activities in three olive cultivars under contrasting water availability regimes. *Environ. Exp. Bot.* 67: 345–352.

Ben-Gal, A.; Dag, A.; Yermiyahu, U.; Zipori, Y.; Presnov, E.; Faingold, I. & Kerem, Z. (2008). Evaluation of irrigation in a converted, rain fed olive orchard: the transition year. *Acta Hortic.* 792: 99–106.

Ben Mimoun, M.; Loumi O.; Ghrab, M.; Latiri, K. & Hellali, R. (2004). Foliar Potassium Application On Olive Tree M. *IPI regional workshop on Potassium and Fertigation development in West Asia and North Africa; Rabat, Morocco, 24-28 November, 2004.*

Bini, G.; Sapia, L.; Briccoli Bati, C. & Toscano P. (1997). Irrigazione dell'olivo con sistemi a microportata: risultati sulla cv Carolea in Calabria. *Inf. Agrario*, LIII, 22: 37-44.

Blank, H. (1932-1939). *Handbuch der Bodenlehre*. 7 volumes, Springer Verlag, Berlin.

Bonciarelli, F. (1981). *Agronomia*. Edagricole (Ed.), Bologna, Italy.

Bouat, A. (1968). Physiologie de l' olivier et analyse des feuilles. *Inf. Oleic. Int.* 42: 73-93.

Briccoli Bati, C.; Rinaldi, R. & Sirianni, T. (1992). Prime osservazioni sulla presenza di micorrize di tipo VAM in oliveti dell'Italia meridionale. In Atti *Giornate Scientifiche SOI-1992.* Ravello, Italy, pp. 46-47.

Briccoli Bati, C.; Tocci, C.; Toscano, P. & Gazea, F. (1995). Influenza di potatura ed irrigazione sullo stato nutrizionale dell'olivo. *Conv. "Tecniche, Norme e Qualità in olivicoltura".* Potenza, Italy, pp. 305-316.

Briccoli Bati, C.; Toscano, P.; Antonuccio, S. & Failla, O. (2002). Effetto dell'inerbimento sullo stato vegeto-produttivo e nutrizionale dell'olivo. *Inf. Agrario, anno LVIII.* Vol 31: 43-46.

Briccoli Bati, C.; Toscano, P. & Perrotta, M.L. (2006). La raccolta meccanica in diverse tipologie olivicole calabresi . *Conv. Naz. "Maturazione e raccolta delle olive: strategie e tecnologie per aumentare la competitività in olivicoltura".* Alanno (PE), Italy. pp. 95-99.

California Fertilizer Association, (1998). *Western Fertilizer Handbook*, Second horticultural edition, Sacramento, CA. pp. 362.

Calvente, R.; Cano, C.; Ferrol, N.; Azcón-Aguilar, C.& Barea, J.M. (2004). Analysing natural diversity of arbuscular mycorrhizal fungi in olive tree *(Olea europaea* L.) plantations and assessment of the effectiveness of native fungal isolates as inoculants for commercial cultivars of olive plantlets. *Applied Soil Ecology*, 26: 11-19.

Chalmers, D.J.; Mitchell, P.D. & Vanheek, L. (1981). Control of peach tree growth and productivity by regulated water supply, tree density and summer pruning. *J. Am. Soc. Hortic. Sci.* 106: 307–312.

Cimato, A.; Marranci, M. & Tattini, M. (1990). The use of foliar fertilization to modify sinks competition and to increase yield in olive (Olea europaea cv Frantoio). Acta Hort 286: 175-178.

Cimato, A.; Marranci, M.; Tattini, M. & Sani, G. (1991). La concimazione fogliare come mezzo per incrementare la produzione nell'olivo. *Inf. Agrar.* 14: 71-73.

Cimato, A.; Sani, G.; Marzi, L. & Marranci, M. (1994). Efficienza e qualità della produzione in olivo: riflessi della concimazione fogliare con urea. *Olivae* n. 54: 48-55.

Cimato, A.; Marranci, M.; Marzi, L. & Sani, G. (1995). Riflessi della concimazione fogliare sulla produttività dell'olivo. Atti conv. su: *"Tecniche, norme e qualità in olivicoltura"*. Potenza (Italy). pp. 137-148.

Connor, D.J. & Fereres, E. (2005). The physiology of adaptation and yield expression in olive. *Hortic. Rev.* 31: 155-229.

d'Andria, R.; Lavini, A.; Morelli, G.; Patumi, M.; Terenziani, S.; Calandrelli, D. & Fragnito, F. (2004). Effects of water regimes on five pickling and double aptitude olive cultivars (*Olea europaea* L.). *J. Hortic. Sci. Biotech.* 79: 18–25.

Dag, A; Ben-Gal, A; Yermiyahu, U.; Basheer, L.; Yogev, N. & Kerem, Z. (2008). The effect of irrigation level and harvest mechanization on virgin olive oil quality in a traditional rain-fed "Souri" olive orchard converted to irrigation. *J. Sci. Food Agric.* 88: 1524–1528.

Dale, J.E. (1988). The control of leaf expansion. *Annual Rev. Plant Physiol.* 39: 267–295.

De Simone, G. & Tombesi, A. (2006). Influenza delle caratteristiche strutturali della pianta di olivo sulle rese di raccolta con l'uso di vibratori di tronco. *Atti Conv. Naz. "Maturazione e raccolta delle olive: strategie e tecnologie per aumenrtare la competitività in olivicoltura"*. Alanno (PE), Italy pp. 63-68.

Dexter, A.R. (1988). Advances in characterization of soil structure. *Soil Tillage Research* 11: 199-239.

Dias, A.B.; Peça, J.; Pinheiro, A.; Santos, L., Morais, N. & Pereira, A.G. (2008). The influence of mechanical pruning on olive production and shaker efficiency. *Acta Horticulturae* 791: 307-313.

Dichio, B.; Romano, M.; Nuzzo, V. & Xiloyannis, C. (2002). Soil water availability and relationship between canopy and roots in young olive trees (cv. Coratina). *Acta Horticulturae*, 586: 255-258.

European Soil Bureau, (1999). *Database georeferenziato dei suoli europei. Manuale delle procedure.* ESB, Comitato Scientifico. Versione 1.1. Versione italiana. E. A.C. Costantini (ed.). JRC, Ispra (VA), pp.170.

Failla, O.; Scienza, A.; Stringari, G.; Porro, D.; Tarducci, S.; Bazzanti, N. & Toma, M. (1997). Diagnostica fogliare per l'olivicoltura toscana. *Informatore agrario*, n. 39: 63-71.

Farinelli, D.; Boco, M.; Ruffolo, M. & Tombesi, A. (2006). Ottimizzazione del periodo di raccolta per aspetti agronomici, meccanici e di qualità dell'olio. Atti Conv. Naz. "Maturazione e raccolta delle olive: strategie e tecnologie per aumentare la competitività in olivicoltura". Alanno (PE), Italy, pp. 175-182.

Farinelli, D.; Ruffolo, M. & Tombesi, S. (2009). Influenza della potatura meccanica sulla produzione e sulla qualità dell'olio di olivi allevati a vaso. *Rivista di Frutticoltura e di Ortofloricoltura* 11: 76-81.

Ferguson, L.; Heraclio, R. & Metheney, P. (1999). Mechanical harvesting and hedging of California black ripe *(Olea europea* L.) cv. 'Manzanillo' table olive. *Acta Hort.* 474: 193-196.

Ferguson, L.; Krueger, W.H.; Reyes, H. & Metheney, P. (2002). Effect of mechanical pruning on *(Olea europea* L.) California black ripe cv. 'Manzanillo' table olive. *Acta Hort.,* 586, pp. 281-284.

Fernández, J.E.; Moreno, F.; Cabrera, F.; Arrue, J.L. & Martin-Aranda, J. (1991). Drip irrigation, soil characteristics and the root distribution and root activity of olive trees. Plant Soil 133: 239-251.

Fernández-Escobar, R. (1999). Fertilización. In: Barranco, D., Fernández-Escobar, R., Rallo, L. (Eds.), El cultivo del olivo. Mundi-Prensa, Madrid, pp. 247-265.

Fontanazza, G. (1988). Olivicoltura la ripresa si fonda su concimazione ed irrigazione adeguate. Giornale di Agric. 15: 36-39.

Freeman, M.; Uriu, K. & Hartmann, H.T. (1994). Diagnosing and correcting nutrient problems. In. Ferguson, L., Sibbett, G.S., Martin, G.C. (Eds.) *Olive production manual.* University of California, USA, pubb. 3353: 77-86.

Giametta, G. & Zimbalatti, G. (1997). Mechanical pruning in new olive - groves. *J. Agr. Eng.Res.* 68: 15-20.

Giametta, G. & Pipitone, F. (2004). La raccolta delle olive. Innovazioni tecnologiche. *Conv. UNACOMA e Accademia Nazionale di Agricoltura: Innovazione tecnologica e ricerca per lo sviluppo di una moderna meccanizzazione agricola,* Bologna, Italy, pp. 79-93.

Giardini, L. (1986). Agronomia generale. Pàtron (Ed.), Bologna, Italy.

Giordano, A. (1999). Pedologia. UTET, Torino, Italy

Gisotti, G. (1988). Principi di geopedologia. Calderini, Bologna, Italy.

Goldhamer, D.; Dunai, J.; Ferguson, L.; Lavee, S. & Klein, I. (1994). Irrigation requirements of olives trees and responces to sustained deficit irrigation. *Acta Hort.* 356: 172-175.

Gómez-Rico, A.; Salvador, M.D.; La Greca, M. & Fregapane, G. (2006). Phenolic and volatile compounds of extra virgin olive oil *(Olea europaea* L. Cv. Cornicabra) with regards to fruit ripening and irrigation management. *Agric. Food Chem.* 54: 7130-7136.

Gómez-Rico, A.; Salvador, M.D.; Moriana, A.; Pérez, D.; Olmedilla, N.; Ribas, F. & Fregapane, G. (2007). Influence of Different Irrigation Strategies in a Traditional Cornicabra cv. Olive Orchard on Virgin Olive Oil Composition and Quality. *Food Chem.* 100: 568-578.

Gucci, R.; Lodolini, E.M. & Rapoport H.F. (2007). Productivity of olive trees with different water status and crop load. *J. of Hort. Sci. and Biotechnology,* 82(4): 648-656.

Hakansson, I.; Voorhees, W.B. & Riley, H. (1988). Weather and other environmental factors influencing crop responses to tillage and traffic. *Soil Tillage Research* 11: 239-282.

Hao, X.; Ball, B.C.; Culley, J.L.B.; Carter, M.R. & Parkin, G.W. (2008). Soil density and porosity. In: Carter, M.R. & Gregorich, E.G. (Eds.) *Soil sampling and methods of analysis.* CRC Press.

Hartmann, H.T. (1958). Some responses of the olive to nitrogen fertilzers. *Proc. Am. Soc. Hort. Sci.* 72: 257–266.

Hartmann, H.T. & Reed, W. (1975). Mechanical harvesting of olives. *California Agriculture, Jou.,* 75: 4-6.

Hendershot, W.H.; Lalande, H. & Duquette, M. (2008a). Ion exchange and exchangeable cations. In: Carter M.R. and Gregorich E.G. (Eds.) *Soil sampling and methods of analysis*. CRC Press.

Hendershot, W.H.; Lalande, H.& Duquette, M. (2008b). Soil reaction and exchangeable acidity. In: Carter M.R. and Gregorich E.G. (Eds.) *Soil sampling and methods of analysis*. CRC Press.

Inglese, P.; Barone, E. & Gullo, G. (1996). The effect of complementary irrigation on fruit growth, ripening pattern and oil characteristic of olive (*Olea europaea* L.) cv. Carolea. *J. Hort. Sci.* 71(2): 257-263.

Inglese, P.; Gullo, G. & Pace, L.S. (1999). Summer drought effects on fruit growth, ripening and accumulation and composition of 'Carolea' olive oil. *Acta Hortic*. 474: 269–273.

Inglese, P.; Gullo, G. & Pace,L.S. (2002). Fruit growth and olive quality in relation to foliar nutrition and time of application. *Acta Hort*. 586: 507-509.

Jeffries, P. & Barea, J.M. (2001). Arbuscular mycorrhiza—a key component of sustainable plant-soil ecosystems. In: Hock, B. (Ed.) *The Mycota IX Fungal Associations*. Springer, Berlin, pp. 95–113.

Jones, J.B. (2003). *Agronomic handbook*. CRC Press.

Kabata-Pendias, A. & Mukherjee, A.B. (2007). *Trace elements from soil to human*. Springer-Verlag, Berlin Heidelberg.

Kay, B. (1990). Rates of change of soil structure under different cropping systems. *Adv. Soil Sci*. 12: 1-41.

Lavee, S. & Woodner, M. (1991). Factor affecting the nature of oil accumulation in fruit of olive (*Olea europaea* L.) cultivars. *J. of Hort. Sci*. 66: 583-591.

Lavee, S. (2006). Biennial bearing in olive (*Olea europaea* L.). In FAO (Ed.) *Olea*. FAOOlive Netw 25: 5–13.

Lavee, S.; Hanoch, E.; Wodner, M. & Abramowitch, H. (2007). The effect of predetermined water deficit on the performance of cv. Muhasan olives (*Olea europaea* L.) in the eastern coastal plain of Israel. *Scientia Hort*. 112: 156–163.

Lavee, S. (2010). Integrated mechanical, chemical and horticultural methodologies for harvesting of oil olives and the potential interaction with different growing systems. A general review. *Avd. Hort. Sci*. 24(1): 5-15.

Lovato, P.E.; Gianinazzi-Pearson, V.; Trouvelot, A. & Gianinazzi, S. (1996). The state of art of mycorrhizas and micropropagation. *Adv. Hortic. Sci*. 10, 46–52.

Massoud, M.A.; Scrimshaw, M.D. & Lester, J.N. (2003). Qualitative assessment cf the effectiveness of the Mediterranean action plan: wastewater management in the Mediterranean region. *Ocean Coast. Manage*. 46: 875–899.

Michelakis, N.I.C.; Vouyoucalou, E. & Clapaki, G. (1994). Plant growth and yield response of the olive tree cv Kalamon, to different levels of soil water potential and methcds of irrigation. *Acta Hortic*. 356: 205–210.

Miller, J.J. & Curtin, D. (2008). Electrical conductivity and soluble ions. In Carter, M.R. & Gregorich, E.G. (Eds), 2008. *Soil sampling and methods of analysis*. CRC Press.

Moriana, A.; Pérez-López, D.; Gómez-Rico, A.; de los Desamparados Salvador, M.; Olmedilla, N.; Ribas, F. & Fregapane, G. (2007). Irrigation scheduling for traditional, low-density olive orchards: water relations and influence on oil characteristics. *Agric. Water Manage.* 87: 171–179.

Natali, S. (1993). Cure colturali dell'oliveto. *Inform. Agrar.* n.23: 67-71.

Nuzzo, V.; Xiloyannis, C.; Dichio, B.; Montanaro, G. & Celano, G. (1997). Growth and yield in irrigated and non-irrigated olive trees cultivar Coratina over four years after planting. *Acta Hortic.*, 449, pp. 75–82.

Palese, A.M.; Celano, G.; Dichio, B.; Nuzzo, V. & Xiloyannis, C. (1997). Esigenze nutrizionali dell'olivo in fase di allevamento. *Informatore Agrario*, 44, pp. 35-40.

Pandolfi, A.; Proietti, P.; Famiani, F.; Guelfi, P.; Farinelli, D.; Sisani, G.& Tombesi A. (2000). Effetti del tipo di gestione del suolo e dell'irrigazione sull'attività vegeto produttiva dell'olivo in centro Italia. *V giornate scientifiche SOI 2000*, Sirmione, Italy pp. 329-330.

Pannelli, G.; Selvaggini, R.; Servili M.; Baldioli M. & Montedoro, G.F.(1996). La produzione e la composizione dell'olio in relazione alla fisiologia dello stress idrico in olivo (*Olea europaea* L.) cv Leccino. In: Atti convegno *'Olivicoltura mediterranea: stato e prospettive della coltura e della ricerca'*. Cosenza, Italy, pp. 701-721.

Pascuzzi, S.; Toscano, P. & Guarella, A. (2007). La meccanizzazione della potatura di riforma degli oliveti - Atti Conv. Naz. - *Tecnologie innovative nelle filiere:orticola, vitivinicola, e olivicola-olearia* – Pisa-Volterra, Italy, Vol. III: 77-80

Pastor, M. (1989). Incidenza delle male erbe sull'evoluzione del contenuto d'acqua del suolo in oliveto di seccagno. *Olivae* 28: 32-37.

Patumi, M.; D'Andria, R.; Fontanazza, G.; Morelli, J.; Giorgo, P. & Sorrentino, G. (1999). Yield and oil quality on intensively trained trees of three cultivars of olive (*Olea europea* L.) under different irrigation regimes. *J. Hortic Sci Biotechnol.* 74: 729-737.

Patumi, M.; D'Andria, R.; Marsilo, V.; Fontanazza, G.; Morelli, G. & Lanza B. (2002). Olive and olive oil quality after intensive monocone olive growing (*Olea europaea* L., cv. Kalamata) in different irrigation regimes. *Food Chem.*, 77: 27–34.

Peça, J.O.; Dias, A.B.; Pinheiro, A.C.; Santos, L.; Morais, N.; Pereira, A.G. & De Souza, D.R. (2002). Mechanical pruning of olive trees as an alternative to manual pruning. *Acta Hort.* 586: 295-299.

Pefica, S.; Bellaloui, N.; Greve, C.; Hu, H. & Brown, P.H. (2001): Boron Transport and Soluble Carbohydrate Concentrations in Olive. *J. Amer. Soc. Hort. Sci.* 126 (3): 291-296.

Perica, S.; Goreta, S. & Selak, G.V. (2008). Growth, biomass al location and leaf ion concentration of seven olive (*Olea europaea* L.) cultivars under increased salinity. *Scientia Hort.*, 117, pp. 124-129.

Petruccioli, G. & Parlati, M. V. (1983). La concimazione dell'olivo. *Terra e vita. anno XXIV*, 3: 55 - 59.

Raglione, M.; Toscano, P.; Angelini, R.; Briccoli Bati, C.; Spadoni, M.; De Simone, C. & Lorenzoni, P. (2000). Olive yield and soil loss in hilly environment of Calabia (Southern Italy). Influence of permanent cover crop and ploughing. *VI International meeting on soils with mediterranean type of climate (IMSMTC)*. Barcellona (Catalonia), 4-9 Jul 1999, pp. 1038-1040.

Ranalli, A.; De Mattia, G.; Ferrante, M.L. & Giansante, L. (1997). Incidence of olive cultivation area on the analytical characteristics of the oil. Note 1. *Riv. Ital. Sost. Grasse* 74: 501–508.

Rasiah, V. & Kay, B.D. (1995). Runoff and soil loss as influenced by selected stability parameters and cropping and tillage practices. *Geoderma*, 68: 321-329.

Roldán-Fajardo, B.E. & Barea, J.M. (1985). Mycorrhizal dependency in the olive tree (*Olea europaea* L.). In: Gianinazzi-Pearson, V., Gianinazzi, S. (Eds.) *Physiological and Genetical Aspects of Mycorrhizal*. INRA, Paris, pp. 323–326.

Servili, M.; Esposto, S.; Lodolini, E.; Selvaggini, R.; Taticchi, A.; Urbani, S.; Montedoro, G.; Serravalle, M. & Gucci, R. (2007). Irrigation effects on quality, phenolic composition and selected volatiles of virgin olive oils cv Leccino. *J. Agric. Food Chem.* 55: 6609–6618.

Soyergin, S.; Moltay, I. ;Genç, Ç.; Fidan, A.E. & Sütçü, A.R. (2000). Nutrient status of olives grown in the Marmara region. *Acta Horticulturae* 586: 375-380.

Stein, O.R.; Neibling, W.H.; Logan, T.J. & Moldenhauer, W.C. (1986). Runoff and soil loss as influenced by tillage and residue cover. *Soil Sci. Soc. Am. J.* 50: 1527-1531.

Tognetti, R.; D'Andria, R.; Morelli, G. & Alvino, A. (2005). The effect of deficit irrigation on seasonal variations of plant water use in *Olea europaea* L. *Plant Soil* 273: 139-155.

Tombesi, A.; Michelakis, N. & Pastor, M. (1996). Raccomandazioni del gruppo di lavoro di tecniche di produzione in olivicoltura e produttività. *Olivae*, 63: 38-51.

Tombesi, A.; Pilli, M. & Boco, M. (2002). Prime valutazioni su intensità e periodicità della potatura nell'olivo. *Rivista di Frutticoltura* 10:71-76.

Tombesi, A. 2006. Planting systems, canopy management and mechanical harvesting. *OliveBioTecq, Recent advances in olive industry*. Mazara del Vallo, Italy: 307-316.

Tombesi, A.; Guarella, P.; Di Vaio, C. & Toscano, P. (2007). Innovazioni nella meccanizzazione della raccolta e della potatura e ristrutturazione degli impianti olivicoli. *Atti Convegno internazionale Ricerca ed innovazione per la filiera olivicolo-olearia dei Paesi del Mediterraneo*. AGRILEVANTE, Bari. Pgt RIOM, Tomo I, Olivicoltura, Guido Ed. Rende,pp, 175-190.

Tombesi, A.; Farinelli, D.; Ruffolo, M. & Sforna, A. (2008). Primi risultati sulla meccanizzazione agevolata ed integrale della potatura dell'olivo. In: *Atti Con. COM.SI.OL "Competività del Sistema Olivo in Italia"*, Spoleto, Italy pp. 141-147.

Tombesi, A. & Tombesi, S. (2007). Olive harvesting and mechanization. In International Olive Council (Ed.) *"Production Techniques in Olive Growing"*. Madrid, Spain pp. 317-346.

Toscano, P. (1998). MANOLI: *Ipermanuale di olivicoltura*. Progetto IIRA (Ipertesti, Internet e Ricerca Agraria), INEA(Ed.), Roma, Italy.

Toscano, P.; Briccoli Bati, C. & Trombino, T. (1999). Grass-cover effects on the vegetative and productive state of a young hilly olive-grove. *Acta Horticolturae*, 474: 181-184.

Toscano, P. (2000). Gestione del suolo: Ciclo delle lavorazioni, inerbimenti e controllo della flora spontanea degli oliveti. *Conv. Rassegna olio qualità ambiente. Metodi e sistemi innovativi dell'olivicoltura biologica e sostenibile. Stato della ricerca e della sperimentazione*. ISO Rende, Italy pp. 73-83.

Toscano, P.; Briccoli Bati, C. & Sirianni, R. (2000). Tecnica colturale: effetti delle concimazioni fogliari in olivicoltura. *V Giornate Scientifiche S.O.I., Vol II*, Sirmione (BS) Italy pp. 583-584.

Toscano, P.; Madeo, A. & Godino, G. (2002a). Risposte di tre cultivar di olivo a differenti tipi di concimazione in un ambiente litoraneo del Sud Italia. *Conv. Internaz. di Olivicoltura. Centenario I.S.Ol. Spoleto* (PG), Italy, pp. 310-315.

Toscano, P.; Godino, G.; Belfiore, T. & Briccoli Bati C. (2002b). Foliar fertilization: a valid alternative for olive cultivar. *Acta Horticolturae*, 594, pp. 191-195.

Toscano, P.; Briccoli Bati, C.; Godino, G.; De Simone, C.; Raglione, M.; Lorenzoni, P.; Angelini, R. & Antonuccio, S. (2004a). Effetti Agronomici e pedologici di due diverse tecniche di gestione del suolo in un oliveto collinare del Meridione d'Italia. *Olivae* 102: 21-26.

Toscano, P.; Scazziota, B. & Briccoli-Bati, C. (2004b). Strategie di gestione eco-compatibile per il recupero di competitività in sistemi olivicoli marginali: esperienze sull'inerbimento in suolo declive ed effetti sugli aspetti vegeto-produttivi su cv Carolea. In atti Conv. *"Il futuro dei sistemi olivicoli in aree marginali: aspetti socio-economici, gestione delle risorse naturali e produzioni di qualità"*. Matera, Italy, pp. 399-407

Toscano, P. (2005). Foliar fertilization on olive growing: first results of a specifics foliar fertilizers applied on some cultivars in different environments of Southern Italy. *FAO Land And Water Bulletin*, 10, pp. 127-130.

Toscano, P. & Casacchia, T. (2006). La raccolta meccanica dell'olivo: effetti dell'Olive HarvestVant su Leccino in Calabria. In I.T.A.S. Cuppari (Ed.). Atti Conv. Naz. *Maturazione e raccolta delle olive: strategie e tecnologie per aumentare la competitività in olivicoltura.* - - Alanno (PE), Italy,pp. 101-105.

Toscano, P.; Casacchia, T.; Briccoli Bati, C. & Scazziota, B. (2006). Inerbimento temporaneo e permanente in olivicoltura. *Inf. Agrario*, 21, pp. 36-39.

Toscano, P. (2007). Come organizzare i cantieri per l'olivicoltura calabrese. *Olivo&Olio* 11-12/07:60-64.

Toscano, P. (2008). Concimazione fogliare insieme a quella al terreno. *Olivo&Olio* 3: 50-53.

Toscano, P. (2009). The controlled grass-cover soil management in southern Italy olive orchard environments. *18th Symposium Of The International Scientific Centre Of Fertilizers - More Sustainability In Agriculture: New Fertilizers And Fertilization Management*. Rome, Italy, pp. 381-386.

Toscano, P. (2010a). Potatura meccanica solo ad anni alterni. *Olivo&Olio*, 3: 44-48.

Toscano, P. (2010b). Raccolta delle olive efficiente con macchine e cantieri adeguati. *Inf. Agrario*, 3: 82-84.

Toscano, P. & Godino, G. (2010). Concimare le foglie per aumentare la resa. *Olivo&Olio*, 3: 40-43.

Von, A. (1998). State of commercial use of AMF-inoculum in Germany. In: Gianinazzi, S.; Schuepp, H. (eds.) *Arbuscular mycorrhizas in sustainable soil–plant systems*. Report of 1997 activities. Iceland: Cost Action 8.21, pp. 153.

Wollny, E. (1898). Untersuchungen über den Einflub der mechanischen Bearbeitung auf die Fruchtbarkeit des Bodens. *Forsch. Geb. Agrik. Phys.* 20: 231-290.

Xiloyannis, C. & Palese, A.M. (2001). Efficienza dell'uso dell'acqua nella coltivazione dell'olivo. In COI; Regione Campania; CNR Ist. Irrigazione (Eds) *"Gestione dell'acqua e del territorio per un olivicoltura sostenibile"*, Corso Internazionale di aggiornamento tecnico scientifico, Napoli, Italy.

Xiloyannis, C. & Dichio, B. (2006). L'irrigazione sostenibile in frutticoltura. *Ital. J. Agron.* 3: 507-517.

Chapter 5

Susceptibility of Cultivars to Biotic Stresses

Nino Iannotta and Stefano Scalercio

Additional information is available at the end of the chapter

1. Introduction

The Italian history of crop protection has been dominated by the use of agrochemical, generally having a negative impact on the environment and residues in final products. The approach based on the use of genotypes with low susceptibility to biotic stresses searching for sources of resistances was largely uninvestigated for long time. Only recently researchers focussed their studies on this subject. Studies devoted to this subject are very difficult mainly for permanent crop such as the olive tree because time consuming and because it is very hard to compare the behaviour of a large germplasm in the same pedoclimatic condition. Furthermore, it is very hard to design tests of resistances under controlled conditions because of the difficulty due to the hard tolerance of captivity of pests such as olive fly and many others. In any case, recently research activities on this field has been greatly improved and several research projects are supported by many institutions.

On the other hand, abiotic stresses has been sufficiently studied, mainly for those concerning the cold tolerance. The CRA-OLI of Rende studied from fifteen years the behaviour of several cultivars in respect to the main biotic stresses utilising two large varietal collections planted in two experimental fields, the largest on the ionian coast of Calabria (Mirto-Crosia, Cosenza) and the smallest near Rende (Cosenza). Results showed a high behavioural diversity of observed cultivars showing different degree of susceptbility to main pathogens and phytofagous of the olive tree. Some cultivars has been deeply investigated searching for the mechanisms determining such behavioural differences.

2. Susceptibility to biotic stresses

Many studies reported in this paragraph have been carried out in two experimental olive groves 20-years old located in Calabria, South Italy, where hundreds of cultivars coming from around the world grow in the same pedoclimatic conditions and permitted us to produce useful data for comparing susceptibility to biotic stresses.

2.1. Susceptibility to *Bactrocera oleae*

It is well known that cultivars have a differentiated susceptibility to olive fly infestations. In this paragraph are chronologically arranged some of papers devoted to the comparison of susceptibility of cultivars to the major insect pest of the olive. The cultivars Carboncella di Pianacce, Gentile, Bardhi i Tirana, Kokermadh i Berat and Nociara showed an infestation level significantly lower (less than 10%) than the cultivars Carolea, Cassanese, Cucco, Giarraffa, Intosso, Kalinjot, Nocellara del Belice, Picholine and Santa Caterina (more than 20 %) during two years of observations (1997-1998) (Iannotta et al., 2001) (Table 1). Other cultivars investigated by the same authors (Dritta di Moscufo, Leccino, Maiatica di Ferrandina and Mixan) showed intermediate infestation levels. Among these cultivars has been observed that cultivars showing a low infestation level had a higher percentage of sterile oviposition stings.

Iannotta et al. (2001) also underlined that the cultivars having a high amount of oleuropein in the pulp of drupes are those with the lowest level of infestation in the period considered optimal for the harvesting (end of October-beginning of November) (Table 1). In fact, the cultivars Carboncella di Pianacce, Gentile di Chieti, Bardhi i Tirana, Kokermadh i Berat and Nociara have an amount of oleuropein higher of 30g/kg of fresh pulp, while the cultivars with higher infestation level have an amount of oleuropein lower than 20g/kg of fresh pulp. From these data emerge a correlation between olive fly infestation and oleuropein content of drupes (Fig. 1).

Cultivar	Sterile Oviposition Stings (%)		Infestation (%)		Oleuropein (g/Kg f.p.)
Bhardi i Tirana	27.5	C	8.5	A	36.60
Carboncella di Pianacce	26.0	C	9.5	A	34.09
Carolea	13.0	AB	22.5	CDE	20.21
Cassanese	17.0	ABC	24.5	E	16.12
Cucco	23.5	ABC	26.5	E	19.11
Dritta di Moscufo	26.0	C	11.0	AB	18.51
Gentile di Chieti	26.0	C	9.5	A	31.37
Giarraffa	16.5	AB	23.5	DE	10.81
Intosso	18.0	ABC	31.0	E	24.04
Kalinjot	12.5	A	23.0	DE	9.29
Kokermadh i Berat	26.5	C	10.0	AB	31.18
Leccino	21.5	ABC	20.0	BCDE	29.01
Maiatica di Ferrandina	19.5	ABC	12.5	ABCD	27.88
Mixan	19.0	ABC	11.5	ABC	19.80
Nocellara del Belice	18.5	ABC	23.0	DE	16.47
Nociara	25.0	BC	9.5	A	32.73
Picholine	17.0	ABC	24.0	E	19.58
Santa Caterina	16.5	ABC	23.5	DE	16.83

Table 1. Percentages of olive fly infestation showed by olive cultivars growing in the same pedoclimatic conditions. Letters indicate significant statistical differences (P<0.01; ANOVA test) (from Iannotta et al., 2001, modified).

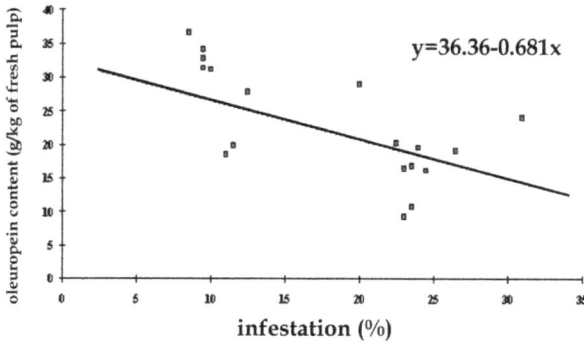

Figure 1. Correlation between oleuropein content of drupes and olive fly infestation (from Iannotta et al., 2001, modified).

Studies carried out in Sicily (Iannotta et al., 2002) highlighted the low susceptibility to olive fly infestations of the cv. Turdunazza antimosca compared to cvs. Tonda Iblea, Moresca and Verdese in the same olive grove. In this case, the low suceptibility of the cultivar Turdunazza antimosca seems to be related to the repellent action against the olive fly of that cultivar.

Iannotta et al. (2006a) observed a very low active infestation (percentage of drupes with living stages of the olive fly, such as eggs, larvae or pupae) rate along the ripening season for cvs. Cellina di Nardò and Cima di Mola, strongly reducing drupes damages and avoiding any kind of pesticide use for the production of a high quality olive oil (Table 2). The cv. Ogliarola del Vulture showed a low infestation level only until the end of October and this lead to an anticipated harvesting for producing a high quality olive oil without field pesticide applications. The cv. Leccino registered a high infestation only during the first period of the olive ripening, while during October and November infestation was lower than the 20% (Table 2), i.e. tollerable for obtaining high quality olive oil. Other cultivars, mainly cv. Maurino, Moraiolo and Grossa di Spagna, showed a high susceptibility to olive fly attacks, showing active infestation levels higher than the 20%. For last cultivars the field applications of pesticides are needed. Percentages of sterile oviposition stings were higher at the end of September for cultivars with low susceptibility, when the content of oleuropein within drupes is the highest of the ripening season. These results confirmed previous observations that demonstrated the role of oleuropein for increasing the mortality of eggs and reducing the hatching rate of young larvae (Iannotta et al., 2002).

These data underline that increasing studies on cultivar susceptibility could effectively produce significant results under a perspective of olive growing sustainability

Iannotta et al. (2006b) obtained results concerning 16 Italian cultivars (Table 3), displaying data concerning different development stages of the olive fly, active and total (active infestation plus exit holes of adults) infestation levels, detected amounts of oleuropein and cyanidine and weight of 100 drupes.

Cultivar	First ripening period (%)		Second ripening period (%)		Third ripening period (%)	
Cellina di Nardò	16,50	D	6,50	D	14,50	CD
Cima di Mola	22,00	BCD	15,50	CD	13,00	D
Coratina	49,00	ABC	79,50	A	51,50	ABCD
Dolce Agogia	41,00	ABCD	20,50	CD	23,00	BCD
Frantoio	39,50	ABCD	35,00	BCD	45,00	ABCD
Grossa di Spagna	21,50	CD	16,50	CD	74,50	AB
Leccino	44,50	ABCD	15,00	CD	17,00	CD
Maurino	26,00	BCD	38,50	BCD	77,50	A
Moraiolo	56,00	A	49,50	ABC	65,00	ABC
Ogliarola Barese	32,00	ABCD	41,00	BCD	38,50	ABCD
Ogliarola Vulture	25,00	BCD	16,50	CD	71,00	AB
Peranzana	51,00	AB	66,00	AB	70,50	AB
Pisciottana	24,00	BCD	43,50	BC	53,00	ABCD

Table 2. Active infestation trend. Letters indicate significant statistical differences (P<0.01; ANOVA test), from Iannotta et al. (2006a), modified.

The same table shows that cvs. Ascolana tenera and Nostrana di Brisighella turn out to be significatively the less infested cultivars, both for active and total infestation in all observed ripening times. Cellina di Nardò shows the lowest susceptibility to olive fly attacks (9.83% of active infestation and 17.67% of total infestation). Also cvs. Nera di Cantinelle, Frantoio, Tonda di Strongoli, Nolca, Cima di Melfi and Termite di Bitetto exhibit low susceptibility (active infestation lower than 13%), while cvs. Dolce Agogia, Dolce di Rossano, Nostrale di Fiano Romano, Ogliarola del Bradano, Ogliarola garganica and Ogliarola del Vulture display intermediate susceptibility values. The active and total infestations, obtained as mean values for any cultivars concerning the different ripening times, turn out to be increasing during the season (Table 4).

The results obtained for 9 non-Italian cultivars show cvs. Gordal sevillana and Hojiblanca with the lowest level of active infestation (<15%), cv. Konservolia the most infested and cvs. Arbequina, Kalamata, Koroneiki, Lucques, Manzanilla and Picual register intermediate infestation percentages (Table 5).

The results of Iannotta et al. (2006b) confirm a different behavior of olive cultivars concerning their susceptibility to olive fly attack. Investigated genotypes in the area of observation displayed a contained percentage of attack lower than 13% in the Italian cvs. Cellina di Nardò, Nera di Cantinelle, Frantoio, Tonda di Strongoli, Nolca, Cima di Melfi and Termite di Bitetto and lower than 15% in non-Italian cvs. Gordal sevillana and Hojiblanca.

Active percentage within the limit of 15% is compatible with a high qualitative product (olive oil) avoiding the use of pesticides. It allows the achievement of the fixed aims, consisting in ecosystem and biocoenotic balances safeguard, which make economically positive the ecocultivation (organic and integrated farming).

Weight 100 drupes (g)	Eggs Fertile (Aborted)	Sterile oviposition stings	Larvae (pupae)	Exit holes	Active (total)	Oleuropein (ppm)	Cyanidine (ppm)
Ascolana tenera							
676.1	19.0 (2.5)	10.0	7.5 (3.5)	5.5	31.0 (49.0)	7974.1	0.0
725.3	10.0 (2.0)	8.0	12.0 (6.0)	15.0	30.0 (55.0)	3595.4	17.1
842.7	13.5 (2.5)	7.5	24.5 (4.0)	13.0	42.0 (65.0)	1419.0	129.3
Cellina di Nardò							
125.7	3.5 (1.0)	4.5	0.0 (0.0)	0.0	3.5 (9.0)	8165.2	79 0
131.1	5.5 (0.5)	5.0	2.5 (1.0)	0.0	9.0 (14.5)	1482.1	1019.7
143.3	2.0 (2.5)	3.5	12.5 (2.5)	6.5	17.0 (29.5)	0.0	1690.0
Cima di Melfi							
247.4	9.5 (1.0)	8.5	0.0 (2.0)	3.0	11.5 (24.0)	-	-
271.0	11.5 (5.0)	17.0	0.0 (0.0)	4.5	11.5 (38.0)	2469.8	17 4
260.8	5.5 (4.0)	10.0	4.5 (3.5)	6.0	13.5 (33.5)	0.0	9.9
Dolce Agogia							
181.2	8.0 (2.5)	7.5	4.5 (0.0)	2.0	12.5 (24.5)	8919.3	27.3
209.5	3.0 (0.5)	11.0	16.0 (2.5)	7.5	26.5 (45.5)	4889.6	0.0
227.2	7.0 (1.0)	16.5	2.5 (7.5)	17.5	17.0 (52.0)	42.9	0.0
Dolce di Rossano							
133.7	7.0 (2.0)	13.5	3.0 (2.0)	0.0	12.0 (27.5)	11860.3	0.0
186.6	4.0 (0.5)	6.5	8.0 (0.5)	3.5	12.5 (23.0)	681.7	96.7
175.5	7.5 (2.0)	5.5	15.5 (4.5)	14.5	27.5 (49.5)	149.2	503.1
Frantoio							
157.6	6.5 (3.0)	9.0	0.0 (0.0)	0.0	6.5 (18.5)	7831.2	12.7
175.3	6.0 (0.0)	9.5	3.0 (0.0)	1.0	9.0 (19.5)	1292.0	0.0
221.4	5.0 (0.5)	3.5	7.5 (4.5)	2.0	17.0 (23.0)	561.4	0.0
Moraiolo							
163.6	8.0 (2.5)	5.5	3.5 (0.0)	0.5	11.5 (20.0)	11810.7	0.0
183.0	13.0 (4.0)	5.5	0.5 (3.0)	1.0	16.5 (27.0)	5162.9	33.9
231.2	5.5 (0.5)	8.5	12.0 (4.5)	5.0	22.5 (36.5)	1771.3	337.6
Nera di Cantinelle							
226.2	3.0 (0.0)	9.0	1.5 (0.0)	0.5	4.5 (14.0)	8727.3	10.5
253.3	4.5 (0.0)	11.0	9.0 (0.0)	6.5	13.5 (31.0)	666.9	56.6

Weight 100 drupes (g)	Eggs Fertile (Aborted)	Sterile oviposition stings	Larvae (pupae)	Exit holes	Infestation Active (total)	Oleuropein (ppm)	Cyanidine (ppm)
270.9	4.0 (0.0)	6.5	7.5 (2.5)	16.5	14.0 (37.0)	39.1	210.1
Nolca							
250.2	1.0 (0.0)	13.0	2.5 (0.5)	2.5	4.0 (19.5)	155.9	91.1
327.0	2.0 (2.0)	6.5	3.5 (6.0)	7.0	11.5 (27.0)	0.0	601.8
325.1	2.0 (2.0)	8.0	13.5 (5.0)	10.0	21.0 (41.0)	0.0	1120.3
Nostrale di Fiano Romano							
255.5	10.5 (2.0)	10.5	3.5 (3.5)	12.0	17.5 (42.0)	3174.0	34.0
262.4	7.0 (1.5)	8.0	6.0 (3.5)	8.5	19.0 (37.0)	463.4	25.4
342.3	7.5 (0.0)	16.0	3.0 (4.5)	15.5	15.0 (46.5)	93.8	13.8
Nostrana di Brisighella							
481.7	13.5 (2.5)	10.5	12.5 (8.0)	5.0	36.0 (54.0)	6444.0	0.0
801.2	3.5 (0.5)	0.0	5.0 (6.5)	19.5	22.0 (42.0)	5002.3	17.2
674.9	15.5 (2.0)	6.5	15.5 (8.0)	29.5	39.0 (77.0)	454.0	0.0
Ogliarola del Bradano							
153.9	10.0 (1.0)	6.0	2.5 (1.0)	1.5	13.5 (22.0)	3830.2	0.0
171.6	13.5 (3.5)	6.0	4.5 (2.0)	3.0	20.0 (32.5)	2238.0	0.0
235.4	0.0 (0.0)	5.0	11.0 (5.0)	7.5	16.0 (28.5)	88.8	0.0
Ogliarola Garganica							
261.2	8.5 (4.5)	16.5	1.0 (0.0)	1.0	9.5 (31.5)	5387.2	71.6
238.2	8.5 (1.0)	13.5	1.5 (3.5)	5.0	13.5 (33.0)	641.6	46.3
250.6	2.5 (1.0)	4.0	10.0 (5.0)	13.5	20.5 (39.0)	444.0	705.6
Ogliarola del Vulture							
195.4	4.5 (2.0)	11.5	1.5 (0.0)	2.5	6.0 (22.0)	9082.3	36.9
188.8	3.5 (1.0)	6.0	7.0 (2.5)	3.0	13.0 (23.0)	6669.3	21.4
265.4	12.0 (3.5)	21.0	12.5 (3.0)	13.5	27.5 (65.5)	588.0	0.0
Termite di Bitetto							
228.0	4.5 (0.0)	8.0	4.5 (0.0)	1.5	9.0 (18.5)	1654.4	34.0
268.8	2.0 (0.0)	6.5	10.0 (0.0)	14.0	12.0 (32.5)	345.6	513.5
382.9	2.5 (1.0)	11.0	5.5 (5.5)	10.0	17.5 (39.5)	229.4	1647.1
Tonda di Strongoli							
389.7	7.5 (1.0)	7.0	1.0 (1.0)	2.0	9.5 (19.5)	8299.6	0.0
448.7	8.5 (0.0)	11.0	3.0 (0.0)	1.0	11.5 (23.5)	4153.3	0.0
453.3	2.5 (0.5)	6.5	7.0 (4.0)	10.5	13.5 (31.0)	557.1	15.5

Table 3. Detailed data obtained in the different theses and ripening times concerning olive fly infestation and oleuropein and cyanidine drupe contents. For any cultivar, the observation was performed in three different ripening times (from Iannotta et al., 2006b, modified).

Ripening time	Active infestation	Tukey test	Total infestation	Tukey test	
26 Sept. 2006	12.38	B	28.81	c	C
26 Oct. 2006	15.69	B	35.88	b	B
29 Nov. 2006	21.28	A	49.06	a	A

Table 4. Data concerning the comparison among the different investigated ripening times referred to Italian cultivars (from Iannotta et al., 2006b, modified). Letters indicate significant statistical differences (Capital letters: P<0.01; small letters: P<0.05; ANOVA test).

Weight drupes (g)	Eggs Fertile (Aborted)	Sterile oviposition stings	Larvae (pupae)	Emergency holes	Infestation Active (total)	Oleuropein (ppm)	Cyanidine (ppm)
Arbequina							
149.6	11.5 (1.0)	9.5	2.5 (0.0)	1.0	14.0 (25.5)	508.9	11.5
195.1	6.0 (1.0)	4.5	9.5 (1.5)	2.5	18.0 (26.0)	101.8	23.2
191.0	5.5 (0.0)	5.0	7.5 (3.0)	14.0	16.0 (35.0)	1956.4	0.0
Gordal sevillana							
506.9	10.0 (0.5)	12.0	0.0 (0.0)	0.0	10.0 (22.5)	3994.2	17.6
803.6	10.5 (1.0)	7.5	4.0 (1.5)	4.0	16.0 (28.5)	1764.8	77.7
922.3	4.5 (2.0)	9.5	10 (2.5)	7.0	17.0 (35.5)	35.4	140.7
Hojiblanca							
			1.5 (0.0) 15.0				
219.6	2.0 (3.0)	11.5	(2.0)	0.5	3.5 (18.5)	5808.2	0 0
290.8	3.5 (0.0)	8.0	15.0	5.0	20.5 (33.5)	5424.5	187.5
322.6	4.5 (0.0)	5.0	(1.5)	18.0	21.0 (44.0)	-	-
Kalamata							
			2.0 (0.0)				
197.9	11.5 (3.5)	14.0	6.5 (3.5)	1.0	13.5 (32.0)	10571.0	0.0
298.3	5.0 (1.0)	0.0	11.5	2.5	15.0 (18.5)	3644.6	22.0
301.0	11.0 (2.0)	1.0	(6.5)	3.0	29.0 (35.0)	240.7	0.0
Konservolia							
			2.5 (0.0)				
322.9	13.5 (3.5)	8.5	5.5 (2.5)	2.5	16.0 (30.5)	5761.2	187.5
465.3	8.5 (1.5)	10.5	16.0	9.0	16.5 (37.5)	3203.6	122.0
508.4	12.0 (1.5)	7.5	(3.0)	3.5	31.0 (43.5)	234.2	36.9
Koroneiki							
			3.5 (0.0) 10.0				
93.0	3.5 (0.0)	10.0	(2.5)	2.5	7.0 (19.5)	6987.9	0.0
87.3	5.5 (0.5)	4.5	14.0	4.5	18.0 (27.5)	2029.9	0.0
89.7	9.5 (1.5)	7.5	(5.0)	5.0	28.5 (42.5)	5774.4	0.0

Weight drupes (g)	Eggs Fertile (Aborted)	Sterile oviposition stings	Larvae (pupae)	Emergency holes	Infestation Active (total)	Oleuropein (ppm)	Cyanidine (ppm)
Lucques							
			6.5 (3.0)				
515.7	6.0 (1.5)	0.0	6.5 (5.5)	5.0	15.5 (22.0)	3557.7	12.6
623.9	2.0 (0.0)	12.0	12.5	6.5	14.0 (32.5)	2119.3	96.4
853.3	5.0 (0.0)	5.5	(3.0)	13.5	20.5 (39.5)	0.0	12.4
Manzanilla							
			1.5 (0.0)				
376.3	11.5 (2.5)	5.5	7.5 (2.5)	3.0	13.0 (24.0)	10517.0	85.6
353.2	5.0 (1.0)	9.0	18.0	6.0	15.0 (31.0)	6028.2	742.6
423.8	5.5 (1.5)	55	(2.0)	8.5	25.5 (41.0)	519.1	186.1
Picual							
			3.5 (0.0)				
			10.0				
364.5	8.0 (1.5)	14.0	(5.0)	0.0	11.5 (27.0)	8915.3	21.81
435.1	6.5 (2.5)	1.0	17.5	6.0	21.5 (31.0)	5663.4	178.5
507.1	4.5 (2.0)	3.5	(1.5)	7.0	23.5 (36.0)	2596.7	1695.6

Table 5. Detailed percentages obtained in the different theses and ripening times concerning *B. oleae* infestation and oleuropein and cyanidine drupe contents. For any non-Italian cultivar, the observation was performed in three different ripening times (from Iannotta et al., 2006b, modified).

The susceptibility to *B. oleae* of other ten cultivars has been investigated by Iannotta et al. (2007a). Observations were carried out detecting the percentages of sterile oviposition stings, active infestation (presence of pre-imago stages: eggs, larvae and pupae) and total infestation (emergence holes, feeding tunnels and pre-imago stages) on olive fruit samples, 200 drupes per cultivar. Samples were collected in three different times during fruit ripening (Table 6).

Weight drupes (g)	Eggs Fertile (Aborted)	Sterile stings	Larvae (pupae)	Emergence holes	Reinfested by Larvae (pupae)
Bardhi Tirana					
309.69	9.5 (2.0)	9.0	0.5 (0.0)	0.5	0.0 (0.0)
358.91	2.5 (0.5)	9.0	4.5 (0.0)	4.5	0.0 (0.0)
415.48	7.5 (1.0)	6.0	12.5 (1.0)	7.0	2.5 (0.0)
Carboncella di Pianacce					
100.86	11.0 (0.5)	16.5	1.5 (0.0)	5.0	0.0 (0.0)
128.96	9.5 (0.5)	9.5	11.0 (5.5)	9.5	0.0 (0.0)
129.91	9.5 (1.5)	5.5	17.5 (3.5)	10.5	1.0 (1.0)
Carolea					
321.34	11.0 (1.0)	11.0	8.5 (2.0)	3.5	0.0 (0.0)

Weight drupes (g)	Eggs Fertile (Aborted)	Sterile stings	Larvae (pupae)	Emergence holes	Reinfested by Larvae (pupae)
390.70	4.0 (1.0)	1.0	22.0 (4.0)	11.0	0.0 (2.5)
463.46	1.0 (1.0)	5.5	18.0 (5.5)	21.5	9.0 (2.0)
Cassanese					
228.09	5.0 (0.0)	5.5	2.5 (0.0)	0.5	0.0 (0.0)
285.48	8.5 (0.5)	3.0	13.5 (0.0)	7.0	0.5 (1.0)
312.36	7.0 (2.5)	1.0	19.0 (2.0)	17.0	7.5 (0.0)
Gentile di Chieti					
511.35	14.0 (0.5)	19.0	4.0 (0.0)	5.0	0.0 (0.0)
495.03	10.0 (3.0)	5.5	12.5 (3.0)	9.5	2.5 (0.0)
524.06	6.5 (0.0)	7.0	14.5 (1.0)	12.0	4.0 (0.0)
Giarraffa					
606.61	5.0 (0.5)	15.0	6.0 (0.0)	8.0	0.0 (0.0)
711.97	5.0 (1.0)	6.5	20.0 (4.0)	8.5	5.0 (0.5)
771.70	2.5 (0.0)	2.0	23.0 (6.0)	17.5	12.0 (0.0)
Nocellara del Belice					
349.88	5.0 (1.0)	10.0	3.0 (0.5)	4.0	0.0 (0.0)
408.33	0.5 (1.0)	3.0	20.5 (5.0)	10.0	8.0 (2.5)
496.77	7.5 (1.5)	4.0	19.0 (4.0)	9.0	4.0 (1.5)
Nociara					
163.15	12.0 (6.5)	18.5	5.0 (0.0)	2.5	0.0 (0.0)
196.58	5.5 (0.0)	5.5	20.0 (3.0)	10.0	4.0 (0.0)
211.31	7.5 (2.5)	4.5	16.0 (5.0)	11.0	5.5 (1.0)
Picholine					
416.12	5.0 (0.5)	14.0	1.0 (1.5)	2.5	0.0 (0.0)
422.25	6.5 (0.5)	5.5	13.0 (1.5)	13.5	4.0 (0.0)
466.44	10.0 (2.0)	1.0	25.0 (5.0)	16.0	2.0 (0.0)
Tonda nera dolce					
242.92	2.0 (0.5)	3.5	0.0 (0.0)	0.0	0.0 (0.0)
301.60	5.0 (2.5)	10.5	3.5 (0.0)	3.0	0.0 (0.0)
298.97	4.0 (0.5)	12.5	4.0 (0.0)	3.5	0.0 (0.0)

Table 6. Detailed percentages obtained in the different cultivars and ripening times concerning olive fly infestation (from Iannotta et al., 2007a). Observations were performed in three different ripening times (03rd October, 04th November and 5th December) for any investigated cultivar. Reported values are referred to 100 drupes.

Drupes weight increased during the season, according to the physiological processes involved in fruit maturation. Fertile eggs were more abundant than aborted ones, which did not exhibit a trend related to ripening times. For many observed cultivars, sterile oviposition stings were much more abundant in the first ripening time. Few pupae were registered within drupes in respect to larvae. As expected, emergence holes increased during the season showing a low value in the first ripening time. Very low value of reinfestation were observed according to female egg laying behavior.

Infestation values due to preimago stages seem to be generally related to the investigated ripening time rather than to the single investigated cultivar. More evident trends were determined by the olive plant phenology, however some cultivars showed an interesting and peculiar behavior in relation to attack levels of olive fly.

Results show that the less susceptible cultivars to *B. oleae* attacks are cvs. Tonda nera dolce and Bardhi Tirana while cvs. Carolea, Carboncella di Pianacce, Gentile di Chieti, Giarraffa, Nocellara del Belice, Nociara and Picholine displayed a considerable susceptibility (Table 6). The low susceptibility observed for cvs. Tonda nera dolce and Bardhi Tirana could be attributed to different causes. The high amount of oleuropein present in the drupes of Bardhi Tirana, according to previous studies (Iannotta et al., 2001; 2002; 2006a), could be involved in processes determining a low incidence of olive fly attack. It hasn't been observed an high presence of the glycoside in cv. Tonda nera dolce while a considerable amount of cyanidine in the drupes was registered. The high value of cyanidine in cv. Tonda nera dolce could determine the observed low susceptibility. Probably, the dark color given by anthocyanins, achieved ever since in the early ripening stages, could confuse female olive flies in drupe recognition with a consequent decreased ovideposition.

The results obtained in the present research confirm those ones achieved in previous investigations, proving the need to explore the existent olive germplasm to search genetic resistance sources. It suggests the utility to achieve these results both to transfer directly to farmers' world and to emphasize ecosystem health and biodiversity conservation.

2.2. Susceptibility to *Pseudomonas savastanoi*

To keep under control the olive knot disease, the use of preventive measures turns out to be essential. Among these ones, the use of less susceptible cultivars emerged to be promising. Behavioural investigations showed a large variability in severities of olive disease caused by *Pseudomonas savastanoi* (Smith). A study carried out in the spring of 2005 by performing a large-scale investigations on the different responses to the pathogen of 262 Italian and 43 non-Italian cultivars, in the germplasm conservation field, where plants are cultivated under the same environmental and growing conditions (Iannotta et al., 2006c). The response to pathogen was evaluated by examining the symptomatology on the basis of the quantity of tubercles present on branches, arranged in classes of infection. During this time, several adverse meteorological events took place, including record minimum temperatures which influence the onset of the disease. Results displayed a different behaviour of olive cultivars to *P. savastanoi* in relation to their different susceptibility to the pathogen. Among observed Italian cultivars, 61% showed an infection's percentage ranging from 0 to 20%, 22.5% ranging from 20 to 40%, 11.1% ranging from 40 to 60%, 5% ranging from 60 to 80% and 0.4% ranging from 80 and 100%. Among investigated cultivars, 86 displayed no symptom of disease (Abunara, Aitana, Ascolana dura, Ascolana semitenera, Aurina, Bianchera,

Cacaredda, Capolga, Caprina Casalanguida, Caprina vastese, Carbonchia, Carpinetana, Cavalieri, Cellina Rotello, Colombina, Corneglia, Cornia, Corniola, Corniolo, Correggiolo, Dolce Andria, Fosco, Gentile Larino, Gentile Colletorto, Giusta, Gragnaro, Gragnan, Grappolo, Grossa Venafro, Grossale, I/77, Laurina, Lavagnina, Mantonica, Marina pugliese, Marzio, Minna di vacca, Morfa, Morchiaio, Morellona Grecia, Morinello, Nasitana, Nebbio Chieti, Nocellara etnea ovale, Ogliara, Ogliastro grande, Oliva grossa, Olivastra seggicnese, Olivastro Bucchinico, Olivo da olio, Olivo della Madonna, Orbetana, Ortice, Pampagliosa, Pennulara, Piangente, Piantone Moiano, Pignola, Posola, Posolella, Precoce, Puntella, Racioppella, Rastellina, Raza, Razzo, Remugnana, Resciola Venafro, Romanella molisana, Rosciola coltodino, Rosciola Rotello, Rustica, Saligna, Sammartinara, Sammartinenga, San Benedetto, San Francesco, Santa Maria, Sperone di gallo, Tombarello, Tonda Alife, Tonda dolce, Tonda dolce Partanna, Tunnulidda, Vicio, Zinzifarica) and no cultivar with a percentage of infection equivalent to 100% was observed, emphasizing the presence of genetic resources in Italian germplasm for olive knot disease prevention. Among analysed non-Italian cultivars, 41.9% showed a percentage of infection included between 0 and 20%, with 5 cultivars showing no sign of attack (Bardhi, Chetani, Hojiblanca, Lucques, Salonenque); 23.3% included between 20 and 40%; 4.7% included between 40 and 60%; 4.7% between 60 and 80%; 23.3% included between 80 and 100%. Cultivars Drobnica, Koroneiki and Vasilikada showed an infection equivalent to 100%. This different susceptibility evaluated under the same agro-environmental conditions, confirms a different response to the pathogen in relation to the ratio plant/parasite and appears tightly dependent from cultivars tolerance to low temperatures.

2.3. Susceptibility to *Spilocaea oleagina*

The different susceptibility of olive cultivars to knot disease has been observed by several authors. Iannotta and Monardo (2004) observed that both percentage of the number of leaves infected and surface occupied on leaves by the symptoms of the disease are significantly different in observed cultivars. These authors observed that out of 35 cultivars studied cultivars Bardhi i Tirana, Carboncella di Pianacce, Cassanese, Dritta di Moscufo, Gentile di Chieti, Kalinjot and Leccino did not show any kind of symptoms of the disease, while cultivars Bosana, Carolea, Nocellara del Belice, Nera di Villacidro, Maiatica di Ferrandina, Itrana and Tonda di Cagliari were clearly infected by *Spilocaea oleagina*. The others observed cultivars showed intermediate values of the disease incidence (Table 7). The same authors related the cultivar susceptibility to the oleuropein content of leaves showing a direct relation between oleuropein content and cultivar susceptibility. In fact, low susceptible cultivars are those with the highest oleuropein content in leaves. Iannotta & Monardo (2004) affirmed that oleuropein could play an important role in determining the development inhibition of the fungus within the leaves.

Results are similar to those obtained by Iannotta et al. (2001) concerning the relation among oleuropein content of drupes and susceptibility to olive fly infestation, demonstrating that studies devoted to the exploration of genetic variability of olive cultivars is a focal subject, to date not sufficiently developped but of great importance.

Cultivar	FI%		SI%		Oleuropeina (mg/g)	
Ascolana	0.80	BC	0.40	B-E	59.64	B-F
Bardhi i Tirana	0.00	C	0.00	E	107.62	AB
Bosana	39.20	A	13.60	A-C	43.69	DEF
Buscionetto	2.50	A-C	1.20	B-E	42.38	DEF
Carboncella	0.00	C	0.00	E	86.31	ABC
Carolea	8.30	A-C	12.80	A-D	44.62	DEF
Cassanese	0.00	C	0.00	E	60.45	B-F
Coratina	1.70	A-C	2.70	A-E	54.13	C-F
Cucco	2.50	A-C	1.50	A-E	68.07	B-F
Dritta di Moscufo	0.00	C	0.00	E	112.40	A
Gentile di Chieti	0.00	C	0.00	E	79.82	A-D
Giarraffa	4.20	A-C	1.90	A-E	31.99	EF
Grossa di Spagna	9.20	A-C	3.30	A-E	68.67	A-E
Intosso	4.20	A-C	1.90	A-E	66.02	B-F
Itrana	26.70	A-C	17.10	A	54.99	C-F
Kalinjot	0.00	C	0.00	E	86.85	A-D
Leccino	0.00	C	0.00	E	103.67	ABC
Maiatica di Ferran.	27.50	A-C	18.30	A-B	90.76	ABC
Mixan	2.50	A-C	0.50	B-E	74.73	A-E
Morghetana	0.80	BC	0.10	C-E	47.60	DEF
Nera di Gonnos	10.00	A-C	2.80	A-E	37.65	DEF
Nera di Villacidro	29.20	A-C	16.00	A-D	48.67	DEF
Noc. del Belice	32.50	AB	15.00	A-B	63.94	A-F
Noc. etnea	3.30	A-C	1.70	A-E	45.43	DEF
Noc. messinese	4.20	A-C	1.10	B-E	37.60	DEF
Nostrale di Rigali	13.30	A-C	7.90	B-E	40.22	DEF
Passulunara	4.20	A-C	1.20	B-E	63.76	B-E
Picholine	4.20	A-C	2.10	A-E	65.78	A-F
Piddicuddara	5.00	A-C	2.70	A-E	70.11	B-F
Pizz'e Carroga	11.70	A-C	4.40	A-E	73.77	B-E
S.Agostino	5.00	A-C	3.20	A-E	38.63	DEF
S.Caterina	0.80	BC	2.10	A-E	42.33	DEF
Santagatese	1.70	A-C	1.70	A-E	80.82	A-E
Tonda di Cagliari	22.50	A-C	11.70	A-E	47.87	DEF
Tonda iblea	3.30	A-C	1.10	B-E	61.26	B-F

Table 7. Percentage of infected leaves (FI%) and percentage of surface occupied by symptoms on leave's surface (SI%) of 35 cultivars (from Iannotta and Monardo, 2004). Letters indicate significant statistical differences (P<0.01; ANOVA test).

2.4. Susceptibility to *Camarosporium dalmaticum*

Iannotta et al. (2006d) found a different level of susceptibility among cultivars for fruit rot disease. In particular, a low susceptibility was observed for cvs. Frantoio, Tonda di Strongoli and Dolce di Rossano. On the contrary, cvs. Ascolana tenera and Nostrana di Brisighella showed a high susceptibility (Table 8). Since the cultivars displaying the lowest susceptibility to the fungus are the same which show the lowest susceptibility to olive fly attacks, a direct correlation between could be hypotised. Also data concerning non-Italian cultivars show a different behaviour. In fact, cvs. Arbequina, Hojiblanca and Picual are the less infected and cv. Gordal sevillana the most affected by mycosis (Table 9). The study proves the utility of further investigations in order to characterize the different cultivars behaviour in relation to their parasites, so as to define their specific susceptibility.

Cultivar	C. dalmaticum	
	Infection	LSD test
Ascolana tenera	12.00	A
Cellina di Nardò	3.33	BC
Cima di Melfi	4.67	BC
Dolce Agogia	3.17	BC
Dolce di Rossano	1.67	BC
Frantoio	0.67	C
Moraiolo	4.00	BC
Nera di Cantinelle	3.17	BC
Nolca	4.33	BC
Nostrale di Fiano Romano	3.67	BC
Nostrana di Brisighella	11.17	A
Ogliarola del Bradano	4.50	BC
Ogliarola garganica	2.83	BC
Ogliarola del Vulture	6.00	B
Termite di Bitetto	3.67	BC
Tonda di Strongoli	1.50	BC

Table 8. Mean values concerning *C. dalmaticum* infection referred to each Italian cultivars. Letters indicate significant statistical differences (P<0.01; ANOVA test). (from Iannotta et al., 2006d, modified).

Cultivar	C. dalmaticum	
	Infection	LSD test
Arbequina	3.33	b
Gordal sevillana	7.50	a
Hojiblanca	3.83	b
Kalamata	5.00	ab
Konservolia	5.50	ab
Koroneiki	4.50	ab
Lucques	5.50	ab
Manzanilla	5.00	ab
Picual	3.00	b

Table 9. Mean values concerning *C. dalmaticum* infection referred to each non-Italian cultivars. Letters indicate significant statistical differences (P<0.01; ANOVA test). (from Iannotta et al., 2006d, modified).

Trials performed by Iannotta et al. (2007a) in the collection field of the CRA OLI compared susceptibility to *Camarosporium dalmaticum* of cultivars planted in the same environmental and agronomic conditions (Fig. 2). The investigation has been performed in 2005 by analyzing 10 cultivars, in the experimental olive plantation made up by a cultivar collection consisting in 20-years old plants. Observations were carried out detecting on 200 drupes per cultivar the infection level (%) of *C. dalmaticum* by direct observation of drupes. Samples were collected in three different period during fruit ripening (03rd October, 04th November and 5th December).

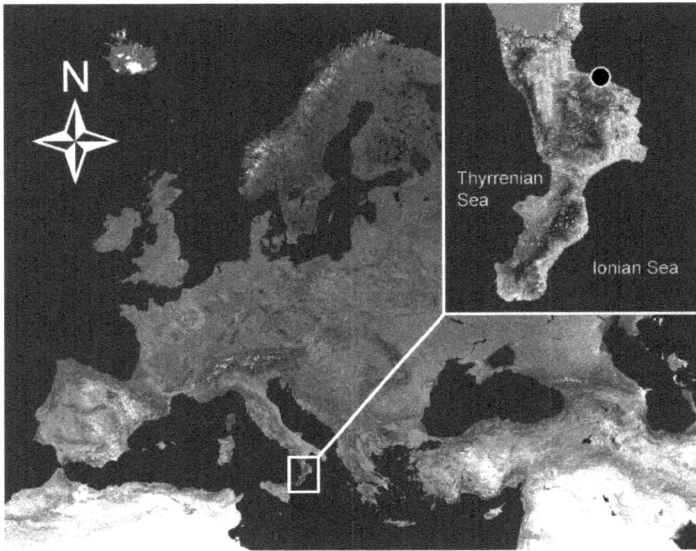

Figure 2. Location of CRA-OLI experimental field, where several cultivars are planted under the same agronomical and climatic conditions, on the ionian coast of Calabria (Mirto-Crosia, Cosenza, Italy).

Cultivar	Infection	
	Mean*	SD
Bardhi Tirana	1.67^{DE}	1.53
Carboncella di Pianacce	1.33^{DE}	2.31
Carolea	4.67^{AB}	1.26
Cassanese	2.17^{BCDE}	1.89
Gentile di Chieti	4.33^{ABC}	1.61
Giarraffa	5.67^{A}	1.04
Nocellara del Belice	2.00^{CDE}	2.29
Nociara	1.17^{DE}	1.26
Picholine	3.17^{ABCD}	1.53
Tonda nera dolce	0.58^{E}	0.43

Table 10. Mean infected drupes concerning *Camarosporium dalmaticum* infection. Letters indicate significant statistical differences (P<0.01; ANOVA test). (From Iannotta et al., 2007a, modified).

Results display a different susceptibility of investigated genotypes in relation to the fungal infection, as indicated by significativity letters concerning analysis of variance (Table 10). The lowest susceptibility in relation to fungal infection has been observed for cv. Tonda nera dolce. Also cvs. Bardhi Tirana, Carboncella di Pianacce and Nociara showed a low susceptibility in relation to the pathogenic fungus infection. The cv. Giarraffa turn out to be the most susceptible cultivar while cvs. Carolea, Cassanese, Gentile di Chieti, Nocellara del Belice and Picholine display an intermediate susceptibility value. Comparison among investigated ripening times for pathogenic fungus emphasise an increase of infection percentages through the season.

2.5. Resistance to *Verticillium dahliae*

Verticillium wilt, caused by the fungus *Verticillium dahliae* Kleb., is a vascular wilt with a very large host range. Verticillium wilt is increasing in several Mediterranean countries, and it is very difficult to reduce its incidence because is not easy to apply in the field chemical compounds such as fosetyl-Al, directly inoculated within vascular system. Very little is known about biochemical and molecular mechanisms of olive resistance to the Verticillium wilt. Among investigated aspects, it seems to be very important the role of phenolic methabolism (Baidez et al., 2007; Markakis et al., 2010). Vizzarri et al. (2011) investigated the susceptibility of cvs. Arbequina, Arbosana, Frantoio, Ottobratica, Sant'Agostino and Urano by inoculating them with different isolates of *Verticillium dahliae*. Plantlets of 18 months and 160 cm high, has been inoculated by applying a small square of agar with sporulating fungal hyphae on wounded wood. After inoculation plantlets were observed for one year in a greenhouse. The severity of the disese was evaluated by utilising 5 classes of the percentage of damaged plant and the intensity of symptoms: 0%, healthy plant; 20%, plant with large clorosis and moderate foliar symptoms; 50%, severe foliar symptoms and desiccation of vegetative apex; 75%, desiccation of more than half of the plant; 100% dead plant, without defoliation (Colella et al., 2008). Furthermore, Vizzarri et al. (2011) defined a molecular protocol for studying the expression of gene involved in defense mechanisms of the olive. They utilised plants of 14 months of the cv. Leccino, some artificially inoculated as described above and some non inoculated as negative control. Results showed that genetic resistance of tested cultivars, inoculated with different Verticillium isolates, is subordinated to the virulence of the inoculated isolate (Table 11).

Cultivar / Isolato	Arbequina				Arbosana				Urano				S. Agostino				Ottobratica				Frantoio			
	20%	50%	75%	100%	20%	50%	75%	100%	20%	50%	75%	100%	20%	50%	75%	100%	20%	50%	75%	100%	20%	50%	75%	100%
ND (Sicilia)																2		1	1					
ND M (Puglia)		2	2								1					1		1	1					
D (Puglia)			3													6			2					
Mirto(Calabria)																								
ND(Umbria)																								

Table 11. Results of the genotypic susceptibility assay. Number of plants showing a given range of symptoms. Cultivars and isolates used (10 replicates for each cultivar). (from Vizzarri et al., 2011).

The cv. Frantoio and Urano were the most resistant with no symptoms on tested plants, while cvs. Ottobratica and Sant'Agostino were higly susceptible. Further studies are needed for assessing the behaviour of cvs. Arbequina and Arbosana, largely utilised in superintensive olive plantations, that showed a differentiated susceptibility depending on the utilised isolate for inoculation.

Vizzarri et al. (2011) also tested a method for evaluating the expression of the genes (PAL and CHS) involved in defense mechanisms of olive. The level of trascript of these genes showed significant increments after plantlets wounding. Attaining the highest value 9 hours after plantlets wounding. Afterward, the level of transcript of both genes decrease, more evidently for the gene PAL after 24 hours, in any case maintaining high expression levels until 30 days after wounding (Fig. 3).

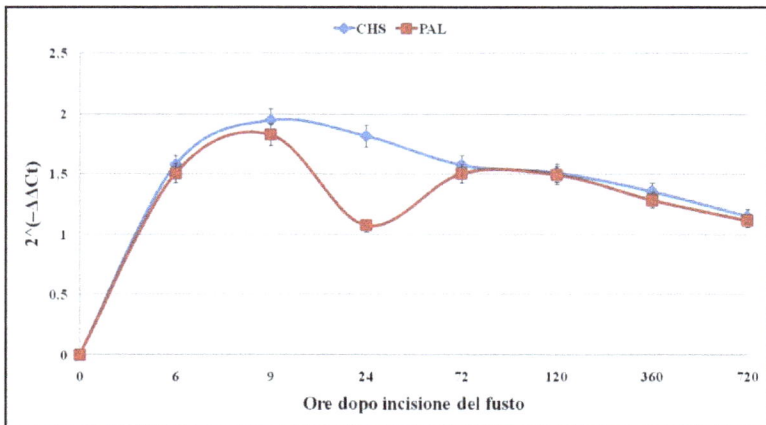

Figure 3. Time course of the relative transcript level of PAL and CHS genes in the leaves of stem-wounded olive plantlets (cv. Leccino), as compared to unwounded plantlets (from Vizzarri et al., 2011).

The use oh this method permit to verify that the response of plant to the injury is quite rapid, more or less 9-12 hours. Vizzarri et al. (2011) hypothised that the evaluation of the expression level of genes PAL e CHS for cultivars with different resistance could be important for verifying the role of phenolic methabolism in olive resistance to pathogenes. In fact, recent papers demonstrated that phenolic compounds are very important for modulating resistance/susceptibility of olive cultivars to verticillium wilt. Phenolic response to verticillium wilt is very different in resistant and susceptible cultivars. The resistant cv. Koroneiki showed higher increasing of phenols than susceptible cv. Amfissis when inoculated with verticillium wilt (Markakis et al., 2010). Genes PAL e CHS play an important role in the biosynthesis of phenolic compounds, then a role in determing resistance of olive cultivars to verticillium wilt could be hypothesised.

3. Factors affecting cultivar susceptibility

Morphological parameters such as dimensions and coulor are known to be important in determining susceptibility of cultivars affecting female choice for laying eggs. Recently, a great effort is done for determining biochemical factors involved in resistance/susceptibility of olive cultivars to biotic stresses.

3.1. The role of phenolic compounds

Differences in the response of olive cultivars to olive fly infestations have been observed (Gümusay et al., 1990; Iannotta et al., 1999, 2006a, 2007a, 2007b; Pereira et al., 2004; Basile et al., 2006; Rizzo and Caleca, 2006; Daane and Johnson, 2010). Two phenolic compounds, oleuropein and cyanidine, were assessed to determine their role in the reduced susceptibility of certain olive cultivars to olive fly (Iannotta et al., 2006a, 2007a, 2007b). In addition, a positive correlation has been observed between the drupe oleuropein content and a low susceptibility of olive cultivars to olive fly damage such that when the drupe oleuropein content is high, the olive fruit is less susceptible to attack (Iannotta et al., 2006a, 2007a, 2007b).

3.1.1. Oleuropein

Iannotta et al. (2001) investigated the huge olive showing a low susceptibility to olive fly infestations of some cultivars due to the high content of oleuropein within drupes. That cultivars became particularly interesting in respect to cultivars having a low oleuropein content within drupes also when planted in the same environmental and agronomical conditions. Although a correlation between high oleuropein content and low susceptibility of olive cultivars to olive fly infestations is generally accepted, it is nopt clear the mechanism of action of this compound. Some authors hypothesised a mechanism of action against eggs and young larvae of olive fly explicated by oleuropein and their methabolites within the tissue of drupes, causing a reduction of the preimaginal population of this pest. Iannotta et al. (2001) evaluated the amount and the localisation of oleuropein within drupes of ten cultivars selected among them known as low-susceptible and high-susceptible ones. Furthermore, absolute oleuropein has been applied directly on the oviposition sting in order to evaluate its efficacy to control egg hatchling and the following larval development. Results confirm the different behaviour of tested cultivars with cvs. Bardhi i Tirana, Carboncella di Pianacce, Gentile di Chieti and Nociara less susceptible than cvs. Carolea, Nocellara del Belice, Giarraffa, Cucco, Picholine and Cassanese (Table 12). Susceptibility of cultivars is correlated to the amount of oleuropein within drupes. Furthermore, the amount of oleuropein is higher where female lays eggs. The higher amount of this compound in the epicarp found for the low susceptible cultivars seems to be related to genetic characteristics of cultivars more than to phisiological response to olive fle attacks, as demonstrated by comparing the distribution of oleuropein in healthy and infested drupes.

Cultivar	Active infestation (%)	Sterile oviposition stings (%)	Oleuropein (drupe) (mg/g)	Oleuropein (epicarp) (mg/g)	Oleuropeina (mesocarp) (mg/g)
Bardhi i Tirana	8,5A	27,5C	29,60cd	32,89	26,48
Carboncella di Pianacce	9,5A	26,0C	60,04b	70,54	49,55
Gentile di Chieti	9,6A	26,4C	38,82bc	37,69	39,96
Nociara	8,9A	25,2BC	91,91a	141,37	42,45
Carolea	22,7CDE	13,8AB	2,25cd	23,90	20,61
Nocellara del Belice	23,1DE	18,6ABC	40,52bc	48,93	32,12
Giarraffa	23,6DE	16,5AB	11,41d	13,05	9,78
Cucco	26,8E	23,5ABC	21,10cd	23,24	18,87
Picholine	24,1E	17,1ABC	18,80cd	17,36	20,25
Cassanese	27,4E	16,9ABC	14,30d	17,24	11,36

Table 12. Detailed percentages obtained in the different cultivars concerning olive fly infestation (from Iannotta et al., 2001, modified). Reported values are referred to 100 drupes. Letters indicate significant statistical differences (P<0.01; ANOVA test).

The use of oleuropein directly on oviposition stings confirm the role of control agent of this compound. After ten days from the oleuropein application, within treated sample only the 31% of olive were infested, while within the untreated sample the infested olive were the 65%.

Phenolic composition and concentration are related to genetic features of a given olive cultivar. These genetic features can be used as varietal markers and as indicators of fruit maturation (Esti et al., 1998). Furthermore, a correlation between olive fruit size and oleuropein content has been shown. Small-fruit cultivars are characterised by high oleuropein content (Amiot et al. 1986). Phenolic compounds are important for the defence of plants against pathogens and insect infestations (Haukioja et al., 1985; Hudgins et al., 2003). The antimicrobial activity of phenolic compounds is well documented (Bisignano et al. 1999; Rauha et al., 2000; Proestos et al., 2005; Pereira et al., 2006, 2007).

It has been shown that oleuropein and cyanidine contents are inversely related. During olive fruit maturation the oleuropein content decreases rapidly (Limiroli et al., 1995) while flavonoid content as cyanidine increases (Amiot et al., 1989). In detail, olive fruit maturation consists of three phases: the growth, green maturation and black maturation (Amiot et al., 1989). While in the growth phase an accumulation of oleuropein occurs, in the green maturation phase it decreases. The black maturation phase is characterized by the appearance of anthocyanins and by the progressive decrease of oleuropein levels (Amiot et al., 1989). In Iannotta et al. (2006a) the mean content of both phenolic compounds appears genetically determined. Similar results were observed by Iannotta et al. (2007a, 2007b) confirming a different olive genotype behavior which depends on the genetically determined content of phenolic compounds (Esti et al., 1998). Moreover, the oleuropein content of drupes is not affected by *B. oleae* attacks, as it is not moved to the damage site. No differences in the oleuropein content were observed between non-infested and infested drupes belonging to the same cultivar (Iannotta et al., 2002).

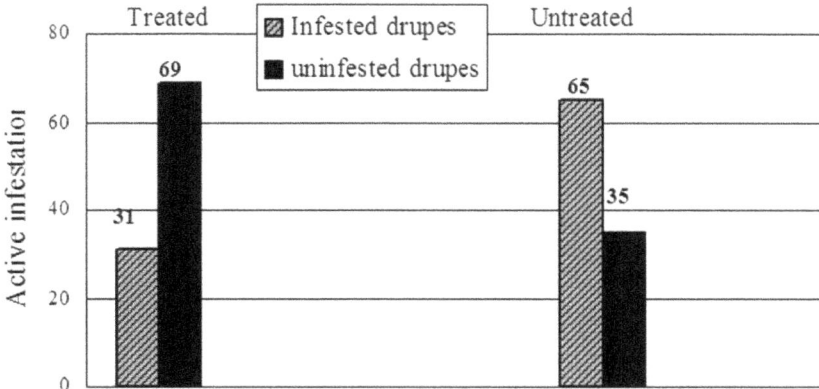

Figure 4. Results of treatment test with oleuropein of oviposition stings (from Iannotta et al., 2001 modified).

In previous studies, it has been established that the drupe oleuropein content is genetically determined since it hasn't been observed a statically significative difference between oleuropein content in non infested and infested drupes by *B. oleae* belonging to the same genotype (Iannotta et al., 2001). Moreover, it has been proved that the differences in oleuropein amounts are correlated to the different behavior of the cultivars in relation to olive fly attacks (Iannotta et al., 2001, 2002, 2006b).

Oleuropein was first isolated from olive leaves (Panizzi et al., 1960) where it is present in high levels (Le Tutour and Guedon, 1992). In addition, it occurs throughout the tree and in any constituent part of the fruit (Servili et al., 1999). Oleuropein confers resistance to diseases and to insect infestation of the olive tree (Soler-Rivas et al., 2000). The bactericidal and bacteriostatic activities of oleuropein and its degradation products against many pathogenic microorganisms have been investigated (Hirschman, 1972; Federici and Bongi, 1983) and its in vitro activity has been detected in relation to several bacteria, fungi, viruses, and parasitic protozoans (Hirschman, 1972; Walter et al., 1973; Gourama and Bullerman, 1987; Tassou et al., 1991; Tranter et al., 1993; Tassou and Nychas, 1994, 1995). Oleuropein can also interfere with the synthesis of virus amino acids, prevent viral shedding, budding or assembly at the cell membrane, inhibit viral replication and, in the case of retroviruses, neutralize the production of reverse transcriptase and proteases. Oleuropein is also able to stimulate phagocytosis, as a response of the immune system against pathogenic microorganisms (Hirschman, 1972). A strong chemotactile repulsion exerted by oleuropein in the oviposition of olive fly eggs has been described (Soler-Rivas et al., 2000). Small droplets of olive sap exuded just after oviposition prevent other females from ovipositing on the same fruit (Girolami et al., 1981; Lo Scalzo et al., 1994). Oleuropein acts by inhibiting the development of olive fly immature stages, especially eggs and first instar larvae during the early ripening period (Iannotta et al., 2002). The higher concentration of oleuropein in the epicarp than in the mesocarp may be due to the biological function of oleuropein in drupe protection against pests (Soler-Rivas et al., 2000). In fact, the epicarp is the interface between the outer environment and the inner olive fruit. Therefore, high levels of oleuropein in the epicarp protect the fruit against olive fly ovideposition (Iannotta et al., 2002).

Moreover, the defence response of fruits damaged both by pathogens and mechanical means, is mediated by β-glucosidase; this enzyme hydrolyses the oleuropein, producing highly reactive aldehyde molecules. Olive cultivars with different levels of enzyme activity have differing degrees of susceptibility to the olive fly. This may be related to the ability of the β-glucosidase to produce highly reactive aldehyde molecules in damaged tissues. A strong peroxidase activity is thereafter detected as a consequence of damage (Spadafora et al., 2008). Results obtained by Iannotta et al. (2001) showed that five cultivars (Bardhi i Tirana, Carboncella di Pianacce, Gentile di Chieti, Kokermadh i Berat, and Nociara) with high levels of drupe oleuropein (31.18 – 36.60 g kg^{-1}) had low levels of infestation (lower than 10%). When oleuropein content decreases, a corresponding increase in the amount of damage caused by olive flies occurs. In the same cultivars, Iannotta et al. (2001) found that the percentage of sterile oviposition stings ranged from 25.0 to 27.5%. Similar results were also observed for cultivars Sant'Agostino, Leccino, and partially Frantoio (Basile et al., 2006). Sterile sting numbers and oleuropein content are inversely proportional to infestation (Iannotta et al., 2001). The role of oleuropein in the inhibition of the development of olive fly immature stages has been shown by performing a comparison between untreated olive samples and samples treated with oleuropein belonging to the cv. Carolea. The cultivar Carolea was chosen because it is susceptible to the olive fly. After ten days, infestation levels were 31% and 65%, respectively, in the oleuropein-treated and non-treated samples (Iannotta et al. 2002). The concentration of oleuropein is greater in the epicarp rather than in the mesocarp during the entire ripening process, except in the case of cvs. Gentile di Chieti and Picholine (Iannotta et al., 2002, 2007a). In these varieties, there is a slightly lower content of oleuropein in the epicarp during the early ripening period. In another study no correlation was observed between infestation and oleuropein content (Iannotta et al., 2006a). In fact, olive fly infestation may be different on the same olive cultivar under different environmental conditions (Fontanazza, 2000) inasmuch as the oleuropein content might be affected by climatic trend (Iannotta et al., 2006a). In a study performed in 2005 in an experimental field located on the Ionian coast of Calabria (Southern Italy), it was observed that cv Cellina di Nardò was the least infested by the olive fly in terms of total infestation (17.67%). In contrast, cvs. Ascolana tenera and Nostrana di Brisighella were the most damaged attaining percentages of total infestation at 56.33% and 57.67%, respectively (Iannotta et al., 2006a). This difference is presumably related also to fruit size (Daane and Johnson, 2010). In fact, Cellina di Nardò has relatively small fruits compared to Ascolana tenera and Nostrana di Brisighella. In addition, it has been shown that small-fruit cultivars are characterised by high oleuropein content (Amiot et al., 1986), playing a synergic role in determining low susceptibility.

Interestingly, it has also been observed that many cultivars characterized by low susceptibility to olive fly attacks showed low susceptibility to the fungal pathogen *Spilocaea oleagina* (Cast.) Hugh. and a negative correlation between oleuropein content in olive leaves and fungal infection has been found (Iannotta and Monardo, 2004). In addition, a correlation between *B. oleae* infestation and *Camarosporium dalmaticum* (Thüm.) Zachos & Tzav.-Klon. infection has been established (Iannotta et al., 2007d). Since the same cultivars showed low levels of susceptibility to both parasites, it could be assumed that high levels of oleuropein may play a role also in determining low cultivar susceptibility to fungal disease (Iannotta et al., 2006c, 2007a).

3.1.2. Cyanidine

Cyanidine occurs in olive fruits (Servili et al., 1999) and an increase of cyanidine content at the end of the maturation stages of the olive fruit, as a consequence of hydrolytic processes, was found (Vinha et al., 2005). On the reasons of different genotype behavior concerning the susceptibility to olive fly attacks, the direct influence of cyanidine in the drupes could be, in effect, supposed. It is evident in cvs. Cellina di Nardò, Nolca and Termite di Bitetto which register high value of cyanidine, increasing during the season (Iannotta et al., 2006b). When investigated genotypes are cultivated in the same pedoclimatic conditions and samples obtained from them are collected in the same ripening times, it is possible attribute the differences, concerning cyanidine amount, to a strong influence of the different investigated genotypes genetic diversity. It has been observed that the completely pigmented drupes are not very recognizable by *B. oleae* females determining considerable difficulties for their ovideposition (Caleca, pers. comm.).

A role played by cyanidine in resistance to herbivores was additionally assessed (Harborne and Williams, 1998). Significant differences were found among cultivars in relation to active and total infestations and cyanidine content (Iannotta et al., 2006a). Cultivars Ascolana tenera and Nostrana di Brisighella had the highest level of active infestation (34.33% and 32.33%, respectively) while cv. Cellina di Nardò was the least infested (9.83%). In addition, cvs. Frantoio, Gordal sevillana, Koroneiki, Nera di Cantinelle, Nolca, Ogliarola garganica, and Tonda di Strongoli showed low levels of susceptibility to olive fly (lower than 15%). Cultivars Cellina di Nardò, Nolca, and Termite di Bitetto had higher levels of cyanidine than other cultivars in the study and had low levels of infestation.

In a study undertaken in 2005 in an experimental field located on the Ionian coast of Calabria (Southern Italy), Iannotta et al. (2007a) found the lowest susceptibility to olive fly attack for cvs Tonda nera dolce and Bardhi i Tirana (6.67% and 13.50%, respectively). On the contrary, cvs. Carolea, Cassanese, Carboncella di Pianacce, Gentile di Chieti, Giarraffa, Nocellara del Belice, Nociara and Picholine were susceptible with a mean percentage of active infestation ranging from 22.17 to 29.83%. The presence of cyanidine in the first ripening period only for cv. Tonda nera dolce suggests a possible role in determining the lowest level of active infestation observed.

Studies on cultivars Bardhi i Tirana, Carolea and Tonda nera dolce (Iannotta et al., 2007a) corroborate results obtained in previous investigations. Cultivars Bardhi i Tirana and Carolea were selected because they have low and high levels of susceptibility, respectively, to olive fly whilst cultivar Tonda nera dolce was selected because it shows high levels of cyanidine. Differences were found among the three cultivars in relation to active and total infestations and oleuropein and cyanidine contents. The lowest percentages of active and total infestations were observed on cv. Tonda nera dolce (8.62% and 20.12, respectively), while cv. Carolea had the most damage (29.00% and 49.38%, respectively). Cultivar Bardhi i Tirana showed intermediate values of active and total infestations. The low susceptibility found for cv Bardhi i Tirana, although greater than the susceptibility showed by cv Tonda nera dolce, might be due to the highest oleuropein content observed. The content of cyanidine in cv. Tonda nera dolce might be responsible for the lowest susceptibility found.

Cyanidine probably acts by giving olive fruits a dark colour during the early ripening stages. This may confuse female olive flies in drupe recognition resulting in a decrease in oviposition. This hypothesis is supported by the evidence that in herbivorous diurnal insects, visual cues may play an important role in the location of host plants and essential resources, such as food, mating, and oviposition sites (Prokopy and Owens 1983). This situation is very common for frugivorous Tephritid flies that feed and oviposit on fruits (Katsoyannos, 1989; Fletcher and Prokopy, 1991; Diaz-Fleischer et al., 2000; Prokopy and Papaj, 2000). A laboratory experiment investigating the effect of fruit colour on attracting olive fruit fly females was carried out by Katsoyannos et al. (1985). In this study, females were left to select for oviposition sites among hollow, hemispheric, ceresin wax domes of different colours. Yellow and orange domes were preferred for oviposition compared to domes of other colours. Red, blue, black, and white domes were the least preferred. Red, blue, and black correspond to the colour of ripening olives reached during the maturation stages. Fruit colour is genetically determined in some cultivars and is related to the content of anthocyanins. Olive fly females prefer green olives for oviposition compared to red and black olives (Cirio, 1971; Rizzo and Caleca, 2006).

The high content of phenolic compounds found in several cultivars is a resource in olive germplasm. Field researches demonstrated that a high content of phenolic compounds is related to low susceptibility to olive fly attacks and other parasites. Therefore, the planting of genotypes containing high amounts of these compounds may greatly contribute to a significant reduction of pesticides inputs. In addition, the presence of phenolic compounds in olive fruits is also associated with various benefits for human health deriving from high quality olive oil intake. Therefore, the conservation of olive intraspecific biodiversity preserves sources of genetic resistance to various pests. To preserve olive germplasm biodiversity in accordance with CAP directives and minimize pesticide use diversifying agronomic practices are strongly related. Strategic B. oleae control is thus a priority for safeguarding both environmental integrity and consumer health. The identification of genetic resistance sources may represent an effective means for olive crop management. In fact, using olive cultivars with low susceptibility to olive fly may represent an effective strategy for organic and integrated pest management, eliminating or decreasing pesticides inputs.

Author details

Nino Iannotta and Stefano Scalercio
Agricultural Research Council - Olive Growing and Oil Industry Research Centre, Rende (CS), Italy

Acknowledgement

We thanks Veronica Vizzarri, Tiziana Belfiore, Maria Elena Noce, Luigi Perri and all the other colleagues involved in field and laboratory works and that allow us to write this contribute. Financial support was provided by the Italian Ministry of Agriculture, Food and Forestry Policy through the project GERMOLI "Salvaguardia e valorizzazione del GERMoplasma OLIvicolo delle collezioni del CRA-OLI"

4. References

Amiot, M.J.; Fleuriet, A. & Macheix, J.J. (1986). Importance and evolution of phenolic compounds in olive during growth and maturation. *Journal of Agricultural and Food Chemistry*, Vol.34, pp. 823–826.

Amiot, M.-J.; Fleuriet, A. & Macheix, J.-J. (1989). Accumulation of oleuropein derivatives during olive maturation. *Phytochemistry*, Vol.28, pp. 67–69.

Baidez, A.G.; Gomez, P.; Del Rio, J.A. & Ortuno, A. (2007). Disfunctionality of the xylem in *Olea europaea* L. plants associated with the infection process by *Verticillium dahliae* Kleb. role of phenolic compounds in plant defense mechanism. *Journal of Agricultural and Food Chemistry*, Vol.55, pp. 3373-3377.

Basile, B.; Romano, R.; Garonna, A.P.; Forlani, M. & Rao, R. (2006). Preliminary study of the susceptibility of different olive cultivars to olive fruit fly [Bactrocera oleae (Gmel.)]. *Proceedings of the Second International Seminar on "Biotechnology and Quality of Olive tree Products around the Mediterranean basin", Olivebioteq, Marsala-Mazara del Valle 5-10 November 2006*, Vol. II: 319-322.

Bisignano, G.; Tomaino, A.; Lo Cascio, R.; Crisafi, G.; Uccella, N. & Saija, A. (1999). On the in vitro antimicrobial activity of oleuropein and hydroxytyrosol. *J. Pharm. Phrmacol.*, Vol.51, pp. 971-974.

Cirio, U. (1971). Reperti sul meccanismo stimolo-risposta nell'ovideposizione del *Dacus oleae* Gmelin (Diptera, trypetidae). *Redia*, Vol.52, pp. 577-600.

Colella, C.; Miacola, C.; Amenduni, M.; D'Amico, M.; Bubici, G. & Cirulli, M. (2008). Sources of verticillium wilt resistence in wild olive germplasm from the Mediterranean region. *Plant Pathology*, Vol. 57, pp. 533-539.

Daane, K.M. & Johnson, M.W. (2010). Olive fruit fly: managing an ancient pest in modern times. *Annual Review of Entomology*, Vol.55, pp. 151-169.

Díaz-Fleischer, F., Papaj, D.R., Prokopy, R.J., Norrbom, A.L. & Aluja, M. (2000). Evolution of fruit fly oviposition behavior, *in* M. Aluja & A.L. Norrbom (eds.), *Fruit flies (Diptera: Tephritidae): Phylogeny and evolution of behavior*, CRC Press, Boca Raton, pp. 811-841.

Esti, M.; Cinquanta, L. & La Notte, E. (1998). Phenolic compounds in different olive varieties. *Journal of Agricultural and Food Chemistry*, Vol.46, pp.32–35.

Federici, E. & Bongi, G. (1983). Improved method for isolation of bacterial inhibitors from oleuropein hydrolysis. *Appl. Environ. Microbiol.*, Vol.46, pp. 509–510.

Fletcher, B.S. & Prokopy, R.J. (1991). *Host location oviposition in tephritid fruit flies, in* A.A. VV., *Reproductive behaviour of Insects: Individuals and Populations*, Chapman and Hall, London, pp. 139-171.

Fontanazza, G. (2000). Olivicoltura intensiva meccanizzata, Edagricole, Bologna, Italy.

Girolami, V.; Vianello, A.; Strapazzon, A.; Ragazzi, E. & Veronese, G. (1981). Ovipositional deterrents in *Dacus oleae*. *Ent. Exp. Appl.*, Vol.29, pp. 177-188.

Gourama, H. & Bullerman, L.B. (1987). Effects of oleuropein on growth and aflatoxin production by *Aspergillus parasiticus*. *Lebensm. –Wiss. u. Techn.*, Vol.23, p. 226.

Gümusay, B.; Özilbey, U.; Ertem, G. & Oktar, A. (1990). Studies on the susceptibility of some important table and oil olive cultivars of Aegean region to olive fly (*Dacus oleaea* Gmel.) in Turkey. *Acta Horticulturae*, Vol.286, pp. 359-362.

Harborne, J.B. & Williams, C.A. (1998). Anthocyanins and other flavonoids. *Natural Product Reports*, Vol.15, pp. 631-652.

Haukioja, E.; Suomela, J. & Neuvonen, S. (1985). Long-term inducible resistance in birch foliage: triggering cues and efficacy on a defoliator. *Oecologia*, Vol.65, pp. 363-369

Hirschman, S.Z. (1972). Inactivation of DNA polymerases of murine leukaemia viruses by calcium elenolate. *Nat. New Biol.*, Vol.238, pp. 277–279.

Hudgins, J.W.; Christiansen, E. & Franceschi, V.R. (2003). Methyl jasmonate induces changes mimicking anatomical defenses in diverse members of the Pinaceae. *Tree Physiol.*, Vol.23, pp. 361–371.

Iannotta, N.; Perri, L.; Tocci, C. & Zaffina, F. (1999). The behaviour of different olive cultivars following attacks by *Bactrocera oleae* (Gmel.). *Acta Horticulturae*, Vol.474, pp. 545-548

Iannotta, N.; Monardo, D.; Perri, E. & Perri, L. (2001). Comportamento di diverse cultivar di olivo nei confronti degli attacchi di *Bactrocera oleae* (Gmel.) e correlazione con la quantità di oleuropeina presente nelle drupe, *Atti Convegno "Biodiversità e sistemi ecocompatibili", Caserta, 2001*, pp. 649-653.

Iannotta, N.; Monardo, D. & Perri, L. (2002). Relazione tra contenuto e localizzazione dell'oleuropeina nella drupa e attacco di *Bactrocera oleae* (Gmel.), *Atti Convegno Internazionale di Olivicoltura. Spoleto, 2002*, pp. 361-366.

Iannotta, N. & Monardo, D. (2004). Suscettibilità di cultivar di olivo a *Spilocaea oleagina* (Cast.) Hugh. e correlazione con il contenuto di oleuropeina nelle foglie. *Conv. "Germoplasma Olivicoli e tipicità dell'olio"*, Perugia 5 dicembre 2003, pp. 216-220.

Iannotta, N.; Condello, L.; Perri, L. & Belfiore, T. (2006a). Valutazione di suscettibilità di genotipi di olivo nei confronti di *Bactrocera oleae* (Gmel.). *Italus Hortus*, Vol.13 (2), pp. 242-245.

Iannotta, N.; Macchione, B.; Noce, M.E.; Perri, E. & Scalercio, S. (2006b). Olive genotypes susceptibility to the *Bactrocera oleae* (Gmel.) infestation. *Proceedings of the Second International*

Seminar on "Biotechnology and Quality of Olive tree Products around the Mediterranean basin", Olivebioteq, Marsala-Mazara del Vallo 5-10 November 2006, Vol. II: 261-266.

Iannotta, N.; Noce, M.E.; Scalercio, S. & Vizzarri, V. (2006c). Behaviour of olive cultivars towards the knot disease caused by *Pseudomonas savastanoi*. *Journal of Plant Pathology*, 88 (3, Special Issue), S45.

Iannotta, N.; Noce, M.E.; Perri, L.; Scalercio, S. & Vizzarri, V. (2006d). Susceptibility of olive cultivars to the *Camarosporium dalmaticum* (Thüm) infections. Proceedings of the *Second International Seminar on "Biotechnology and Quality of Olive tree Products around the Mediterranean basin", Olivebioteq, Marsala-Mazara del Vallo 5-10 November 2006*, Vol. II: 311-314.

Iannotta, N.; Noce, M.E.; Ripa, V.; Scalercio, S. & Vizzarri, V. (2007a). Assessment of susceptibility of olive cultivars to the *Bactrocera oleae* (Gmel.) and *Camarosporium dalmaticum* (Thüm.) Zachos & Tzav.-Klon. attacks in Calabria. *Journal of Environmental Science and Health, Part B*, Vol.42, pp. 789-793.

Iannotta, N.; Belfiore, T.; Monardo, D.; Noce, M.E.; Scalercio, S. & Vizzarri, V. (2007b). Indagine nel germoplasma dell'olivo sul comportamento di numerosi genotipi in relazione alla loro suscettibilità agli attacchi parassitari. *Acta Biologica*, Vol.83, pp. 215-220.

Katsoyannos, B.I. (1989). Field responses of Mediterranean fruit flies to spheres of different color patterns and to yellow crossed panels, *in* R. Cavalloro (ed.), *Fruit flies of Economic Importance*, Balkema, Rome, Italy, pp. 393-400.

Katsoyannos, B.I.; Patsouras, G. & Vrekoussi, M. (1985). Effect of colour hue and brightness of artificial oviposition substrates on the selection of oviposition sites of *Dacus oleae*. *Entomol. Exp. Appl.*, Vol.38(3), pp. 205-214.

Le Tutour, B. & Guedon, D. (1992). Antioxidative activities of *Olea europea* leaves and related phenolic compounds. Phytochemistry, Vol.31, pp. 1173-1178.

Limiroli, R.; Consonni, R.; Ottolina, G.; Marsilio, V.; Bianchi, G. & Zetta, L. (1995). 1H and 13C NMR Characterization of new Oleuropein Aglycones. *J. Chem. Soc., Perkin Trans. 1*, Vol.1, pp. 1519-1523.

Lo Scalzo, R.; Scarpati, M.L.; Verzengnassi, B. & Vita, G. (1994). *Olea europaea* chemical repellent to *Dacus oleae* females. *J. Chem. Ecol.*, Vol.20, pp. 1813–1823.

Markakis, E.A.; Tjamos, S.E.; Antoniou, P.P.; Roussos, P.A.; Paplomatas, E.J. & Tjamos, E.C. (2010) . Phenolic responses of resistant and susceptible olive cultivars induced by defoliating and nondefoliating *Verticillium dahliae* pathotypes. *Plant Disease*, Vol.94 (9), pp. 1156-1162.

Panizzi, L.; Scarpati, M.L. & Oriente, E.G. (1960). Structure of oleuropein, bitter glycoside with hypotensive action of olive oil. *Note II. Gazz. Chim. Ital.*, Vol.90, pp. 1449-1485.

Pereira, J.A.; Alves, M.R.; Casal, S. & Oliveira, M.B.P.P. (2004). Effect of olive fruit fly infestation on the quality of olive oil from cultivars Cobrançosa, Madural and Verdeal Transmontana. *Ital. J. Food Sci.*, Vol.16, pp. 355-365.

Pereira, J.A.; Pereira, A.P.G.; Ferreira, I.C.F.R.; Valentão, P.; Andrade, P.B.; Seabra, R.; Estevinho, L. & Bento, A. (2006). Table olives from Portugal: phenolic compounds, antioxidant potential and antimicrobial activity. *J. Agric. Food Chem.*, Vol.54, pp. 8425-8431.

Pereira, J.A.; Oliveira, I.; Sousa, A.; Valentao, P.; Andrade, P.B.; Ferreira, I.C.F.R.; Ferreres, F.; Bento, A.; Seabra, R. & Estevinho, L. (2007). Walnut (Juglans regia L.) leaves: Phenolic compounds, antibacterial activity and antioxidant potential of different cultivars. Food and Chemical Toxicology, Vol.45, pp. 2287–2295.

Proestos, C.; Chorianopoulos, N., Nychas, G.J. & Komaitis, M. (2005). RP-HPLC analysis of the phenolic compounds of plant extracts investigation of their antioxidant capacity and antimicrobial activity. *J. Agric. Food Chem.*, Vol.53, pp. 1190-1195.

Prokopy, R.J. and Owens, E.D. (1983). Visual detection of plants by herbivorous insects. *Ann. Rev. Entomol.*, Vol.28, pp. 337-64.

Prokopy, R.J. and Papaj, D.R. (2000). Behavior of flies of the genera Rhagoletis, Zonosemata, and Carpomya (Trypetinae: Carpomyina), *in* M. Aluja and A.L. Norrbom (eds.), *Fruit flies (Tephritidae): phylogeny and evolution of behavior*, CRC, Boca Raton, FL., pp. 219–252.

Rauha, J.P.; Remes, S.; Heinonen, M.; Hopia, A.; Kähkönen, M.; Kujala, T.; Pihlaja, K.; Vuorela, H. & Vuorela, P. (2000). Antimicrobial effects of Finnish plant extracts containing flavonoids and other phenolic compounds. *Int. J. Food Microbiol.*, Vol.56, pp. 3-12.

Rizzo, R. & Caleca, V. (2006). Resistance to the attack of Bactrocera oleae (Gmelin) of some Sicilian olive cultivars. *Proceedings of the Second International Seminar on "Biotechnology and Quality of Olive tree Products around the Mediterranean basin", Olivebioteq, Marsala-Mazara del Vallo 5-10 November 2006*, Vol. II: 291-298.

Servili, M.; Baldioli, M.; Mariotti, F. & Montedoro, GF. (1999). Phenolic composition of olive fruit and virgin olive oil: distribution in the constitutive parts of fruit and evolution during oil mechanical extraction process. *Acta Horticulturae*, Vol.474, pp. 609-619.

Soler-Rivas, C.; Espin, J.C. & Wichers, H.J. (2000). An easy and fast test to compare total free radical scavenger capacity of foodstuffs. *Phytochem. Anal.*, Vol.11, pp. 330-338.

Spadafora, A.; Mazzuca, S.; Chiappetta, F.; Parise, A.; Perri, E. & Innocenti, A.M. (2008). Oleuropein-specific-b-glucosidase marks early response of olive fruit (*Olea europaea*) to mimed insect attack. *Agricultural sciences in China*, Vol.7, pp. 703-712.

Tassou, C.; Nychas, G.J.E. & Board, R.G. (1991). Effect of phenolic compound and oleuropein on the germination of *Bacillus cereus* T-spores. *Biotechnol. Appl. Biochem.*, Vol.13, pp. 231–237.

Tassou, C. & Nychas, G.J.E. (1994). Inhibition of Staphylococcus aureus by Olive phenolics in broth and in Food Model System. *Journal of Food Protection*, Vol.57, pp. 120-124.

Tassou, C. & Nychas G.J.E. (1995). Inhibition of *Salmonella enteritidis* by oleuropein in broth and in a model Food system. *Letters in Applied Microbiology*, Vol.20, pp. 120-124

Tranter, H.S.; Tassou, S.C. & Nychas, G.J. (1993). The effect of the olive phenolic compound, oleuropein, on growth and enterotoxin B production by *Staphilococcus aureus. J. Appl. Bacteriol.*, Vol.74, pp. 253-259.

Vinha, A.F.; Ferres, F.; Silva, B.M.; Valentão, P.; Gonçalves, A.; Pereira, J.A.; Oliveira, M.B.; Seabra, R.M. & Andrade, P.B. (2005). Phenolic profiles of Portuguese olive fruits (Olea europaea L.): Influences of cultivar and geographical origin. Food Chemistry, Vol.89, pp. 561-568.

Vizzarri, V.; Ferrara, M.; Salimonti, A.; Zelasco, S.; Iannotta, N.; Santilli, E.; Perri, E. &Nigro, F. (2011). Studio della resistenza di genotipi di olivo a *Verticillium dahliae* kleb. e relativa risposta della pianta. *Acta Italus Hortus*, in press.

Walter, W.M.Jr.; Fleming, H.P. & Etchells, J.L. (1973). Preparation of antimicrobial compounds by hydrolysis of oleuropein from green olives. *Appl. Microbiol.*, Vol.26, pp. 773-776.

Chapter 6

Phytosanitary Certification

Giuliana Albanese, Maria Saponari and Francesco Faggioli

Additional information is available at the end of the chapter

1. Introduction

Olive plants are among the most ancient cultivated fruit trees. Over the centuries, propagation occurred mainly vegetatively. The longevity of trees and the latency of most of the virus infections allowed the dissemination through the propagative material of hidden viruses, which were not detected until recently, when the advent of novel diagnostic tools surprisingly revealed that virus infections are in fact widespread.

In the past, the selection of high value olive germplasm has been mainly based on the agronomic and pomological traits of the plants and on the quality and yield of the olive-derived product (oil). Specifically, investigation on the sanitary status of the selected ecotypes were done mainly by visually inspections. This fact leaded to the propagation and spread of systemic pathogens harbored either in a latent form or in the form of specific symptoms that initially have been confused with the phenotypic expression of the plant (as in the case of symptoms caused by infection of *Strawberry latent ringspot virus* (SLRSV) in the cultivar 'Raggiola')

In order to prevent the spread of dangerous pathogens (*Verticillium dahliae, Pseudomonas savastanoi* pv. *savastanoi)* and viruses in particular, remedies rely mainly on preventive measures such as the use of pathogen-tested propagative material. The main approach used to obtain, propagate and commercialize plants free from harmful pathogens is through phytosanitary selection and certification programs, which also encompass pomological selection for trueness to type and superior quality traits. In order to obtain pathogen-free material from infected trees, sanitation treatments such as heat therapy, meristem tip culture and micrografting, although still limited for their application for virus elimination in olive plants, can be applied.

A certification program is a procedure whereby single well-analysed candidate mother plants (nuclear stock plants) are used as sources of propagation material with a process of filiation. In this way, it is possible to provide growers with high quality (genetic and sanitary) material. The certification scheme in general, and phytosanitary in particular, can be adopted either for worldwide spread varieties or for those locally distributed.

Each step of the propagation (descendent filiation) (Pre-basic, Basic and Certified material) must comply with the requirements that are intended to produce and maintain the selected material in the best growing conditions as specified by the enforced phytosanitary regulations. In particular, the sanitary status must be assessed following the officially recognized technical procedures, regarding the list of the target pathogens, type of sample, period of sampling and protocol for testing.

Phytosanitary selection requires the use of appropriate diagnostics protocols for pathogen detection. The difficulty in recognising and/or diagnosing virus-infected olive trees during field surveys imposes the use of laboratory tests in order to assess the absence of the target pathogens. Due to the lack of indicators for the biological assays and the unreliability of the ELISA test in olive, the application of molecular diagnostic techniques for viruses, fungi and bacteria detection became, in the recent past, critical for the assessment of the sanitary status of a given selected ecotype. These sensitive and reliable methods are absolutely necessary as they are at the basis of efficient and valid certification programs.

The increasing international demand for olive products, and therefore the expansion of olive crops, is stimulating the exchange of olive germplasm in new areas of the world, prompting for the adoption at European and International level of harmonised Certification Programs that reduce risks of pathogen dissemination and ensure the commercialisation of high quality propagative material and, consequently, guarantee high quality olive productions.

2. Systemic pathogens transmissible with the propagative material

To date, 8 virus-like diseases have been described, and fifteen different viruses (Tab. 1) and five phytoplasmas have been identified in olive plants. The actual Italian olive certification law (DM 20/11/2006) imposes the absence, in the propagation material, of some of the abovementioned pathogens as well as the most dangerous fungus and the most widespread bacterium in olive crops (Tab. 2). An appraisal of other ways of transmission than vegetative propagation if known, the susceptible hosts, effects and diseases caused by these pathogens and Countries where they have been detected in olive is reported below.

Arabis mosaic virus (ArMV) is a member of the genus *Nepovirus*, family *Secoviridae* (Sanfaçon et al., 2011). It is transmitted by the longidorid nematode *Xiphinema diversicaudatum,* but there is no evidence of its transmission to olive plants by this vector. The main hosts of this virus are strawberry, hop, *Vitis* spp., raspberry (*Rubus idaeus*), *Rheum* spp., *Sambucus nigra*, sugarbeet, celery, gladiolus, horseradish and lettuce. The most common symptoms induced by ArMV are leaf mottling and flecking, stunting and several forms of deformation including enations. Because of the serious damages caused on some crops this virus is inserted among the "harmful organisms known to occur in the community and relevant for the entire community" in Directive 2000/29/EC and its absence must be determined on plant material of *Fragaria* and *Rubus*. The symptoms vary depending on the host plant but also on the virus isolate, cultivar, season and year. Many plant species infected with ArMV, including olive trees, do not show any symptoms (Martelli et al., 2002). The virus has been reported in olive trees from Italy (Savino et al., 1979), Portugal (Martelli, 2011) Egypt, USA (Saponari & Savino, 2003), Turkey (Çağlayan et al., 2004), Syria (Alabdullah et al., 2005) and Lebanon (Fadel et al., 2005).

Cherry leaf roll virus (CLRV) belongs to the family *Secoviridae*, genus *Nepovirus*, subgroup c. Even if it is classified as a *Nepovirus*, its transmission by nematodes has not yet been demonstrated to date, whereas it effectively occurs by pollen and, in some hosts, very efficiently by seed too. In olive plants, its transmission by means of pollen has not been demonstrated, but has been ascertained by seeds at the rate of 41% (Saponari et al., 2002). CLRV infects many herbaceous, shrubs and woody plants of genera: *Betula, Celtis, Cornus, Fagus, Juglans, Ligustrum, Olea, Populus, Ulmus, Rubus, Sambucus* and *Rheum*. The virus often induces symptoms in ash, birch, cherry, elderberry and walnut including delayed leaf development, chlorotic leaf streaks or spots, as well as dieback of branches or whole trees but it is symptomless in olive trees (Savino & Gallitelli, 1981). Its presence in olive trees was reported in Italy, Portugal, Spain (Martelli, 1999), then in Egypt, USA (Saponari & Savino, 2003), Turkey (Çağlayan et al., 2004), Syria (Alabdullah et al., 2005), Lebanon (Fadel et al., 2005) and recently in Croatia (Luigi et al., 2011), where it has been shown to have a negative impact on olive fruit and virgin oil quality (Godena et al., 2012).

Strawberry latent ringspot virus (SLRSV) is an unassigned species in the *Secoviridae* family. It is transmitted by the nematode *X. diversicaudatum* and by seed in several species (Cooper, 1986), but in olive plants, these kinds of means of transmission have not been demonstrated. SLRSV infects strawberry and raspberry, mostly without symptoms but resulting in various degrees of mottle and decline in some cultivars. The virus was isolated for the first time from olive in cv. 'Corregiolo' in Italy (Savino et al., 1979) and later in Portugal (Henriques et al., 1992), Spain (Bertolini et al., 1998), Egypt, USA (Saponari & Savino, 2003), Turkey (Çağlayan et al., 2004), Lebanon (Fadel et al., 2005), Syria (Alabdullah et al., 2005), Croatia (Bjelis et al., 2007), Tunisia (Martelli, 2011) and Albania (Luigi et al., 2009). Small, pear-shaped, puckered fruits with deformed kernels (bumpy fruits), narrow and twisted leaves, bushy growth and reduced crop were described in olive trees of cv. 'Ascolana tenera' affected by SLRSV (Marte et al., 1986). Similar symptoms were observed in cvs 'Negrinha' and 'Galega' in Portugal, associated with a severe reduced rooting ability of the cuttings (Henriques et al., 1992). Among 15 different olive cultivars reporting plants being affected by SLRSV in Portugal, only some showed symptoms (Henriques et al., 1992) in agreement with what was observed in Italy (Savino et al., 1979; Marte et al., 1986); no symptoms are apparently associated with SLRSV infections in Spain (Bertolini et al., 1998). Very interesting is the fact that previously, the 'Raggiola' and 'Frantoio' were considered different olive varieties due to morphological and agronomical dissimilarities. A relatively recent study showed that the two cultivars are genetically identical and that their differentiations are due to the constant presence of SLRSV in 'Raggiola' and the repeated SLRSV absence in 'Frantoio' (Fig. 1) (Ferreti et al., 2002). Rooting trials conducted to compare SLRSV-infected 'Raggiola' with virus-free 'Frantoio' showed that the virus does not influence the rooting rate of olive cuttings (Roschetti et al., 2009), contrary to what had been previously reported in Portugal.

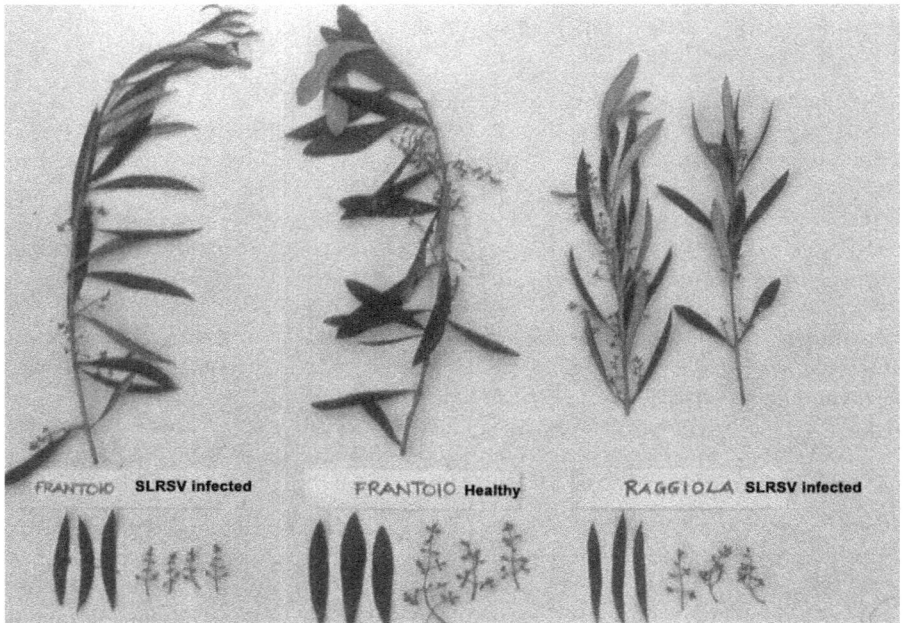

Figure 1. Phenotypic expression of olive cultivars 'Frantoio' SLRSV-affected, healthy 'Frantoio' and 'Raggiola' SLRSV-affected. See as morphological aspect of 'Frantoio' SLRSV-affected and 'Raggiola' SLRSV-affected are identical (narrow leaves and small inflorescences)

Cucumber mosaic virus (CMV) belongs to the genus *Cucumovirus*, family *Bromoviridae*. It is one of the most dangerous virus affecting vegetable plants (about 800 wild and cultivated plant species are its hosts). CMV induces important vegetative and productive reductions (up to 100% in plants such as tomato and pepper). When this virus affects herbaceous plants, it is transmitted very efficiently by 75 different aphid species and with varying efficiency by seed. CMV infection in olive is symptomless and its transmission by aphid vectors to/from olive has not yet been proven. It was isolated the first time from olive trees by Savino and Gallitelli (1983) in Italy. This report was confirmed in Portugal by Rei et al. (1993) who detected CMV alone, as well as together with SLRSV. CMV in olive trees was also found in Spain (Bertolini et al., 1998), Turkey (Çağlayan et al., 2004), Syria (Alabdullah et al., 2005), Croatia (Bjelis et al., 2007), Tunisia (Martelli, 2011) and recently in California (Al Rwahnih et al., 2011).

Olive latent virus 1 (OLV-1) is a member of the genus *Necrovirus*, family *Tombusviridae*. The virus is one of the few viruses detected in olive trees that is transmitted by seed (at a rate of 82%) (Saponari et al., 2002). It was detected in olive trees in Italy (Gallitelli & Savino, 1985),

Jordan (Martelli et al., 1995), Portugal (Felix & Clara, 2002), Egypt, USA (Saponari & Savino, 2003), Lebanon (Fadel et al., 2005), Syria (Alabdullah et al., 2005), Tunisia (Martelli, 2011) and Turkey (Serce et al., 2007). Several OLV-1 isolates have been obtained from symptomless or weakened trees. Since this virus has also been isolated from citrus in Turkey and Italy (Martelli et al., 1996) and from tulips in Japan (Kanematsu et al., 2001) it is reasonable to assume that it, as well as other olive viruses, may also have a larger host range.

Olive latent virus 2 (OLV-2) is the type species of the monotypic genus *Oleavirus,* family *Bromoviridae* (Grieco & Martelli, 1997). OLV-2 was isolated by mechanical inoculation from symptomless olive trees in Apulia, Southern Italy (Savino et al., 1984). It has subsequently been identified in Lebanon (Fadel et al., 2005), Syria (Alabdullah et al., 2005), Croatia (Bjelis et al., 2007) and Tunisia (Martelli, 2011) from symptomless olive cultivars. The host range of OLV-2 was limited to olive trees until castor beans (*Ricinus communis* L.), showing yellowish vein netting and systemic mottling on leaves, were reported in Greece to be infected with this virus (Grieco et al., 2002).

Olive latent virus 3 (OLV-3) is classified as a tentative member of the genus *Marafivirus,* family *Tymoviridae*. The virus is not mechanically transmitted. Search of possible vectors, *Euphyllura olivina* and *Saissetia oleae,* was not successful even if OLV- 3 was detected by RT-PCR in the psyllid. A survey conducted in the Mediterranean region showed the OLV-3 presence in Italy, Syria, Malta, Tunisia, Portugal, Turkey, Lebanon and Greece with an infection rate average of 30% always in symptomless olive trees (Alabdullah et al., 2010).

Olive latent ringspot virus (OLRSV) is an approved species of the genus *Nepovirus,* family *Secoviridae*. The virus is transmitted by mechanical inoculation and the existence of a natural vector is unknown. OLRSV is latent in olive trees, but it causes some symptoms on diagnostically susceptible hosts, such as apical necrosis on *Chenopodium quinoa* and *C. amaranticolor,* and red-rimmed local lesions and malformation on tip leaves of *Gomphrena globosa*. The virus was isolated from asymptomatic olive trees in Lazio, Central Italy (Savino et al., 1983), in Portugal in 1990, in Syria (Alabdullah et al., 2005) and then in Tunisia (Martelli, 2011).

Olive leaf yellowing associated virus (OLYaV) is an unassigned species in the family *Closteroviridae*. Various studies have been published (Sabanadzovic et al., 1999; Essakhi et al., 2006; Luigi et al., 2010) and are still in progress to define its taxonomic position. OLYaV presence in psyllid *E. olivina* and unidentified mealybugs of genus *Pseudococcus* gave the indication that transmission by these vectors could be possible (Sabanazdovic et al., 1999). The olive leaf yellowing (OLY) disease was recorded for the first time in Italy on cv. 'Biancolilla' (Savino et al., 1996) and it is characterized by a bright leaf yellow discoloration (Fig. 2). A survey conducted in Italy showed that, in old OLYaV-affected olive trees, leaf yellowing symptom is frequently absent (Albanese et al., 2003). The OLY syndrome, consisting of poor fruit set, bright yellow discoloration of the foliage, mottling, necrosis, extensive defoliation and dieback, has been associated to other viruses such as *Olive vein yellowing associated virus (OVYaV)* (Faggioli & Barba, 1995) and *Olive yellow mottle and decline associated virus (OYMDaV)*

(Savino et al., 1996), but their presence on olive trees was very rare. On the other hand, OLYaV seems to be one of the most widespread olive viruses: in Italy it infects more than 60% of southern Italy olive cultivars (Faggioli et al., 2005) and it has also been reported in high percentages in Israel (Martelli, 2011), Egypt, USA (Saponari & Savino, 2003), Lebanon (Fadel et al., 2005), Spain (Martelli, 2011), Syria (Alabdullah et al., 2005), Albania (Luigi et al., 2009), Croatia (Bjelis et al., 2007), Tunisia (Martelli, 2011) and California (Al Rwahnih *et al.*, 2011). A study on the rooting and grafting capacity of OLYaV-infected 'Carolea' and its respective healthy controls showed that the virus does not influence the rate of rooting of the cuttings and does not interfere with the grafting success rate; positive significant effects in grafting ability were observed on infected material only during a temperature stress, probably due to the reduced water need of infected shoots (Roschetti et al., 2009). Significant difference in vegetative growth was observed between virus-free and OLYaV-infected young olive plants, demonstrating negative OLYaV interference (Cutuli et al., 2011). To date, no other hosts have been found for this virus.

(a)   (b)

Figure 2. Yellowing symptoms in an olive cultivar 'Carolea' (a) which tested positive for OLYaV and detail of yellow leaves of the same tree (b)

Olive semilatent virus (OSLV) is still unclassified. It was transmitted mechanically from olive tree to *Nicotiana benthamiana* (Materazzi *et al.*, 1996). The main symptom observed in Italy on OSLV-affected olive trees was a very mild chlorotic vein clearing of the leaves, but there is not enough evidence of the etiological involvement of this virus in the disease (Martelli, 1999).

Tobacco mosaic virus (TMV) belongs to the genus *Tobamovirus*, family *Virgaviridae*. Mechanical transmission to herbaceous indicator plants was possible, but not easily. It was isolated in central Italy from olive trees showing vein banding, discolorations along the main veins, severe defoliation and decline (Triolo et al., 1996). However, there is no conclusive evidence that TMV is agent of these symptoms.

Tobacco necrosis virus (TNV). Viruses with properties similar to those of TNV were first detected in symptomless olive trees by Félix & Clara (2002) in Portugal. One isolate was studied further, revealing its identity as TNV-D species (Cardoso et al., 2004). However, further genomic characterization of this isolate led to its classification as a new species in the *Necrovirus* genus named *Olive mild mosaic virus (OMMV)* (Cardoso et al., 2005). To date, it is not clear if TNV-D can be considered among the viruses isolated from olive trees, even if recent data shows the presence of this species in olive trees (Cardoso and colleagues deposited the complete genome sequence of a TNV-D isolate from olive trees in the Gene Bank, accession number FJ666328). Virions are readily transmitted by mechanical inoculation and naturally by the fungus *Olpidium brassicacae*. TNV has a wide host range that includes monocotyledonous and dicotyledonous plants, which frequently cause necrotic lesions on the roots and leaves.

Acronym	Virus species	Genus	Geographical distribution
OLV 1	*Olive latent virus 1*	*Necrovirus*	Italy, Jordan, Portugal, Egypt, USA, Lebanon, Syria, Turkey, Tunisia
OLV 2	*Olive latent virus 2*	*Oleavirus*	Italy, Syria, Croatia, Lebanon, Tunisia
OLV 3	*Olive latent virus 3*	*Marafivirus*	Greece, Italy, Lebanon Malta, Portugal, Syria, Tunisia, Turkey
OLRSV	*Olive latent ringspot virus*	*Nepovirus*	Italy, Portugal, Syria, Tunisia
OVYaV	*Olive vein yellowing associated virus*	*Potexvirus*	Italy
OYMDaV	*Olive yellow mottling and decline associated virus*	Unclassified	Italy
OLYaV	*Olive leaf yellowing associated virus*	*Closteroviridae*, unassigned species	Albania, Croatia, Egypt, Italy, Israel, Lebanon, Spain, Syria, Tunisia, USA
OSLV	*Olive semilatent virus*	Unclassified	Italy
OMMV	*Olive mild mosaic virus*	*Necrovirus*	Portugal
SLRSV	*Strawberry latent ringspot virus*	*Secoviridae*, unassigned species	Croatia, Egypt, Italy, Lebanon, Portugal, Spain, Syria, Tunisia, Turkey, USA
CLRV	*Cherry leafroll virus*	*Nepovirus*	Croatia, Egypt, Italy, Lebanon, Portugal, Spain, Syria, Tunisia, Turkey, USA
ArMV	*Arabis mosaic virus*	*Nepovirus*	Egypt, Italy, Lebanon, Portugal, Syria, Turkey, USA
CMV	*Cucumber mosaic virus*	*Cucumovirus*	Croatia, Italy, Portugal, Spain, Syria, Tunisia, Turkey, USA
TMV	*Tobacco mosaic virus*	*Tobamovirus*	Italy
TNV	*Tobacco necrosis virus*	*Necrovirus*	Portugal

Table 1. Viruses identified in olive trees and their geographical distribution (Martelli, 2011; Çağlayan et al., 2009)

Phytoplasmas constitute a monophyletic clade within the *Mollicutes* class. Their classification has been possible through the use of restriction fragment length polymorphism (RFLP) analysis and sequencing of the conserved 16S rRNA gene (Lee et al., 1998; Semüller et al., 1998). A variable range of symptoms in olive trees such as shoot proliferation, shortening of internodes, witches'-brooms, little leaves (Fig. 3a), leaf rolling and yellowing, leaf bronzing, phyllody, flower abortion, hypertrophied inflorescences (Fig. 3b), fasciation, erect growth, dwarfing, decline and die-back have been frequently associated with the presence of phytoplasma in Spain, Italy and Iran (Ahangara et al., 2006; Bertaccini et al., 2002; Font et al., 1998; Pasquini et al., 2000). Identification of phytoplasmas detected in olive plants showed they were members of the 16S-IB (Aster yellow), 16S-IC (Clover phyllody), 16Sr-III (Peach X disease), 16S-VA (Elm yellow) or 16S-XIIA (Stolbur) groups and subgroups. The failure to detect phytoplasmas in many symptomatic olive trees leads to doubts on whether these type of alterations could be associated with other causes (Barba, 1993; Camele et al., 1999). Nevertheless, phytoplasmas detected in olive plants are agents causing very well known and severe diseases in other hosts. These include aster yellow, clover phyllody, peach X disease, elm yellow and stolbur in solanaceous plants, as well as grapevine yellow (= Bois Noir). Even if their transmission by leaf-hopper vectors has been proven for some of them (among various host plants but not yet in and from olive plants) their presence in olive plants indicates a serious potential threat for other important crops.

(a)   (b)

Figure 3. Shortened internodes, witches'-brooms and little leaves (a); hypertrophied inflorescences (b) on olive trees affected by phytoplasmas

The fungus *Verticillium dahliae* is a soil-borne pathogen that attacks olive trees (as well as over a hundred woody and herbaceous species), particularly when their roots are stressed. It causes the most severe disease suffered by olive plants, named Verticillium wilt that induces yellow leaves, defoliation (Fig. 4) and death due to the fungus attacking the plants' vascular system. Internally, a dark reddish brown streak on the wood occurs in most plants. This is visible on branches when the bark peeled off. If the cross-section of infected branches or trunks is examined, the brown woody coloration may appear as a ring. Although some plants may die quickly, more commonly trees with only a few wilted branches during a growing season become more severely infected the following year. After the first report of Verticillium wilt in Italy, it has later been detected in Algeria, Arizona, California, Egypt, France, Greece, Iran, Malta, Marocco, Syria, Spain and Turkey (Bubici & Cirulli, 2011). Another species, *V. albo-atrum,* may occasionally cause the same disease in olive plants.

Figure 4. Yellow leaves, defoliation and wilt of olive caused by *V. dahliae* (photo by Antonio Ippolito)

The bacterium *Pseudomonas savastanoi* pv. *savastanoi* causes the most frequent disorder occurring in olive plants known as olive knot disease. The disease manifests itself through the growth of tubercles (Fig. 5) , which either appear individually, or in clusters on any part of the plant, but most commonly on twigs, young branches and around wounds on the main trunk. Knots can damage the stem structure and can deform the scaffold of the tree if infection is severe during the early stages of the tree. This may become a serious problem in nurseries that grow olive plantlets for marketing. *P. savastanoi* causes a similar disease in other plants as oleander, ash, jasmine, Japanese privet, *Forsythia* spp., *Phyllirea* spp., *Retama sphaerocarpa, Rhamnus alathernus* and myrtle (Surico & Marchi, 2011). This bacterial disease is present in all areas of the world where olive plants are cultivated. This is due to the ability of its causal agent to colonize the phylloplane of the tree.

Figure 5. Olive knots: rough galls and swellings on twigs and branches caused by *P. savastanoi* pv. *savastanoi*

Harmful organisms	Acronym	Sanitary status	
		Virus-free (VF)	Virus-tested (VT)
VIRUSES:			
Arabis mosaic	ArMV	X	X
Cherry leafroll	CLRV	X	X
Strawberry latent ringspot	SLRSV	X	X
Cucumber mosaic	CMV	X	
Olive latent 1	OLV-1	X	X
Olive latent 2	OLV-2	X	
Olive leaf yellowing associated	OLYaV	X	X
Tobacco necrosis	TNV	X	
PHYTOPLASMAS		X	X
FUNGI:			
Verticillium dahliae		X	X
BACTERIA:			
Pseudomonas savastanoi pv. *savastanoi*		X	X
NEMATODES:			
Meloidogyne incognita		X	X
Meloidogyne javanica		X	X
Pratylenchus vulnus		X	X
Xiphinema diversicaudatum		X	X

X = the absence of this organism must be ascertained

Table 2. Pathogens and pests that must be absent in order to obtain the "virus–free" or " virus–tested" sanitary status according to Italian olive certification law (DM 20/11/2006)

3. Strategies to control invasive olive pathogens and the importance of phytosanitary certification

Pathogens associated with olive propagative material may be systemic (viruses and phytoplasmas and probably *P. savastanoi* pv. *savastanoi*) or associated with the vascular system (*V. dahliae*) and they are unlikely to be eliminated during the vegetative propagation of an infected source. Accordingly, local and long-distance spread of these pathogens through the movement of infected propagative material has caused a highly threatening worldwide distribution of infectious diseases. The symptomless nature of several olive virus infections may also contribute to the inadvertent propagation and distribution of infected material.

To avoid disease and/or pathogen dissemination through vegetative propagation, possible remedies include mainly preventive strategies based on the use and propagation of "healthy" mother plants. In fact, in order to attain sanitary improvements of any crop, a system of preventive, protective and often of sanitation measures has to be established and implemented, encompassing a complex series of interventions currently referred to as "phytosanitary selection and certification".

In the framework of a phytosanitary and clonal improvement program, the main activities include: (i) field surveys for the selection of olive trees with no apparent disease symptoms and fulfilling the pomological traits of the cultivar; (ii) samples collection for laboratory tests, both for pathogen detection and DNA marker analysis; (iii) molecular tests (RT-PCR, dot blot hybridization and dsRNA analysis) for the detection of viruses included in the certification program; (iv) genetic characterisation using SSR markers; (v) sanitation by heat therapy, meristem tip culture and micrografting in case of no healthy trees being detected for one or more cultivars; (vi) propagation of the candidate nuclear stocks under conditions that ensure freedom from re-infections, usually in insect-proof greenhouses.

Field surveys should be carried out in the main olive-growing areas for the specific cultivar undergoing the clonal and sanitary selection program. Usually mature trees are selected (i.e. 25-year-old) based on visual inspection during spring and autumn. Samples for virus testing consist of 10-15 cuttings collected from 1- to 2-year-old twigs or young leaves for DNA extraction and SSR marker analysis.

Despite limited information being available on the application and effectiveness of sanitation protocols on olive plants, *in vivo* and *in vitro* heat therapy, *in vitro* shoot tip culture and micrografting have all been applied in attempts to regenerate OLYaV- and CLRV-free material and some successful results have been obtained (Bottalico et al, 2002). For *in vivo* heat therapy, plants can be grown at 38°C for 3 to 12 months. During heat therapy, 2 to 2.5 cm long shoot tips are excised no earlier than three months from the beginning of the treatment. After surface-sterilization in 0,05% mercuric hydrochloride for 10 min, the shoot tips are placed *in vitro* in petri dishes on different media according to the cultivar [OM (Rugini, 1984); MSM media (Leva et al., 1994)] and grown at 24°C with a 16 h photoperiod.

Regenerated explants are subcultured 3 or 4 times every four weeks using the proliferation medium reported by Rugini (1984), prior to transplanting in jiffy pots. For *in vitro* heat therapy, 2 to 3 cm long in vitro plantlets are exposed for 1 to 3 weeks to 38°C. After heat treatment and subcultures, the surviving plantlets are transplanted in jiffy pots.

For shoot tip culture, apexes are excised from well-established *in vitro* cultures maintained on OM or MSM media. Regenerated apexes are subcultered on the same media prior to being transplanted in jiffy pots.

Olive seedlings for micrografting are recovered from seeds soaked for a few seconds in 70% alcohol before the endocarp removal. Kernels are then soaked for 10 min in 0.05% mercuric hydrochloride solution, rinsed three times in sterile water, and placed on wet sterilized paper in petri dishes at 25° in the dark. After 2-3 months, the regenerated seedlings are cut, leaving about 1.0 cm of the epicotyl. Shoot apexes, excised from *in vitro* grown plantlets, are then grafted on the top of decapitated seedlings. After grafting, plants are cultured in rooting medium (Rugini, 1984). Generally less than 70% of the grafts are successful, and only 10% of the plants survive after transplantation in the soil.

Although phytoplasma diseases may be cured by treatment with certain classes of antibiotics and by heat water therapy, such approaches have not been applied to olive plants in order to obtain sanitation from these pathogens. This is because olive trees affected by phytoplasmas are very rare, and during a phytosanitary selection, phytoplasma-free plants can be easily found. Since their transmission by leaf-hopper vectors is ascertained for other crops, growing nuclear stock plants in insect-proof greenhouses ensures also freedom of infection from phytoplasmas.

The detection of *V. dahliae*-free plants must be carried out with great care. A visual diagnosis is in fact insufficient in guaranteeing the absence of this fungus and have recommended the use of new and sensitive diagnostic tools that are now available (see paragraph 4).

In order to prevent infections by *P. savastanoi* pv. *savastanoi* selected materials must be free from symptoms of the disease, and before propagation material is harvested, mother plants must be sprayed with a copper-based treatment to reduce risk of infections by the epiphytic bacterial population.

The candidate nuclear-stock material obtained through the field selection and/or sanitation treatments describe above can enter the certification program upon official approval (see paragraph 5), and genetically and sanitary certified propagative material will be available to growers.

Demand for olive products is constantly increasing in local and foreign markets, stimulating the expansion of olive crops and encouraging the exchange of olive germplasm at an international level. The activation of a selection and certification program is thus crucial to guarantee the quality of the propagative material and reduce risks for pathogen dissemination.

4. Identification of olive pathogens: updates on diagnostic tools

Sanitary certification programs require reliable and sensitive diagnostic tests in order to allow for the identification of pathogen-free trees and the assessment of their overall plant production processes. Due to the latency of several infections caused for example by viruses, visual inspections are not reliable and laboratory tests must be performed to certify virus-free or virus-tested materials. Biological tests and serological assays, widely used to detect pathogens affecting other crops like stone fruits, grapes, pome fruits, result ineffective in olive plants due to the absence of differential woody indicators for the bioassays and the low viral titre and/or to the interference by some contaminants. All these factors have made olive tree virus diagnosis very problematic. Luckily, in the last decade several molecular approaches have been developed and improved to detect olive viruses, bacteria and fungi in the propagating materials. Different molecular techniques such as RT-PCR in single/double step or nested, PCR, real time PCR, dot blot hybridization and dsRNA analysis, have been implemented in recent years and drastically improved sensitivity and specificity of olive-infecting pathogens' diagnosis. Recently, molecular technology has been successfully applied for routine and large scale detection and could easily be transferred to those Countries that intend to develop their olive crops through production, maintenance and distribution of healthy (virus-free or virus-tested) planting material.

Concerning viruses, RT-PCR assay has proved to be the most rapid, sensitive and reliable technique for detecting an RNA target in infected plants, and in recent years, different protocols have been developed for olive viruses detection (Grieco et al., 2000; Bertolini et al., 2001a, 2003; Pantaleo et al., 2001; Faggioli et al., 2002, 2005). Recently, a one step RT-PCR protocol has been set up and validated in an inter-laboratory ring test (Loconsole et al., 2010) for the diagnosis of the eight most important olive viruses. This should be a starting point for anyone wishing to approach the sanitary selection of olive plants. New and improved diagnostic techniques (e.g. Real Time RT-PCR, multiplex RT-PCR, polyprobe for molecular hybridization) will be continuously developed as the knowledge on the genetics and biology of olive-infecting pathogens advances.

Phytoplasma detection is now accomplished through nested-PCR on total DNA extracted from olive plants using the protocol of Barba et al. (1998). Gene amplification is performed using a direct PCR with primers P1/P7 (Deng & Hiruki, 1991; Schneider et al., 1995), followed by a nested-PCR with primers R16F2/R2 (Lee et al., 1993). The use of this analysis allows to determine whether plants are affected by phytoplasma, but does not give information about the identity of the pathogen. Identification of phytoplasma can be achieved through restriction fragment length polymorphism (RFLP) analysis, but it is not required for sanitary certification since the certified olive material must be free from all phytoplasmas. In recent years, the diagnostic technique has also been developed and improved for phytoplasmas. Real time PCR protocols for the identification and group characterization of phytoplasmas are now available. Whilst not yet applied to the diagnosis of olive phytoplasmas, these techniques have all the necessary features for this purpose, and there is therefore potential for their use in the near future (Christensen et al., 2004; Hodgetts et al., 2009).

Diagnosis of *V. dahliae* is preliminarily performed through an accurate search for foliar symptoms and vascular browning. Foliar chlorosis and necrosis could be due to other causes such as root rot diseases, whereas browning on cross section of stems was sometimes not found. Conclusive detection is attempted by isolating the fungus on agar media from olive tissues and possibly using PCR or nested PCR with *V. dahliae* specific primers (Nigro et al., 2002; Mercado-Blanco et al., 2002) or by Real Time Scorpion PCR (Schena et al., 2004).

Identification of *P. savastanoi* pv. *savastanoi* is very easy when the typical knots are present on plants. Nevertheless, its presence in latent and systemic form has been reported by Penyalver et al., (2006). The presence of *P. savastanoi* pv. *savastanoi* both as epiphytic and entophytic agent makes its control in the certified material absolutely compulsory; moreover, for a more sensitive and reliable diagnosis, molecular techniques are needed. Recently, molecular protocols of nested-PCR have been set up to obtain reliable diagnoses of latent infections (Bertolini et al., 2001b). This is also possible (and has been done) simultaneously with four other olive viruses (Bertolini et al., 2003).

5. Phytosanitary certification program: The Italian experience in the last twenty years

The production of healthy, high-quality olive products depend to a large extent on the quality of the plant material used for olive cultivation. In general, the production of "healthy" plants for planting occurs through defined certification procedures by which a particular cultivated selection, whose health status and trueness to type have been officially attested, is propagated following specific requirements. In a typical certification scheme, the certified material is descended by a defined number of propagation steps from individual plants, found to be free from pathogens and pests, maintained and propagated under rigorous conditions that exclude recontamination.

To this end, appropriate regulations are necessary to ensure the production, marketing and movement of certified plant propagation material with high standards and free from harmful pathogens (Annex I). Over the past twenty years, through globalisation and the expansion of several crops in new areas, concerns were raised about new disease emergencies transmitted by infected propagating material. In an attempt to limit the potential impact of the spread of pests and pathogens through the movement of infected plants, specific laws have been issued at regional, national and European levels.

The Council Directive 2008/90/EEC of 29 September 2008 (recast version of Directive 92/34/EEC) on the marketing of fruit plant (including olive) propagating material and fruit plants intended for fruit production established a harmonized Community regime which ensures that growers throughout the Community receive propagating material and fruit plants which are healthy and of good quality. This applies to fruit plant propagating material and fruit plants of genera and species listed in Directive Annex I, which may only be marketed if they are either CAC (*Conformitas Agraria Communitatis*), Pre-basic, Basic or Certified material.

To be classified as such, material must comply with the criteria of quality, plant health, testing methods and procedures, propagation systems and varietal aspects and must have been recognised following official inspections. In addition, propagating material or fruit plants may, in most circumstances, only be marketed by accredited suppliers, whose production methods and facilities meet the requirements of the Directive. Propagating material and fruit plants from Countries outside of the European Union (EU) may only be marketed within the Community if they offer the same guarantees as materials produced in the Community complying with Council Directive 92/34/EEC. Each EU Member State adopts its own enforcement and implementation policies using the EU Directive as a guide.

In Italy, it is compulsory that the production and marketing of olive propagating material fulfils the requirements established by the Italian Decree amended on 14/04/1997 in compliance with the aforementioned EU Directive. According to this law, assessment of true to type and certification of sanitary status of olive propagating materials are compulsory and plants are certified as CAC. In this kind of certification plants must be free from *Euzophera pinguis, Meloidogine* spp., *S. oleae, P. savastanoi* pv. *savastanoi, V. dahliae* and all known viruses (15 to date). However, several aspects concerning for example the procedures for inspections and controls are not well defined, leading to misinterpretations and heterogeneous application of the law in the different Italian regions by the regional Phytosanitary Services. Besides this compulsory system, which relies on the propagation of olive mother plants identified by the single nurseryman and found free from the target pathogens, a voluntary certification system has been activated since 1987 at the national level by the Ministry of Agriculture. During the last 10 years, the legislation has been revised, and in late 2006 the revision process was completed with the publication of 2 decrees concerning the revised organisation of the system (DM 24/7/2003, DM 4/5/2006) and 4 decrees concerning the updated official technical operations for the certification of pommes, stone fruit, olive and strawberry (DM 20/11/06).

The DM 20/11/06 provides detailed guidance on the production of olive trees and rootstocks. Plant material produced according to this certification scheme is derived from nuclear-stock plants (also identified as Primary source) officially recognised and registered in the database of certified accessions by the Ministry of Agriculture. The material deriving from the first multiplication of the nuclear-stock material enters in the certification process as Pre-basic material. Once the nuclear-stock has been registered, the breeders or Institutions or Research Centres responsible for its production and selection must keep the material under conditions that minimise recontamination risks.

The main outlines of the Italian certification scheme (Tab. 3 and Fig. 6) are the following:

a. *Registration of nuclear-stock material*: Breeders or researchers that intend to introduce a new accession for a specific variety in the certification system must provide detailed information about the trueness to type and sanitary controls performed by filling the official forms defined in the DM 20/11/06. The evaluation and eligibility of the registration request is carried out by a technical certification committee authorised by the Ministry of Agriculture.

b. *Maintenance and propagation of Pre-basic material*: Pre-basic olive material derives directly from the propagation of the nuclear-stock; the Pre-basic plants must be maintained in insect-proof green-houses (at the Conservation for Premultiplication Repository) with at least two replications. The plants are grown in *V. dahliae* and *X. diversicaudatum* -free soil mixture and periodically tested for viruses using molecular tools (10% of the plants each year, starting from the 5th year). Molecular tests should be also performed if, after visual inspection, plants show symptoms of *V. dahliae* or phytoplasmas. Cuttings and seeds collected from the Pre-basic material is used to produce, in the same facilities, the Basic plants for the establishment of the Premultiplication Repository.

c. *Maintenance and propagation of Basic material:* Basic olive planting material is the propagation material that is obtained from Pre-basic material, maintained in open field (Repository for the Premultiplication) in a variable replication number (2 minimum) depending on the importance of the cultivar. Premultiplication field plots must be tested and found free from *V. dahliae* and *X. diversicaudatum,* and have a 20 meters non-cultivated border. Basic plants must be periodically inspected and tested as defined for the Pre-basic material. Cuttings and seeds collected from the Basic material are used to produce the certified mother plants for the establishment of the Multiplication mother blocks.

d. *Maintenance and propagation of Certified material:* Certified mother plants obtained from the propagation of the Basic material represent the source for nursery certified olive plant production. Mother plants are grown in open fields in variable replication numbers depending on the market of the specific cultivar. Plants are visually inspected at least once a year, while each plant must undergo laboratory tests for virus detection at least once within a 30-year period.

e. *Certified nursery productions*: Production takes place in authorized nurseries that join the certification program. The nursery production must comply with the requirements established by the DM 20/11/06 in terms of: (i) soil mixture (free from *V. dahliae, X. diversicaudatum, Meloidogyne incognita, M. javanica* and *Pratylenchus vulnus*), (ii) location of the certifiable olive blocks; (iii) maintenance of a farm business registry. The regional phytosanitary service, following visual inspections and examination of the documentation, is in charge of releasing the official certification (blue label) for every single plant or seedling.

In order to facilitate the certification and the availability of certified material for new olive cultivars or clones, the Premultiplication and the Multiplication blocks may be created directly using planting material deriving from the first multiplication of the nuclear-stock. In this way, the timeframe between the approval of a new accession in the system and the availability of certified plants in the nurseries is effectively reduced.

This certification program has been supported until now mainly by public funds that cover the costs for the management of the Conservation for the Premultiplication and Premultiplication repositories; starting in 2012 the program should shift to a self-sustaining system, in which taxes recovered on each released certification label will make up for the costs of the repository management.

Steps	Plant category	Facilities	Current active Repository for olive certified material in Italy	Controls and certification released by
Selection of Nuclear-stock	Nuclear-stock	Screen-houses	Several Research Centers	Regional Phytosaritary Services
Conservation for Premultiplication	Pre-basic	Screen-houses	- CRA-PAV Rome - University of Bari - Azienda Agricola Sperimentale "Improsta"	
Premultiplication	Basic	Open field	- CRA-PAV Rome - CRSA Basile Caramia, Bari - Azienda sperimentale di Santa Paolina, Follonica - CAV Tebano	
Multiplication	Certified	Open field	ConsorzioVivaisticoPugliese	
		Nurseries	--	

Table 3. Organization of the Certification program for olive propagating material

As aforementioned, the Italian voluntary certification program involves several woody crop species. In most cases, the main reason prompting for the certification of such accessions is the presence of pathogens that can cause detrimental effects on the affected plants (i.e. quarantine pests for stone fruit or citrus). Contrastingly, in the case of olive plants, the main aspect that promoted the adoption of this program has been the high level of genetic and phenotypic variability within each cultivar, which could results in heterogeneous plants and misidentification of such cultivars.

The certification scheme adopted in Italy ensures trueness-to-type and uniformity, since the certified plants are obtained through subsequent clonal propagation steps from a single registered accession.

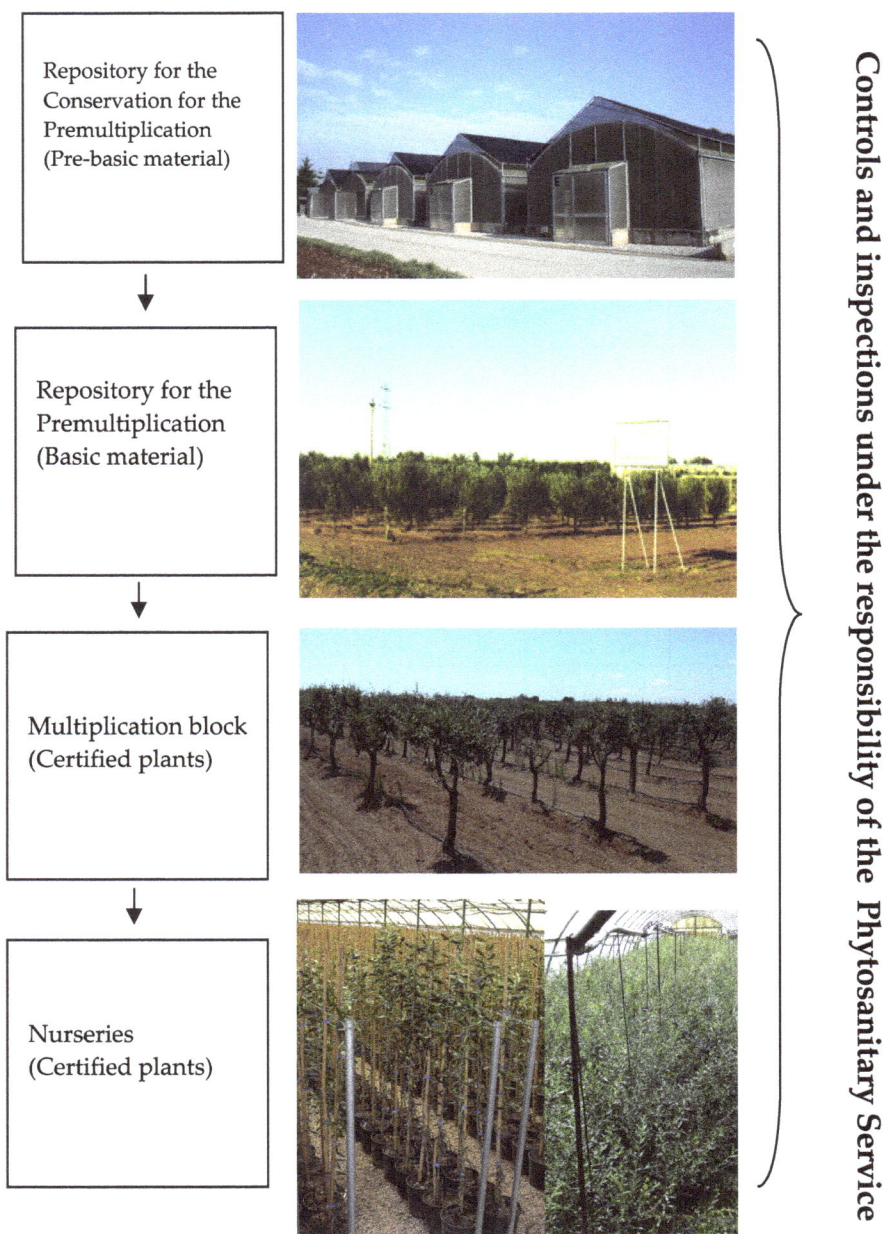

Figure 6. Outline of the general certification steps and facilities

Annex I. – List of the reference European Union (EU) Directives and Italian national regulations

Basic EU Directive

- *Council Directive 92/34/EEC* of 28 April 1992 on the marketing of fruit plant propagating material and fruit plants intended for fruit production
- *Council Directive 2008/90/EC* of 29 September 2008 (Recast version of Directive 92/34/EEC) on the marketing of fruit plant propagating material and fruit plants intended for fruit production

Implementing measures of Directive 92/34/EEC :

- *Commission Directive 93/48/EEC* of 23 June 1993 setting out the schedule indicating the conditions to be met by fruit plant propagating material and fruit plants intended for fruit production, pursuant to Council Directive 92/34/EEC
- *Commission Directive 93/64/EEC* of 5 July 1993 setting out the implementing measures concerning the supervision and monitoring of suppliers and establishments pursuant to Council Directive 92/34/EEC on the marketing of fruit plant propagating material and fruit plants intended for fruit production
- *Commission Directive 93/79/EEC* of 21 September 1993 setting out additional implementing provisions for lists of varieties of fruit plant propagating material and fruit plants, as kept by suppliers under Council Directive 92/34/EEC

Basic Italian Regulations

- *DM 14 aprile 1997* Recepimento delle direttive delle Commissione n. 93/48/CEE del 23 giugno 1993, n. 93/64/CEE del 5 luglio 1993 e n. 93/79/CEE del 21 settembre 1993, relative alle norme tecniche sulla commercializzazione dei materiali di moltiplicazione delle piante da frutto e delle piante da frutto destinate alla produzione di frutto
- *DM 24 luglio 2003* Organizzazione del servizio nazionale di certificazione volontaria del materiale di propagazione vegetale delle piante da frutto
- *DM 4 maggio 2006* Disposizioni generali per la produzione di materiale di moltiplicazione delle specie arbustive ed arboree da frutto, nonché delle specie erbacee a moltiplicazione agamica
- *DM 20 novembre 2006* norme tecniche per la produzione di materiali di moltiplicazione certificati di Agrumi, Fragola, Olivo, Pomoidee, Prunoidee (supplemento ordinario alla Gazzetta Ufficiale n. 141 del 20 giugno 2007)
- *Decreto Legislativo 25 giugno 2010, n.124* Attuazione della direttiva 2008/90 relativa alla commercializzazione dei materiali di moltiplicazione delle piante da frutto destinate alla produzione di frutti (refusione) (pubblicato nella Gazzetta Ufficiale n. 180 del 4 agosto 2010)

6. Conclusion

In this chapter an overview on the olive graft-transmissible pathogens and on the latest phytosanitary directives embodied by the EU and by the Italian Ministry of Agriculture is given.

As remarked, olive has always been considered a very resistant species to diseases caused by different pathogens; however, several pathogens, mainly systemic, can affect the trees and, in some case, invalidate the production. Recent advances in plant pathology and molecular biology, significantly contributed to the discover of new olive pathogens, to characterize their genome, biology and epidemiology.

Italy has been amongst the first Countries to adopt an effective certification system for the production of plant propagation material with high quality standards. After 10 years from its promulgation the Italian Regulation has been revised with the support of a technical committee, in order to improve the program and meet the quality standards amended in the late '90 by the EU, which are mandatory for all member States. In 2006 a revised national Regulation was issued (DM 20/11/2006), updating the list of pathogens that need to be checked and implementing the protocols for their identification. In the last 5 years several valuable virus-free and true-to-type primary sources, belonging to the most widespread or local Italian varieties, have been registered, propagated through the certification system, and made available to the growers. Although, the EU directives and the Italian regulations concerning the production of olive propagation material have been critically revised and implemented, it is necessary to continuously update the list of the pathogens and the diagnostic protocols, including the latest tools for genetic and phytosanitary assessment. It should be considered for example that some specific olive viruses such as OLV-1, OLV-2 and OLRSV are rare, infections are symptomless on olive plants, and there are no evidences about their threat to other crops. OLYaV is currently included in the list of the harmful pathogen for the Italian phytosanitary regulation, but even if OLYaV-infected trees are widespread, there are very few plants showing symptoms of yellowing, and more importantly the association of this virus with the OLY disease has not been clearly demonstrated. Regarding CMV and TNV, although these viruses are polyphagous and very damaging to other crops, they in olive are rare (CMV) or present only in a restricted geographical area (TNV- Portugal). On the basis of these data, the list of the viruses to be included in the phytosanitary certification program could be restricted to the following: SLRSV, CLRV (both are associated to manifest diseases either in olive plants or in other crops), ArMV (one of the harmful pathogens for *Fragaria*, *Rubus* and other crops) and perhaps TNV. Whereas, it is important to ensure that the certified olive material is free from phytoplasmas, *V. dahliae* and *P. savastanoi* pv. *savastanoi*.

Long distance movement of plant propagation material and the expansion of olive crops in new areas impose the use of common and harmonized certification procedures which are crucial to restrict the spread of harmful pathogens and pests.

Author details

Giuliana Albanese
Dipartimento di Gestione dei Sistemi Agrari e Forestali,
Università degli Studi Mediterranea di Reggio Calabria, Reggio Calabria, Italy

Maria Saponari
Istituto di Virologia Vegetale del CNR – Unita' Organizzativa di Supporto di Bari, Bari, Italy

Francesco Faggioli
CRA-Centro di Ricerca per la Patologia Vegetale, Roma, Italy

7. References

Ahangaran, A., Khezri, S., Habibi, M.K, Alizadeh, A. & Mohammadi, G.M. (2006). The first report of detection of a phytoplasma in olive trees in a botanic collection in Iran, *Commun Agric Appl. Biol. Sci.* 7: 1133-1138.

Alabdullah, A., Elbeaino, T., Saponari, M., Hallak, H. & Digiaro, M. (2005). Preliminary evaluation of the status of olive-infecting viruses in Syria, *EPPO Bull.* 35: 249-252.

Alabdullah, A., Minafra, A., Elbeaino, T., Saponari, M.,Savino, V. & Martelli, G.P. (2010). Nucleotide sequenze and menome organization of Olive latent virus 3, a new putative member of the family *Tymoviridae, Virus Research* 152: 10.

Albanese, G., Faggioli, F., Ferretti, L., Sciarroni, R., La Rosa, R. & Barba, M. (2003). Sanitary status evaluation of olive cultivars in Calabria and Sicily, *J. Plant Pathol.* 85: 304.

Al Rwahnih, M., Guo, Y., Daubert, S., Golino D. & Rowhani, A. (2011). Characterization of latent viral infection of olive trees in the National clonal germplasm repository in California, *J. Plant Pathol.* 93 (1): 227-231.

Barba, M. (1993). Viruses and virus-like diseases of olive, *EPPO Bull.* 23: 493-497.

Barba, M., Boccardo, G., Carraio, L., Del Serrone, P., Ermacora, P., Firrao, G., Giunchedi, L., Loi N., Malfitano, M., Marcone, C., Marzachì, C., Musetti, R., Osler, R., Palmano, S., Poggi Pollini, C., Ragozzino, A. (1998). Confronto di differenti tecniche di diagnosi applicate al rilevamento di fitoplasmi in pomacee, *Notiz. protez. piante* 9: 263-278.

Bertaccini, A., Paltrinieri, S., Botti, S., Lugaresi, C. (2002). Malformations associated with phytoplasma presence in olive trees, *Atti Convegno Internazionale di Olivicoltura.*, Spoleto, Italy, 344–349

Bertolini, E., Fadda, Z., Garcia, F., Celada, B., Olmos, A., Gorris, M.T., Del Rio, C., Caballero, J., Duran-Vila, N., & Cambra, M. (1998). Virosis del olivo detectadas en Espana, Nuevos metodos de diagnostico, *Phytoma* 102: 191-193.

Bertolini, E., Olmos, A., Martinez, M. C., Gorris, M. T., & Cambra, M. (2001a). Single-step multiplex RT-PCR for simultaneous and colourimetric detection of six RNA viruses in olive trees., *J.Virol. Methods* 96: 33-41.

Bertolini, E., Cambra, M., Peñalver, R., Ferrer, A., Garcia, A., Del Rio, M.C., Gorris, M.T., Martínez, M.C., Quesada, J.M., García De Oteyza, J., Duran-Vila, N., Caballero, J.L. & López, M.M. (2001b). Métodos serológicos y moleculares de diagnóstico de virus y bacterias de olivo. Evaluación de la sensibilidad varietal y la aplicación a programas de certificación, *Mercacei Magazine* 30: 1-6.

Bertolini, E., Olmos, A., Lopez, M.M., & Cambra, M. (2003). Multiplex nested reverse-trascription polymerase chain reaction in a single tube for sensitive and simultaneous detection of four RNA viruses and *Pseudomonas savastanoi* pv. *savastanoi* in olive trees, *Phytopathology* 93: 286-292.

Bjeliš, M., Loconsole, G. & Saponari, M. (2007). Presence of viruses in Croatian olive groves, *Pomologia Croatica* 13: 165-172.

Bottalico, G., Rodio, M.E., Saponari, M., Savino, V. & Martelli, G.P. (2002). Preliminary results of sanitation trials of viruses-infected olive, *J. Plant Pathol.* 84: 171-200.

Bubici, G. & Cirulli, M. (2011). Verticillium wilt of olives, *in* Schena L., Agosteo G.E. & Cacciola S.O. (ed), *Olive Diseases and Disorders.* Transworld Research Network, Kerala, India, pp. 191-222.

Camele, I., Rana, G.L., Murari, E., Bertaccini, A. (1999). Indagini preliminari su alcune alterazioni morfologiche e cromatiche dell'olivo, *L'Informatore Agrario* 55 (22): 79-81.

Çağlayan, K., Fidan, U., Tarla, G. & Gazel, M., (2004). First report of olive viruses in Turkey. *J. Plant Pathol.* 86: 89-90.

Çağlayan, K., Faggioli, F. & Barba, M. (2009). Virus, phytoplasma and unknown diseases of olive trees, *in* Hadidi, A., Barba, M., Candresse, T. & Jelkmann, W. (ed), *Virus and virus-like diseases of pome and stone fruits*, APS, St. Paul, USA, pp. 289-297.

Cardoso, J.M.S., Felix, M.R., Clara, M.I.E. & Oliveira, S. (2005). The complete genome sequence of a new Necrovirus isolated from *Olea europaea* L., *Arch. Virol.* 150: 815-823.

Cardoso, J.M.S., Felix, M.R., Oliveira, S. & Clara, M.I.E. (2004). A Tobacco necrosis virus D isolate from Olea europaea L.: viral characterization and coat protein sequence analysis, *Arch. Virol.* 149: 1129-1138.

Cooper, J. (1986). Strawberry latent ringspot virus, CMI/AAB Description of Plant Viruses, N° 126.

Christensen, N.M., Nicolaisen, M., Hansen, M. & Schulz, A. (2004). Distribution of phytoplasmas in infected plants as revealed by real-time PCR and bioimaging, *Molecular Plant Microbe Interactions* 17: 1175-1184.

Cutuli, M., Campisi, G., Marra, F.P., Caruso, T. (2011). Vegetative growth and ecophysiological aspects in young olive plants inoculated with olive leaf yellowing associated virus (OLYaV), *Acta Italus Hortus* 1: 356-361.

Deng, S., Hiruki, C. (1991). Amplification of 16SrRNA genes from culturable and nonculturable mollicutes, *J. Microbiol. Methods* 14: 53-61.

Essakhi, S., Elbeaino, T., Digiaro, M., Saponari, M., Martelli, G.P. (2006). Nucleotide sequence variations in the HSP70 gene of Olive leaf yellowing-associated virus, *J. Plant Pathol.* 88: 285-291

Fadel, C., Digiaro, M., Choueiri, E., Elbeaino, T., Saponari, M., Savino, V. & Martelli, G.P. (2005). On the presence and distribution of olive viruses in Lebanon, *EPPO Bull.* 35: 33-36.

Faggioli, F. & Barba, M. (1995). An elongated virus isolated from olive, *Acta Hortic.* 386: 593-600.

Faggioli, F., Ferretti, L., Albanese, G., Sciarroni, R., Pasquini, G., Lumia, V. & Barba, M. (2005). Distribution of olive tree viruses in Italy as revealed by one-step RT-PCR, *J. Plant Pathol*. 87: 49-55.

Faggioli, F., Ferretti, L., Pasquini, G. & Barba, M. (2002). Detection of Strawberry latent ring spot virus in leaves of olive trees in Italy using a one-step RT-PCR, *J.Phytopathol*. 150: 636-639.

Felix, M.R. & Clara, M.I. (2002). Two necrovirus isolates with properties of olive latent virus 1 and of tobacco necrosis virus from olive in Portugal, *Acta Hortic*. 586: 725-728.

Ferreti, L., Faggioli, G., Pasquini, G., Sciarroni, R., Pannelli, G., Baldoni, L. & Barba, M. (2002). Strawberry latent ringspot virus (SLRSV) cause of differentiation among Raggiola and Frantoio olive cultivars, *J. Plant Pathol*. 84: 171-200.

Font, I., Abad, P., Dally, E.L., Davis, R.E. & Jordá, C. (1998). Nueva enfermedad en el olivar español, *Phytoma España* 102: 211-212.

Gallitelli, D., & Savino, V. (1985). Olive latent virus 1. A single-RNA spherical virus isolated from in Apulia (Southern Italy), *Ann. Appl. Biol*. 106: 295-303.

Godena, S., Bendini, A., Giambanelli, E, Cerretani, L., Đermic, D. & Dermic, E. (2012). Cherry leafroll virus: Impact on olive fruit and virgin olive oil quality, *Eur. J. Lipid Sci. Technol*. 114: 535–541.

Grieco, F., Alkowni, R., Saponari, M., Savino, V., & Martelli, G.P. (2000). Molecular detection of olive viruses, *EPPO Bull*. 30: 469-473.

Grieco, F. & Martelli, G.P. (1997). Olive latent virus 2, representative of a putative new genus in the family Bromoviridae, *Phytoparasitica* 25: 1.

Grieco, F., Parrella, G. & Vovlas, C. (2002). An isolate of Olive latent virus 2 infecting castor bean in Greece, *J. Plant Pathol*. 84: 129-131.

Henriques, N.I.C., Rei, F.T., Alit, F.A., Serena, J.F. & Poet, M.F. (1992). Virus diseases in Olea europea cultivars: Immunodiagnosis of Strawberry latent ringspot nepovirus, *Phytopatol. Medit*. 31: 127-132.

Hodgetts, J., Boonham, N., Mumford, R. & Dickinson, M. (2011). Panel of 23S rRNA Gene-Based Real-Time PCR Assays for Improved Universal and Group-Specific Detection of Phytoplasmas, *Applied Environmental Microbiology* 75: 2945-2950.

Kanematsu, S., Taga, Y. & Morikawa, T. (2001). Isolation of Olive latent virus 1 from Tulip in Toyoma Prefecture, *J. Gen. Plant Pathol*. 67: 333-334.

Lee, I.M, Dawn, E, Gundersen-Rindal, D.E, Davis, R.E. & Bartoszyk, M. (1998). Revised classification scheme of phytoplasmas based on RFLP analyses of 16S rRNA and ribosomal protein gene sequences, *Int. J. Syst. Bacteriol*. 48: 1153-1169.

Lee, I.M., Gundersen, D.E., Hammond, R.W., Devis, R.E. (1993). Use of micoplasma like organism (MLO) group specific oligonucleotide primer for nested-PCR assay to detect mixed-MLO infections a single host plant, *Phytopathology* 84: 559-566.

Leva, A.R., Petruccelli, R., Bartolini, G. (1994). Mannitol in vitro culture of *Olea europaea* L. (cv. Maurino), *Acta Hortic*. 356: 43-46.

Loconsole, G., Saponari, M., Faggioli, F., Albanese, G., Bouyahia, H., Elbeaino, T., Materazzi, A., Nuzzaci, M., Prota, V., Romanizzi, G., Trisciuzzi, N. & Savino V. (2010). Inter-laboratory validation of PCR-based protocol for detection of olive viruses, *EPPO Bull.* 40: 423-428.

Luigi, M., Manglli, A., Thomaj, F., Buonaurio, R., Barba, M., Faggioli, F. (2009). Phytosanitary evaluation of olive germplasm in Albania, *Phytopathol. Medit.* 48: 280-284.

Luigi, M., Roschetti, A., Albanese, G., Barba, M., Faggioli F. (2010). Molecular characterization of Olive Leaf Yellowing associated Virus isolates, *Petria* 20 (2): 307.

Luigi, M., Godena, S., Đermić, E., Barba, M., Faggioli, F. (2011). Detection of viruses in olive trees in Croatian Istria, *Phytopathol. Medit.* 50 (1): 150-153.

Marte, M., Gadani, E., Savino, V. & Rugini, E. (1986). Strawberry latent ringspot virus associated with a new disease of olive in Central Italy, *Plant Dis.* 70: 171-172.

Martelli, G.P. (1999). Infectious diseases and certification of olive: an overview, *EPPO Bull.* 29: 127-133.

Martelli, G.P., Sabanadzovic, S., Savino, V., Abu-Zurayk, A.R. & Masannat, M. (1995). Virus-like diseases and viruses of olive in Jordan, *Phytopathol. Medit.* 34: 133-136.

Martelli, G.P., Yılmaz, M.A., Savino, V., Baloğlu, S., Grieco, F., Güldür, M.E., Greco, N. & Lafortezza, R. (1996). Properties of citrus isolate of Olive latent virus 1, a new Necrovirus, *Eur. J. Plant Pathol.* 102: 527-536.

Martelli, G.P., Salerno, M., Savino, V. & Prota, U. (2002). An appraisal of diseases and pathogens of olive, *Acta Hortic.* 586: 701-708.

Martelli, G.P. (2011). Infectious diseases of olive, *in* Schena L., Agosteo G.E. and Cacciola S.O. (ed), *Olive Diseases and Disorders.* Transworld Research Network, Kerala, India, pp. 71-88.

Materazzi, A., Toni, S., Panatroni, A., Osti, M. & Triolo, E. (1996). On the presence of a new isometric virus in *Olea europaea* L., *Atti Convegno Annuale della Societa Italiana di Patologia Vegetale*, Udine, Italy, pp. 57-59.

Mercado-Blanco, J., Rodríguez-Jurado, D., Pérez-Artés, E. & Jiménez-Díaz, R.M. (2002). Detection of the defoliating pathotype of *Verticillium dahliae* in infected olive plants by nested-PCR, *Eur. J. Plant Pathol.* 108: 1–13.

Nigro, F., Schena, L., Gallone, P. (2002). Diagnosi in tempo reale della verticilliosi dell'olivo mediante Scorpion-PCR, *Atti Convegno Internazionale di Olivicoltura*, Spoleto, Italy, pp. 454–461.

Pantaleo, V., Saponari, M. & Galitelli, D. (2001). Development of a nested PCR protocol for detection of olive-infecting viruses in crude extracts, *J. Plant Pathol.* 83: 143-146

Pasquini, G., Marzachì, C., Poggi Pollini C., Faggioli, F., Ragozzino, A., Bissani, R., Vischi, A., Barba, M., Giunchedi, L. & Boccardo, G. (2000). Molecular characterization of phytoplasmas affecting olive trees (*Olea europea* L.) in Italy, *J. Plant Pathol.* 82 : 213-219.

Penyalver, R., García, A., Ferrer, A., Bertolini, E., Quesada, J.M., Salcedo, C.I., Piquer, J., Pérez-Panadés, J., Carbonell, E.A., del Río, C., Caballero, J.M. & López, M.M. (2006).

Factors affecting *Pseudomonas savastanoi* pv. *savastanoi* plant inoculations and their use for evaluation of olive cultivar susceptibility *Phytopathology* 96 (3): 313-319.

Rei, F.T., Henriques, M.I.C., Leitao, F.A., Serrano, J.F. & Potes, M.F. (1993). Immunodiagnosis of Cucumber mosaic cucumovirus in different olive cultivars, *EPPO Bull.* 23: 501-504.

Roschetti, A., Ferretti, L., Muzzalupo, I., Pellegrini, F., Albanese, G. & Faggioli, F. (2009). Evaluation of the possible effect of virus infections on olive propagation, *Petria* 19 (1): 18-28.

Rugini, E. (1984). *In vitro* propagation of some olive (*Olea europaea sativa* L.) cultivars with different root-ability, and medium development using analytical data from developing shoots and embryos, *Scientiae Horticulturae* 24: 123-134.

Sabanadzovic, S., Abou-Ghanem, N., La Notte, P., Savino, V., Scarito, G. & Martelli, G.P. (1999). Partial molecular characterization and RT-PCR detection of a putative closterovirus associated with olive leaf yellowing, *J. Plant Pathol.* 81: 37-45.

Sanfaçon, H., Iwanami, T., Karasev, A.V., van der Vlugt, R., Wellink, J., Wetzel, T. & Yoshikawa, N. (2011). Family Secoviridae. *in* King, A.M.Q., Adams, M.J., Carstens, E.B., Lefkowitz, E.J. (ed), *Virus taxonomy. Ninth Report of the International Committee on Taxonomy of Viruses*, Elsevier-Academic Press, Amsterdam, The Netherlands, pp. 881-899.

Saponari, M., Savino, V. & Martelli, G.P. (2002). Seed transmission in olive of two olive-infecting viruses, *J. Plant Pathol.* 84: 167-168.

Saponari, M. & Savino, V. (2003). Virus and virus-like agents in olive, *Informatore Fitopatologico* 12: 26-29.

Savino, V., Barba, M., Gallitelli, G. & Martelli, G.P. (1979). Two nepoviruses isolated from olive in Italy, *Phytopatol. Medit.* 18: 135-142.

Savino, V. & Gallitelli, D. (1981). Cherry leaf roll virus in olive, *Phytopathol. Medit.* 20 202-203.

Savino, V. & Gallitelli, D. (1983). Isolation of Cucumber mosaic virus from olive in Italy, *Phytopatol. Medit.* 22: 76-77.

Savino, V., Gallitelli, D. & Barba, M. (1983). Olive latent ringspot virus, a newly recognized virus infecting olive in Italy, *Ann. Appl. Biol.* 133: 243-249.

Savino, V., Piazzola, T., Di Franco, A. & Martelli, G.P. (1984). Olive latent virus 2, a newly recognized virus with a differently shaped particle, *Proceeding of the 6th Congress of the Mediterranean Phytopathological Union*, Cairo, Egypt, pp. 24-26.

Savino, V., Sabanadzovic, S., Scarito, G., Laviola, C. & Martelli, G.P. (1996). Two olive yellows of possible viral origin in Sicily, *Informatore Fitopatologico* 46: 55-59.

Schneider, B., Seemüller, E., Smart, C.D., Kirkpatrick, B.C. (1995). Phylogenetic classification of plant pathogenic mycoplasma like organism or phytoplasmas, *in* Razin, S., Tully, J.G. (ed), *Molecular and diagnostic procedures in mycoplasmology*, Academic Press, San Diego, CA, USA, 1, pp. 369-380.

Seemüller, E., Marcone, C., Lauer, U., Ragozzino, A. & Göschl, M. (1998). Current status of molecular classification of the Phytoplasmas, *J.Plant Pathol.* 80: 3-26.

Serce, C. U. , Yalcin, S. , Gazel, M. , Cağlayan, K. & Faggioli, F. (2007). First report of Olive latent virus 1 from olive trees in Turkey, *J. Plant Pathol.* 89: 73.

Schena, L., Nigro, F. & Ippolito, A. (2004). Real-time PCR detection and quantification of soilborne fungal pathogens: the case of *Rosellinia necatrix, Phytophthora nicotianae, P. citrophthora,* and *Verticillium dahlia, Phytopathol. Medit.* 43: 273–280.

Surico, G. & Marchi, G. (2011). Olive knot disease, *in* Schena, L., Agosteo, G.E. & Cacciola, S.O. (ed), *Olive Diseases and Disorders.* Transworld Research Network, Kerala, India, pp. 89-116.

Triolo, E., Materazzi, A. & Toni, S. (1996). An isolate of Tobacco mosaic tobamovirus from *Olea europaea, Advan. Hortic. Sci.* 10: 39-45.

Chapter 7

Floral Biology: Implications for Fruit Characteristics and Yield

Adolfo Rosati, Silvia Caporali and Andrea Paoletti

Additional information is available at the end of the chapter

1. Introduction

Floral biology has important practical implications, in addition to its scientific relevance, given that flower characteristics and bloom affect fruit characteristics and yield. Yield derives from fruit quality (e.g. weight) and quantity (i.e. number), which, in turns, depend on flower quantity and quality: flowers must be suitable to become fruits, and then must be pollinated and fertilized, and must set fruits, which must then grow. Not all flowers can do all of this: some flowers, for instance, have aborted ovaries which are partially developed or absent at bloom, depending on when the abortion occurred. Even when still present, these aborted ovaries are not capable of becoming fruits. Normal pistils, may not be pollinated or fertilized, but also fertilized ovaries may drop after some growth, resulting in fruit drop. From 100 flowers, in olive, all the above phenomena result in one to few fruits (Hartmann, 1950). Because of this low fruit set, it is often believed that cultural practices aimed at improving pollination, increasing fruit set or reducing ovary abortion or fruit drop. may lead to increased olive yields.

In the following paragraphs we will analyze in detail some of these passages (i.e. ovary abortion and fruit set) allowing fruit formation from the flower. We will conclude that, in olive, yield is affected mostly by the nutritional status of the tree and the environmental conditions affecting it, which determine the potential yield. To achieve this potential yield, the plant uses a series of compensating mechanisms which make the yield rather independent of flowering or any single subsequent passage (i.e. ovary abortion, fruit set, drop and size).

Firstly, however, we will begin by describing how fruit characteristics, particularly fruit size, are mostly determined by the ovary characteristics at bloom. In the following paragraphs we will see how these ovary characteristics affect ovary abortion and fruit set. Finally we will discuss the apparent redundant flowering in olive, which produces many more flowers than necessary for the yield it can sustain.

Figure 1. Flower size differs among olive cultivars: Nocellara del Belice, a large-fruited cultivar, on the left, and Koroneiki, a small-fruited cultivar, on the right.

2. Ovary vs. fruit characteristics

Fruit size is important commercially and the study of factors affecting it are of great scientific interest. Fruit size depends mostly on the size and number of fruit cells, though intercellular space may be also important (Bertin *et al.*, 2002; Corelli-Grappardelli, 2004). In olive, fruit size differs greatly among cultivars (Barranco, 1999). Both the endocarp and mesocarp contribute to final fruit size differences among cultivars (Hammami *et al.*, 2011).

Fuit size differences across cultivars are mostly due to cell number, while cell size tends to be similar (Rapoport *et al.*, 2004), despite the fact that fruit growth, from the ovary to the mature fruit, is due in greater part to cell expansion than to cell division (Rapoport *et al.*, 2004; Hammami *et al.*, 2011). In fact, the mature fruit has about 8.5 times the cell number compared to the ovary, while cell size is about 250 times greater (Rapoport *et al.*, 2004). Fruit weight is about 2000 (from 1000 to 4000) times greater than ovary weight at bloom (Rosati *et al.*, 2009). However, from the ovary to the mature fruit, the mesocarp grows much more than the endocarp (Rosati *et al.*, 2012), probably due to its longer growth period (i.e. up to fruit maturity) compared to the endocarp, which stops growing about eight weeks after bloom (Hammami *et al.*, 2011). However, both tissues grow in strict proportion to their initial cell number in the ovary. The greater growth of the mesocarp is related to its initial (i.e. in the ovary at bloom) greater number of cells of smaller size (Rosati *et al.*, 2012). These results agree with the hypothesis that cell number, rather than tissue mass, is related to the sink strength of an organ (Ho, 1992) even though this might be not always the case (Marcelis, 1996), because cell number is only one factor determining sink strength (Gillaspy *et al.*, 1993).

Mature fruit size correlates with flower and, particularly, ovary size at bloom (Rosati *et al.*, 2009). This is true for both the mesocarp and the endocarp independently, though, as stated above, the mesocarp grows proportionally more: the endocarp volume in the fruit is about 4000 times greater than its volume in the ovary, while the same ratio is about 800 for the endocarp (Rosati *et al.*, 2012).

Both the endocarp and mesocarp contribute to the ovary size differences among and within cultivars and even the locular space is proportional to ovary size (Rosati *et al.*, 2012). Similarly, proportionality among the ovary and other flower parts was found both within trees (Cuevas & Polito, 2004) and between cultivars (Rosati *et al.*, 2009). Ovary size depends mostly on cell number rather than size, both across and within cultivars (Rosati *et al.*, 2011a). This is true both for the mesocarp and the endocarp, suggesting that similar mechanisms regulate cell division in different tissues. Cell size does differ among cultivar, but does not correlate with ovary size. Cell size differs greatly among tissues with bigger but fewer cells in the endocarp compared to the mesocarp, suggesting that cell growth patterns differ among tissues (Rosati *et al.*, 2011a). In fact, cell size in the ovary correlates with tissue relative growth (i.e. from bloom to fruit maturity), both across tissues and cultivars, suggesting that cell size at bloom reflects the stage of growth of a given tissue: the larger the cells, the less growth remains to be performed (Rosati *et al.*, 2012). This agrees with the observation that fruit growth is mostly achieved by cell expansion as mentioned above. The larger cell size of the endocarp at bloom suggests that this tissue is at a more advanced stage of growth, compared with the mesocarp and thus, its remaining growth, relative to its size at bloom, is less, at least in terms of cell expansion. This would also explain why the endocarp stops growing earlier than the mesocarp. It is possible that the endocarp might be in a more advanced stage of development at bloom, since it is inside this tissue that vital functions (e.g. fertilisation) take place soon after bloom and thus the endocarp needs relatively more differentiated cells. The mesocarp, on the contrary, does not perform particular activities at bloom, hence, it does not need advanced cell differentiation in this period.

In conclusion, genetic differences in fruit size, among olive cultivars, appear to arise from differences in cell division patterns occurring in the ovary (and probably in the whole flower) before bloom. Similar results have been found in other species where genes coding for cell division before bloom have been found to be responsible for differences in final fruit size. The final fruit size, aside from genetic control, is also related to environmental and endogenous plant conditions that allow the genetic potential growth to be achieved to a varying degree. Analysis the genomics, or of environmental factors affecting fruit size is beyond the scope of this review.

Figure 2. Fruit size differing in olive cultivars: Nocellara del Belice, a large-fruited cultivar, on the right, and Koroneiki, a small-fruited cultivar, on the left at the same date during the fruit growth cycle. The difference was already present among the ovaries at bloom.

3. Ovary (or pistil) abortion

Ovary or pistil abortion indicates the presence of flowers with absent or only partly formed and non-viable ovaries: that is, ovaries incapable of becoming fruits and destined to drop. Flowers with aborted ovaries are often called staminate flowers since only the male organs are complete and functional. Normal flowers with both male and female organs complete and functional are called hermaphrodite flowers. Ovary abortion varies greatly with year, cultivar, individual tree, branch and shoot, and even among and within inflorescences (Morettini, 1939; Bottari, 1951; Badr & Hartmann, 1971; Fabbri *et al.*, 2004; Martin & Sibbett, 2005). Ovary abortion occurs early during flower development, (Pirotta & De Pergola, 1913), mostly 30-40 days before bloom (Uriu, 1959; Cuevas *et al.*, 1999; Reale *et al.*, 2006).

Ovary abortion in olive appears to results mostly from competition among flowers (and ovaries, the future fruits) for resources, which are insufficient for all flowers to develop, given the redundant flowering. The onset of this competition occurs very early and affects both pistil abortion and fruit set, as we will see later (Hartmann, 1950; Uriu, 1959; Cuevas *et al.*, 1994; Perica *et al.*, 2001; Levin & Lavee, 2005). Conditions that affect competition among flowers/fruits or that decrease available resources, usually result in increased abortion and decreased fruit set. Among such conditions are: N deficiency, foliar diseases and low leaf-to-bud ratio, (Petri, 1920; Morettini, 1951; Uriu, 1953, 1959; Fernandez-Escobar et al., 2008); water stress (Melis, 1923; Brooks, 1948; García Gálvez, 2005); insufficient light (Bottari, 1951; Dimassi et al., 1999); adverse climactic conditions and high yield in previous year (Rallo et al., 1981; Rapoport & Rallo, 1991; Lavee, 1996; Cuevas et al., 1994); abundant flowering in current year (Reale et al., 2006); and unfavorable inflorescence position in the canopy (Cuevas & Polito, 2004; Seifi et al., 2008). Aborted flowers do not contain starch, suggesting a link between nutrient availability and pistil abortion (Reale et al., 2009). Removing part of the inflorescences reduces drastically the ovary abortion in olive (Seifi et al., 2008).

On the other hand, pistil abortion is known to be under genetic control, varying with the cultivar (Campbell, 1911; Morettini, 1939, 1951; Magherini, 1971; Lavee, 1996; Lavee et al., 2002). In a recent paper it has been demonstrated that even the genetic component can be interpreted with the competition theory (Rosati et al., 2011b). In fact, as described above, fruit size differences are related to ovary size differences among cultivars. Large-fruited cultivars tend to have bigger flowers, not just ovaries (Rosati et al., 2009). This implies greater energetic costs for the same number of flowers, and thus greater pistil abortion, which in fact tends to be higher in larger-fruited cultivars (Morettini, 1939; Magherini, 1971; Acebedo, 2000; Rosati et al., 2011b). In the previous paragraph we have seen that greater ovary size in large-fruited cultivars results from greater cell number (not greater cell size) and cell number appears to be a key factor in determining the sink strength of an organ (Marcelis, 1996). Larger ovaries with more cells, therefore, might have a greater ability to compete for resources than smaller ovaries, determining both greater abortion and lower fruit set.

4. Fruit set

As mentioned above, fruit set is very low in olive (Hartmann, 1950). It is often believed that positively affecting it (i.e. reducing it) would lead to greater yields. However, many studies show that when flower numbers are artificially reduced, fruit set increases proportionally, leading to similar numbers of set fruits (Suarez et al., 1984, Rallo & Fernandez-Escobar, 1985, Lavee et al., 1996, 1999). This is interpreted as a tendency of the olive tree to set a fixed mass of fruits, that is independent of flower number. This fixed mass is referred to as the fruiting potential (Lavee et al., 1996), and depends on the genotype and environmental factors. Hence, as soon as fruit set reaches such potential, based on tree reserves, the rest of the flowers may drop.

Competition for resources among flowers in olive is reported in many studies (Suarez et al., 1984; Cuevas et al., 1994; Lavee et al., 1999; Seifi et al., 2008; Rapoport & Rallo, 1991; Cuevas et al., 1995). Rugini and Pannelli (1993) showed that fruit set increases when shoot development is mechanically or chemically slowed down, further supporting the hypothesis that competition drives fruit set. Given that resource availability affects fruit set, large-fruited cultivars, having larger ovaries/flowers (Rosati et al., 2009), which need more resources, set fewer fruits, though total fruit mass is similar (Rosati et al., 2010). This suggests compensation between fruit number and size. The genetic effect on fruit set, appears thus to result from different degrees of competition for resources among ovary/flowers of different sizes. This would explain why small-fruited cultivars like Arbequina and Arbosana tend to produce several fruits per inflorescence, while table-olive (i.e. large-fruited) cultivars typically produce only one, but bigger fruit. Nor can it be argued that the small size in small-fruited cultivars is the result of a higher fruit set and consequent source limitations. In fact, aside from the fact that the fruit of these small-fruited cultivars is smaller already at bloom (i.e. the ovary), well before fruit set is decided, thinning fruits to reduce competition increases fruit size minimally, but does not result in fruit size comparable to that of large-fruited cultivars (Rosati *et al.*, 2010). This suggests that differences in fruit size are mainly of genetic origin. Clearly, the ultimate fruit size depends also on resource availability, and varies up to more than 100% within the same cultivar, but among cultivars fruit size may vary much more than that (Barranco, 1999), up to 600% (Rosati *et al.*, 2009). Hence, adjusting fruit load across cultivars with extremely different fruit size needs a wider compensation mechanism than just adjusting fruit size: this mechanism appears to be the adjustment of fruit set.

5. Redundant flowering, andromonecy and fitness

As stated above, the olive produces redundant flowers relative to its yield potential and fruit set is very low. We have seen how ovary abortion increases under conditions that increase competition for resources. In fact, ovary abortion is a means to save resources, balancing the number of ovaries to the resources available (Primack & Lloyd, 1980; Bertin, 1982; Stephenson & Bertin, 1983). Plants, like the olive, that abort part of the ovaries in otherwise hermaphrodite flowers are called andromonoecious (about 4000 species are estimated to be andrommonoecious). From an evolutionary point of view, andromonoecy is considered an intermediate step towards dioecy, and allows saving of resources without affecting the total number of flowers and thus the male function and fitness (Vallejo-Marín & Rausher, 2007). Aborting the ovary does not affect pollen production in olive (Cuevas & Polito, 2004). In this species, large-fruited cultivars have greater ovary abortion, but similar number of flowers (Rosati *et al.*, 2010, 2011b), leaving the male function probably unaltered (although differences in pollen production per flower, among cultivars with different fruit/ovary/flower size have not been studied). Leaving the male function (and fitness) unaltered might explain why only the ovary is aborted instead of the whole flower, which would save more resources, given that the flower is several times bigger than the ovary alone (Rosati *et al.*, 2009; Cuevas & Polito, 2004). It would also explain why the flower production is so redundant compared to the fruiting potential. Producing pollen is less expensive than producing fruits and seeds, while still increasing the plant's fitness, making it more convenient to produce more (male) flowers that fruits. Modeling work demonstrates that fitness is maximized, under limiting resources, when many more flowers than fruits are produced (Morgan, 1993). This author finds that the optimal flower/fruit ratio increases for andromonoecious and monoecious species compared to dioecious ones, and increases even more for wind pollinated species, like the olive.

To increase olive productivity, therefore, it might be possible to breed plants that invest less in male fitness (reduced flowering), thus saving resources, which can be spend in setting and maturing more fruits. This appears reasonable since the energetic cost of flowering is not negligible in olive (Famiani *et al.*, unpublished data).

6. Conclusions

We have seen how competition for available resources among developing flowers and fruits plays a continuous role during the whole developmental cycle of the reproductive organs (Lavee *et al.*, 1999). Hence, pistil abortion first, and fruit set afterwards, appear to be part of a single mechanism that adjusts maternal investments to available resources. Even the genetic component affecting fruit size (i.e. cultivar differences) may be explained with the competition theory, based on the related differences in flower and ovary size (which correlates with fruit size), implying different energetic costs for the development of one fruit. In the absence of catastrophic events that might impair production (e.g. lack of pollination, severe drought or nutrient deficiency, etc.), the olive tree appears to produce fruits according to its potential, independent of the amount of flowers produced. To achieve this potential, the tree uses the different compensatory mechanisms available at any given developmental stage (i.e. pistil abortion, fruit drop, fruit size). Hence, studies reporting positive correlations between abundance of flowering (or pollen in the air) and yield (Moriondo *et al.*, 2001; Fornaciari *et al.*, 2002, Galán *et al.*, 2004), would imply not a causal relationship but simply correlated phenomena: bloom and yield both represent an expression of the tree yield potential. The redundant flowering in olive, may serve the purpose of increasing the mail and the overall fitness, which is advantageous from an evolutionary point of view, but wasteful from an agronomical perspective. For the olive grower, selecting for reduced mail fitness (i.e. fewer flowers) to benefit the female fitness (i.e. yield) might prove beneficial.

Author details

Adolfo Rosati, Silvia Caporali and Andrea Paoletti
Agricultural Research Council - Olive Growing and Oil Industry Research Centre, Spoleto (PG), Italy

Acknowledgment

Financial support for this study was provided by the Italian Ministry of Agriculture, Food and Forestry Policy through the project GERMOLI "Salvaguardia e valorizzazione del GERMoplasma OLIvicolo delle collezioni del CRA-OLI".

7. References

Acebedo, M.M., Cañete, M.L. & Cuevas, J. (2000). Processes affecting fruit distribution and its quality in the canopy of olive trees, *Advances in Horticultural Science* Vol. 14:169–175.

Badr, S.A. & Hartmann, H.T. (1971). Effect of diurnally fluctuating vs. constant temperatures on flower induction and sex expression in olive (Olea europaea L.), *Physiologia Plantarum* Vol. 24:40–45.

Barranco, D. (1999). Variedades y patrones, *in* Barranco, D., Fernàndez-Escobar, R., Rallo, L. (Eds), *El cultivo del olivo*, Ed. Mundi-Prensa, Madrid, pp. 63–89.

Bertin, R.I. (1982). The evolution and maintenance of andromonoecy, *Evolutionary Theory* Vol. 6:25–32.

Bertin, N., Gautier, H. & Roche, C. (2002). Number of cells in tomato fruit depending on fruit position and source-sink balance during plant development, *Plant Growth Regulation* Vol. 36:105–112.

Bottari, V. (1951). Quattro anni di infruttuose osservazioni sull'aborto dell'ovario nel fiore dell'olivo, *Annali della Sperimentazione Agraria* Vol. 5: 359–376.

Brooks, R.M. (1948). Seasonal incidence of perfect and staminate olive flowers, *Proceedings American Society of Horticultural Science* Vol. 52:213–218.

Campbell, C. (1911). L'aborto fiorale nell'olivo, *L'Italia Agricola*, Vol. 16.

Corelli-Grappadelli, L. & Lakso A.N. (2004). Fruit development in deciduous tree crops as affected by physiological factors and environmental conditions, *Acta Horticulturae* Vol. 636:425–441.

Cuevas, J., Rallo, L.& Rapoport, H.F. (1994). Crop load effects on floral quality in olive, *Scientia Horticulturae* Vol. 59:123–130.

Cuevas, J., Rallo, L.& Rapoport, H.F. (1995). Relationships among reproductive processes and fruitlets abscission in Arbequina olive, *Advances in Horticultural Science* Vol. 9:92–96.

Cuevas, J., Pinney, K.& Polito, V.S. (1999). Flower differentiation, pistil development and pistil abortion in olive, *Acta Horticulturae* Vol. 474:293–296.

Cuevas, J.& Polito, V.S. (2004). The role of staminate flowers in the breeding system of Olea europaea (Oleaceae): an andromonoecious, wind-pollinated taxon, *Annals of Botany* Vol. 93:547–553.

Dimassi, K., Therios, I.& Balatsos, A. (1999). The blooming period and self-fruitfulness in twelve Greek and three foreign olive cultivars. *Acta Horticulturae* Vol. 474:275–278.

Fabbri, A., Bartolini, G., Lambardi, M. & Kailis, S.G. (2004). *Olive Propagation Manual*, Landlinks Press, Collingwood, Victoria.

Fernandez-Escobar, R., Ortiz-Urquiza, A., Prado, M. & Rapoport, H.F. (2008). Nitrogen status influence on olive tree flower quality and ovule longevity, *Environmental and Experimental Botany* Vol. 64:113–119.

Fornaciari, M., Pieroni, L., Orlandi, F. & Romano, B. (2002). A new approach to consider the pollen variable in forecasting yield models, *Economic Botany* Vol. 56:66-72.

Galán, C., Vázquez, L., García-Mozo, H. & Domínguez, E. (2004). Forecasting olive (Olea europaea) crop yield based on pollen emission, *Field Crops Research* Vol. 86:43–51.

García Gálvez, G. (2005). Calidad de flor del olivo en relación al régimen de riego y a la salinidad, *Trabajo Professional Fin de Carrera*, E.T.S.I.A.M., University of Cordoba, Spain.

Gillaspy, G., Ben-David, H. & Gruissem, W. (1993). Fruits: A developmental perspective, *Plant Cell* Vol. 5:1439–1451.

Hammami, S.B.M., Manrique, T. & Rapoport, H.F. (2011). Cultivar based fruit size in olive depends on different tissue and cellular processes throughout growth, *Scientia horticulturae* Vol. 130:445–451.

Hartmann, H.T. (1950). The effect of girdling on flower type, fruit set, and yields in the olive, *Proceedings American Society of Horticultural Science* Vol. 56:217–226.

Ho, L.C. (1996). Tomato, *in* Zamki, E., Shaffer, A.A. (Eds.), *Photoassimilate distribution in plant and crops*, Marcel Dekker, Inc., New York, pp. 709–728.

Lavee, S. (1996). Biology and physiology of the olive, *in* IOOC (Eds.), *World Olive Encyclopedia*. International Olive Oil Council, Madrid, Spain, pp. 59–110.

Lavee, S., Rallo, L., Rapoport, H.F.& Troncoso, A. (1996). The floral biology of the olive: effect of flower number, type and distribution on fruitset, *Scientia horticulturae* Vol. 66:149–158.

Lavee, S., Rallo, L., Rapoport, H.F. & Troncoso, A. (1999). The floral biology of the olive II. The effect of inflorescence load and distribution per shoot on fruit set and load, *Scientia horticulturae* Vol. 82:181–192.

Lavee, S., Taryan, J., Levin, J. & Haskal, A. (2002). The significance of crosspollinaticn for various olive cultivars under irrigated intensive growing conditions, *Olivae* Vol. 91:25–36.

Levin, A.G. & Lavee, S. (2005). The influence of girdling on flower type, number, inflorescence density, fruit set, and yields in three different olive cultivars ('Barnea', 'Picual', and 'Souri'), *Australian Journal of Agricultural Research* Vol. 56:827–831.

Magherini, R. (1971). Osservazioni sull'aborto dell'ovario nell'olivo, *L'Agricoltura Italiana* Vol. 71:291–301.

Marcelis, L.F.M. (1996). Sink strenght as a determinant of dry matter partitioning in the whole plant, *Journal of Experimental Botany* Vol. 47:1281–1291.

Martin, G.C. & Sibbett, G.S. (2005). Botany of the olive, *in* Sibbett, G.S., Fergusoɔ, L., Coviello, J.L., Lindstrand, M. (Eds.), *Olive Production Manual*, University of California, Agriculture and Natural Resources, Oakland, California, pp. 15–19.

Melis, A. (1923). Cause di aborto nel gineceo del fiore dell'olivo, *Stazione Sperimentale Agricola Italiana* Vol. 56:302–312.

Morettini, A. (1939). L'aborto dell'ovario nel fiore dell'olivo, *L'Italia Agricola* Vol. 11:815–828.

Morettini, A. (1951). Ulteriore contributo allo studio dell'aborto dell'ovario nel fiore dell'olivo, *Annali della Sperimentazione Agraria* Vol. 5:309–329.

Morgan, M. (1993). Fruit to flower ratio and trade-pffs in size and number, *Evolutionary Ecology* Vol. 7:219–232.

Moriondo, M., Orlandini, S., De Nuntiis, P. & Mandrioli, P.(2001). Effect of agrometeorological parameters on the phenology of pollen emission and production of olive trees (Olea europea L.), *Aerobiologia* Vol. 17:225–232, DOI: 10.1023/A:1011893411266.

Perica, S., Brown, P.H., Connell, J.H., Nyomora, A.M.S., Dordas, C., Hu, H.N.& Stangoulis, J. (2001). Foliar boron application improves flower fertility and fruit set of olive, *HortScience* Vol. 36,714–716.

Petri, L. (1920). Sulle cause di arresto di sviluppo dell'ovario nel fiore dell'olivo, *Rendiconti dell'Accademia Nazionale dei Lincei* Vol. 29, serie 5, sem. I.

Pirotta, R.& De Pergola, D. (1913). Sull'olivo maschio, *Bollettino della Società Botanica Italiana* Vol. 20.

Primack, R.B.& Lloyd, D.G. (1980). Andromonoecy in the New Zealand montane shrub manuka, Leptospermum scoparium (Myrtaceae), *American Journal of Botany* Vol. 67:361–368.

Rallo, L., Martin, G.C. & Lavee, S. (1981). Relationship between abnormal embryo sac development and fruitfulness in olive, *Journal of the American Society for Horticultural Science* Vol. 106:813–817.

Rallo, L. & Fernandez-Escobar, R. (1985). Influence of cultivar and flower thinning within the inflorescence on competition among olive fruit, *Journal of the American Society for Horticultural Science* Vol. 110:303–308.

Rapoport, H.F. & Rallo, L. (1991). Postanthesis flower and fruit abscission in 'Manzanillo' olive, *HortScience* Vol. 116:720–723.

Rapoport, H.F., Manrique T. & Gucci, R. (2004) Cell division and expansion in the olive fruit, *Acta Horticulturae* Vol. 636:461–465.

Reale, L., Sgromo, C., Bonofiglio, T., Orlandi, F., Fornaciari, M., Ferranti, F. & Romano, B. (2006). Reproductive biology of Olive (Olea europaea L.) DOP Umbria cultivars, *Sexual Plant Reproduction* Vol. 19:151–161.

Reale, L., Sgromo, C., Ederli, L., Pasqualini, S., Orlandi, F., Fornaciari, M., Ferranti, F. & Romano, B. (2009). Morphological and cytological development and starch accumulation in hermaphrodite and staminate flowers of olive (Olea europaea L.), *Sexual Plant Reproduction* Vol. 22:109–19.

Rosati, A., Zipančič, M., Caporali, S. & Padula, G. (2009) Fruit weight is related to ovary weight in olive (Olea europaea L.), *Scientia Horticulturae* Vol. 122:399–403.

Rosati, A., Zipančič, M., Caporali, S. & Paoletti, A. (2010). Fruit set is inversely related to flower and fruit weight in olive (Olea europaea L.), *Scientia Horticulturae* Vol. 126:200–204.

Rosati, A., Caporali, S., Hammami, S.B.M., Moreno-Alìas, I., Paoletti, A.& Rapoport, H.F. (2011a). Differences in ovary size among olive (Olea europaea L.) cultivars are mainly related to cell number, not to cell size, *Scientia Horticulturae* Vol. 130:185–190.

Rosati, A., Caporali, S., Paoletti, A. & Famiani, F. (2011 b). Pistil abortion is related to ovary mass in olive (Olea europaea L.), *Scientia Horticulturae* Vol. 127:515–519.

Rosati, A., Caporali, S., Hammami, S.B.M., Moreno-Alías, I., Paoletti, A. & Rapoport, H.F. (2012). Tissue size and cell number in the olive (Olea europaea) ovary determine tissue growth and partitioning in the fruit, *Functional Plant Biology*, in press.

Rugini, E. & Pannelli, G. (1993). Preliminary results on increasing fruit set in olive (Olea europaea L.) by chemical and mechanical treatments, *Acta Horticulturae* Vol. 329:209–210.

Seifi, E., Guerin, J., Kaiser, B. & Sedgley, M. (2008). Inflorescence architecture of olive, *Scientia Horticulturae* Vol. 116:273–279.

Stephenson, A.G. & Bertin, R.I. (1983). Male competition, female choice and sexual selection in plants, *in* Real, L., (ed.), *Pollination Biology*, London, Academic Press, pp. 109–149.

Suarez, M.P., Fernandez-Escobar, R. & Rallo, L. (1984). Competition among fruits in olive II. Influence of inflorescence or fruit thinning and cross-pollination on fruit set components and crop efficiency, *Acta Horticulturae* Vol. 149:131–144.

Uriu, K. (1953). Pistil abortion in the olive (Olea europaea L.) as influenced by certain physiological conditions. *PhD Thesis*, University of California, USA.

Uriu, K. (1959). Periods of pistil abortion in the development of the olive flower, *Proceedings American Society of Horticultural Science* Vol. 73:194–202.

Vallejo-Marín, M. & Rausher, M.D. (2007). The role of male flowers in andromonoecious species: energetic costs and siring success in Solanum carolinense L., *Evolution* Vol. 61:404–412.

Virgin Olive Oil

Chapter 8

Olive Mill By-Products Management

Pietro Toscano and Francesco Montemurro

Additional information is available at the end of the chapter

1. Introduction

Among the Mediterranean countries, Italy results to be the second after the Spain for the olive and olive oil production (FAOSTAT, 2012), with around 1.2 million of olive grove hectares, by 80% displaced in the Italian southern regions (ISTAT, 2010). In the olive oil industry, the oil extraction is carried out in oil mills, which are classified in pressure mills, and in continuous "two" or "three" phases way mills. In all milling typologies, only not over the 20% of processed olives constitute the oil production, while the milling by-products, wastewater and pomace, represents up to 120% of processed olives. These wastes could constitute a problem for their sustainable disposal as well as a resource for soil C stabilization and sequestration, energy generation or production of value-add compounds for the food, pharmaceutical or cosmetic industries.

A great part of husks derived by pressure and three phases mills is still destined to the industry for the extraction of residual oil through solvents (n-exane: $CH_3(CH_2)_4CH_3$). The residual defatted pomace is used as fuel in cogenerative processes (heat and electric power generation) or, as well as the two phases mill husks (fluid pomace) and olive mill wastewaters, disposed on soil as amendments, both raw and after composting. In the Mediterranean countries, where soils have frequently problems of organic matter lack, and active desertification processes, the recycle of the olive wastes as amendments should be better and more important to protect the environment being a valid alternative and a useful solution to the problems both of sustainable utilization of byproducts and soil fertility conservation.

In many experimental trials carried out in many sites, the utilization of olive industry byproducts as organic amendments, raw or stabilized through the aerobic fermentation, frequently showed good agronomic efficiency, in terms of fertility and chemical, physic and microbiological characteristics of the soils as well as crops productivity; generally pointing out some negative effects of fermentable organic matter and better findings using stabilized wastes.

Althought in literature can be found numerous findings on the influence of raw and composted organic materials on soil fertility and crops growth, only few published studies are focused on influences of waste waters and pomace compost on the soil-plant system, for a sustainable crops production.

In this chapter on report the state of the art of the by-products composting, and results of some application of raw and composted olive industry by-products on soil and crops.

2. Olive mill wastes

The olive mill byproducts are classified, according to the different systems of olive oil extraction, as: Olive mill Waste Waters (WW); Virgin Olive Pomace (OP), by pressure mills (OPP), with around 30% moisture; by centrifugal "three phases" mills (OP3), with around 50% moisture; or by centrifugal "two phases" mills (OP2), with moisture more than 60%. The Solid Defatted Pomace (SDP), is the byproduct of the pomace industry, after the extraction with solvents of the residual oil from virgin pomace.

2.1. Waste waters characteristics and treatments

The waste waters constitute the liquid fraction of the olive mill byproducts and are composed by the olives water content and by the waters used in the milling (olives and equipments washing waters and the waters of dilution process in the three phases centrifugal systems). As reported in Table 1, the different extraction systems produce, from 100 kg of processed olives, around 10 liters of waste waters for the two phases system, around 55 liters for the pressure crushers, and more than 100 litres for the continuous three phases system, that they reduce to only 35 litres in case of recycle of water.

Milling technology	H_2O added (%)	Pomace (kg/100 kg olive)	Pomace moisture (%)	Waste waters (kg/100 kg olive)
3 phase continuous	50	55-57	48-54	80-110
Pressure	0-10	30-35	25-30	56-58
2 phase	0-10	75-80	60-70	10
3 phase with water recycle	0-20	56-60	50-52	33-35

Table 1. Mass balance in the olive oil extraction

The main characteristic of WW is the presence of organic compounds such as organic acids, lipids, alcohols and polyphenols that turn it into phytotoxic materials, that might have unfavourable impact on plants (Capasso et al., 1992; Kavdir & Killi, 2008). It also represents a great environmental hazard if not properly managed. However, WW contain valuable resources, such as a high organic matter concentration and some nutrients, especially potassium, that could be usefully used to improve the physic-chemical and biological properties and then soil fertility and productivity; representing a valid option to close the residue-resource cycle (Roig et al., 2006). In Table 2 are reported the averages of main chemical WW characteristics, given by several Authors (Aktas et al., 2001; Filidei et al., 2003; Moreno et al., 1987; Paredes et al., 1999; Piperidou et al., 2000; Saviozzi et al., 1991; Vlyssides et al., 1996; Vlyssides et al., 2004 – as cited in Roig et al., 2006).

Dry matter (%)	6.75
pH	4.84
EC (dS/m)	8.36
O.M. (g l^{-1})	55.80
TOC (g l^{-1})	37.00
TN (g l^{-1})	0.97
P$_2$O$_5$ (g l^{-1})	0.56
K$_2$O (g l^{-1})	4.82
Na (g l^{-1})	0.25
Ca (g l^{-1})	0.35
Mg (mg l^{-1})	121.25
Fe (mg l^{-1})	81.70
Cu (mg l^{-1})	3.15
Mn (mg l^{-1})	6.13
Zn (mg l^{-1})	6.13
d (g cm$^{3\ -1}$)	1.04
Lipids (g l^{-1})	6.38
Poliphenols (g l^{-1})	4.98
Carbohydrates (g l^{-1})	7.16
COD (g l^{-1})	124.67
BOD$_5$ (g l^{-1})	65.00

Table 2. Average composition of olive mill waste waters.

About the microbiological characterization, the results of different analyses performed on different kind of WW, have individualized 130 species of lipolytic microorganisms (56 Fungi, 22 Yeasts, and 52 Bacteria), cellulolytic Bacteria and pectinolytic Fungi, while not are resulting the nitrificants nor the actinomycetes (Pacifico, 1989; Ramos-Cormenzana, 1986).

For the waste waters treatment have been proposed both physic-chemical and biological processes (Amirante & Di Rienzo, 1993; McNamara et al., 2008; Rozzi & Malpei, 1996; Vigo et al., 1990; Vitolo et al., 1999).

Among the first ones, finalized to the volumes reduction, and to the mineralization of organic compounds, they are:

-   distillation, filtering or flocculation and disposal in the water bodies;
-   evaporation and agronomic utilization of concentrated material;
-   raw application on the soil;

-    adoption of two phases mills, and management of the fluid pomaces.

The biological treatments of purification include both aerobic and anaerobic processes. The aerobic biological treatments are based on the microbic degradative activity that transform the decomposable organic matter in not pollutants mineral elements and humus-like substances.

The anaerobic processes are characterized by microbial pools that works in absence of oxygen, converting the organic polluting substances in biogas (methane) and carbon dioxide or in hydrogenated volatile substances (fatty acids and alcohols), (Filidei et al., 2003). Three main physiological groups of microorganisms are involved: fermenting bacteria, organic acid oxidizing bacteria, and methanogenic archaea. Microorganisms degrade organic matter with a sequence of biochemical conversions to methane and carbon dioxide. Syntrophic relationships between hydrogen producers (acetogens) and hydrogen scavengers (homoacetogens, hydrogenotrophic methanogens, etc.) are critical to the process. A wide variety of process applications for biomethanation of wastewaters, slurries, and solid waste have been developed. They vary from the simple WW open-air storage to the treatment in Completely Stirred Tank Reactor (CSTR) through co-digestion with others organic matrixes, up to treatments in special digesters, as UASB reactors (Up-flow Anaerobic Sludge Blanket) and with different process conditions (retention times, loading rates, temperatures, etc.) to maximize the biogas production (Angelidaki et al., 2011).

Even if these treatments are able to demolish the WW polluting power, in the practice they are often not economically sustainable. The seasonality of WW production and the characteristics of biological toxicity, make difficult the management of their treatment in the purification installations of the urban waters. Conversely, the relative low quantity produced by single milling plants do not make economic WW purification in specific installations.

The WW open-air storage is the simplest and economic system of storage and treatment. During the open-air storage by the action of aerobes and anaerobes microorganisms, it will be partially purified and stabilized. However, the WW open-air storage require ample surfaces away from inhabited centers, for the stink development, and exposes to risks of soil and subsoil water contamination, if the basins are not correctly waterproofed (Catalano et al., 1985).

In fact, these wastes can constitute a "secondary raw materials", to be consider as a resource, being of a vegetable origin, which have not undergone chemical manipulations nor received additives, and therefore without pathogenic microorganisms and viruses, as well as pollutants or toxic products. The optimal operative option results therefore to be the agronomic utilization of the WW that, when correctly managed, allows to exploit its fertilizing characteristics with low costs of management, reducing the risks of environmental pollution.

Many studies have been conducted on soil application both on olive orchard and herbaceous crops, studying differing time and doses of WW application. The results showed both the agronomic benefits and the limitations of raw WW disposal. When correctly managed, the WW application on soil generally results increasing the organic matter, phosphorus and potassium content; improving both physical and hydraulic soil properties as well as the crop yield (Alianiello et al., 1998; Andrich et al., 1992; A.R.S.S.A., 2001; Ben Rouina et al., 1999; Bonari et al., 1993; Catalano et al., 1985; Di Giovacchino & Seghetti, 1990; Di Giovacchino et al., 2001; Levi-Minzi et al., 1992; Montemurro et al., 2007, 2011; Saviozzi et al., 1991; Tamburino et al., 1999; Tomati & Galli, 1992).

Another option concerns WW composting through biological processes of aerobic stabilization, on supporting matrixes with adequate physical and mechanical properties (porosity, structure, texture), to allow the reactions of bio-oxidation in the solid phase. In this way it is possible to obtain a not phytotoxic organic material, humus-like partially transformed, that could be used in fertilization plans to restore or maintain soil fertility on partial or total substitution of mineral fertilizers (Benitez et al., 1997; Filippi et al., 2002; Galli et al., 1997; Paredes et al., 2002; Tomati et al., 1995; Vallini et al., 2001).

2.1.1. Waste Waters agronomical value

The WW contain appreciable amount of mineral elements and organic substances, and can be considered as liquid amendments by vegetal origin. WW application to the soil could realize the double purpose to allow a natural chemical and biological degradation and to enrich the soil in organic matter and in mineral elements. This means that the agronomical use of these waters represents one of the most effective systems to decrease their BOD_5. For this reason, the elevated values of WW BOD_5 and COD that make extremely risky WW disposal in surface or ground water bodies, would come to lose importance in the case of soil distribution. The WW poliphenols, held substances of elevated polluting power in the case of disposal in the surface and ground water bodies, do not represent a pollution factor for the soil, being instead precursory in the synthesis of humic substances that represent the most active fractions of soil organic matter with important role in soil organic fertility and protection from synthetic organic contaminating, as pesticides and heavy metals.

For this, the most effective option appears to be the use of WW as liquid organic amendment or fertigant, and the WW recycle in soil could be a valid alternative to other depurative treatments. In fact, WW affect positively crop performances, of olive, grapevine and cereals, when distributed in dates and doses selected through rigorous agronomical criteria.

On the contrary, an incorrect WW management, may cause temporary immobilisation of soil mineral nitrogen and, consequently, crop yield reduction, due to the deficiency of N uptake by the plants; also increasing environmental pollution risks. To avoid these effects, the maximum amount of waste water that can be applied is restricted by the Italian law to 50 and 80 m^3 ha^{-1} yr^{-1}, for the pressure and continuous milling systems, respectively.

Phytotoxic effects can also be avoided applying simple and few expensive WW pretreatment before the application to the soil, using biological or mineral catalysts, or submitting them to suitable composting processes, mixed with other organic by-products.

The stabilization of WW through composting on effect with repeated saturation of supporting materials with suitable physical-mechanics characteristics (porosity, structure, texture) as straws, leaves, sawdust, rapiers, or pomace, to maintain a moisture level useful to the bioxidative reactions (Benitez et al., 1997; Garcia-Gomez et al., 2003; Paredes et al., 2000; Tomati et al., 1995; Vallini et al., 2001).

In this way, on allow on the WW the bioxidative reactions in solid phase, handling to a final product, without phytotoxic effects, useful in the agricultural soils as organic amendment.

2.1.2. Waste Water effects on the herbaceous crops

The effects of the WW distribution on soils destined to herbaceous crops, vary according to the WW composition, the quantity, the distribution dates in relationship to seeding or crop phenologic stage, the rainfall after the WW supply.

The olive mill WW do not generally affect the productivity of the spring-summer crops, when the shedding is effected with an adequate interval time before the seeding. It is also possible to distribute the WW with crop in action, i.e. on autumn-winters cereals, in the full growth phase, limiting the WW doses to not over 40 m^3 ha^{-1} (Bonari et al., 1993; Di Giovacchino & Seghetti, 1990; Di Giovacchino et al., 2001; Marsilio et al., 1990; Montemurro et al., 2007, 2011a; Roig et al., 2006).

Negative effects on grass crops are due to the elevated values of electrical conductivity, that can induce salinity damages, flocculation of soil clay fraction, the phytotoxicity of poliphenols (Della Monica et al., 1978, 1979).

In the last years, because of environmental protection policy, new technologies that minimize these risks were developed. In this framework, the raw olive waste water was differently treated to improve the percentage of recycle of these materials. In particular, different studies indicate that the application of mineral catalyser (MnOx) on WW reduces the level of poliphenols and other pollutants. As a consequence, this treatment decreases both phytotoxicity and temporary immobilization of soil mineral N, thus making the treated WW able to sustain good levels of crops yield and products quality (Montemurro et al., 2007; Vigo et al, 1990).

2.1.3. Waste Water effects on the olive crop

Many experiences have confirmed the possibility to distribute the WW on the olive grove more than over 400 m^3 ha^{-1} without significant variations of vital parameters as the photosynthetic activity, transpiration, stomatic conductance, leaves carbohydrates and chlorophylls content. Conversely, improvements of the productivity on treated plants was frequently observed. A decrease of the vegetative development has been recorded only on young plants raised in pots and treated with higher WW doses (Proietti et al., 1988).

The organic matter content in WW treated soil always results greater in comparison to the untreated control, with greater nitrogen contents and more elevated C/N ratio.

In the soils treated with WW a greater presence of the total microflora was found (Fungi and other microbial groups), as well as an increase of respiratory and enzymatic activities. These findings showed the absence of toxicity of the WW towards the microorganisms, and the improvement of soil fertility. Moreover, the findings of different researches indicate the increase in organic matter and nutrient contents, an improvement in the aggregate stability and, as a consequence, a better physical soil properties.

In order to the possible ground water pollution, other results have shown that also up to 500 m^3 ha^{-1} of WW application do not represent a pollution danger of the surface water in the clay soils. (Andrich et al., 1992; Ben Rouina et al., 1999; Briccoli-Bati & Lombardo, 1990; Lombardo et al., 1995; Palliotti & Proietti, 1992).

2.2. Pomace characteristics and treatments

The olive pomace is composed of fruit matter (olive skins, flesh, seeds and stone fragments), and of different amount of vegetation and process water which contains the water-soluble constituents of the fruits, in order to the extraction system used.

The OPP and OP3 are usually destined to the pomace industry, for the extraction of residual oil by solvents; and then used as fuel, also in cogenerative processes (Molinari & Bonfà, 2005). These by-products could also be used as animal food, or in biodegradative processes to produce ethanol (Ballesteros et al., 2002), or compost for agricultural utilization. Nevertheless the olive pomace, being constituted by vegetable not fermented organic matter, does not contain heavy metals, toxic pollutants or pathogens, and can be considered as a vegetable amendment (Table 3), (Alburquerque, et al. 2004). Therefore, it can be used in the agricultural soils without any treatment, as allowed by the current normative (Law 574/1996).

Parameters	Range
Humidity %	55 - 75
pH (H_2O)	4.8 – 6.5
EC (dS m^{-1})	0.9 – 4.7
Ash (g kg^{-1})	24 - 151
TOC (g kg^{-1})	495 - 539
C/N ratio	28 - 73
Total N (g kg^{-1})	7 - 18
P(g kg^{-1})	0.7 – 2.2
K(g kg^{-1})	7.7 – 29.7
Ca(g kg^{-1})	1.7 – 9.2
Mg(g kg^{-1})	0.7 – 3.8
Na(g kg^{-1})	0.5 – 1.6
Fe (mg kg^{-1})	78 - 1462
Cu(mg kg^{-1})	12 - 29
Mn(mg kg^{-1})	5 - 39
Zn(mg kg^{-1})	10 - 37

Table 3. Main characteristics of the olive pomace (dry weight)

2.2.1. Pomace agronomical value

On the contrary to the WW, no many experiences was carried out on the OP application, because both OP3 and OPP represent a better incoming value when destined to the pomace industry and energetic production. The results of OP application both in olive orchard and herbaceous crop, was different, depending both from doses applied and crop. Against to a remarkable improvement in chemical-physics characteristics of olive orchard soil up to two years after the pomace supply, has no given significant differences in vegetative and productive parameters of treated olive trees. In herbaceous crop a phytotoxic effect was observed on plant growth with highest doses applied, showing that the fresh organic matter distributed in soil will be degraded producing intermediate metabolites which are not compatible with normal plant growth and nutrients availability, with in some cases an increase in the fungal disease (Bing et al., 1994; Bonanomi et al., 2006; Brunetti et al., 2005; Di Giovacchino et al., 2004; Diacono & Montemurro, 2010; Tejada & Gonzales, 2003, 2004).

In the practice, the benefits of olive pomaces recycling in soil occur only if were applied according to best agronomical practices, as taking into account suitable time and specific plan of fertilization, needs of the soil-plant system and the climatic conditions.

Nevertheless, the application of raw olive pomace involves a more difficult management of these biomasses and, also for possible negative aspects, a better management solution is to compost the pomace, as we can see later in this paper.

2.3. Olive mill wastes legislation

According to their nature, the extraction methods and the different phases of management, the olive mill by products have to follow different laws concerning storage, treatment, transport, application to the soil, or other destinations.

2.3.1. Waste waters legislation

The WW legislation is based on their two possible different destinations: agronomic or landfill. In case of agronomic use without any preventive treatment, the reference is the Law n. 574/1996. As established by the Decree of the Ministry of the environment, on the "Individuation of non dangerous wastes submitted to the simplified recovery procedures to the senses of the articles 31 and 33 of the Legislative Decree 05.02.1997 n. 22", the WW can also be recovered through the "Production of fluid fertilizers, as by the Law n. 748 of 19.10.1984: "New norms for the fertilizers discipline".

If the WW cannot be used in agronomical purposes, it is necessary to apply preventive treatment. The products, resulting from the purification process, can be schematized as follows:

-    for the liquid phase, the reference is the Law n. 319, 10.05.1976 (Merli law) with his modifications and integrations, and the Legislative Decree n. 22/1997 (Ronchi Decree) and his modifications and integrations, in which the limits of acceptability are established also for the discharge in surface and ground water, on the soils, or in the subsoils (L.D. 389 of 8.11.1997).

- for the solid phase, the reference, related both to the treatment and discharge phases, is constituted by the Legislative Decree 27.01.1992, n. 99, "Realization of the directive 86/278/CEE regarding the environment protection, particularly of the soil, in the use of the solid phase purification in agriculture", and the Law 19.01.1984, n. 748 with his modifications and integrations "New norms for the discipline of the fertilizers." The art. 8, par. 1, letter "d" of the "Ronchi Decree" establishes that WW are excluded by application, because disciplined by specific dispositions of law "the activities of byproducts treatment that give origin to the fertilizers, selected in reference to the typology and the mode of utilization as by Law 19.10.1984, n. 748 with his modifications and integrations".

Another possible destination of the WW is the agronomic use after an a intermediary phase for production of liquid fertilizers, conforming to the Law 748 19.10.1984, "New norms for the discipline of the fertilizers", according to the norms of the D.M. 05.02.1998 of the Ministry of the environment. As a consequence, WW must be recovered according to the point R3 of the Annex C of Legislative Decree n. 22/1997, article 6, paragraph 1, letter h, ("Recycle/recover of the organic substances not used as solvents, included the composting operations and other biological transformations").

2.3.2. Pomace legislation

The OP storage is disciplined by the Legislative Decree n. 22/1997 with his modifications and integrations, according to which it should be considered as not dangerous special wastes that (ex D.M. 05.02.1998 of the Ministry of the environment) can be submitted to the simplified recovery procedures according to the articles 31 and 33 of the Legislative Decree n. 22/1997 with his modifications and integrations.

About the treatments, in case of distribution on the soil, also the OP2 can be considered as OP ex Law n. 574/1996, for which is not provide any preventive treatment for the agronomic use, because: "To the goals of the application of the present law, the OP2 coming from the olives milling and constituted by the waters and the fibrous part of fruit and the core fragments, can be used as amendments in derogates to the characteristics established by the Law 19.10.1984, n. 748 and modifications". As D.M. 05.02.1998 of the Ministry of the environment, (non dangerous wastes that can be submitted to the simplified recovery procedures as the articles 31 and 33 of the Legislative Decree n. 22/1997 with his modifications and integrations) the pomace could be submitted to processes of anaerobic biostabilization, with production of biogas. This material could also be submitted to the aerobic processes, for the compost production in conformity with the annex to the Law 19.10.1984, n. 748 with modifications.

Other destinations of the OP, with exclusion of the OP2 for elevated moisture, is the conferment to the pomace industry for the extraction with solvents of the residual oil. The defatted pomace that resulted as byproduct of this process, are used as fuel, or as organic amendments after aerobic co-composting.

3. Biological treatments

With this definition, are indicated the processes and the activities on organic materials that, degraded and transformed by various decomposers microorganisms, allow its stabilization in terms of mineralization of the mostly degradable components, and the hygienization of the biomass by pasteurization.

The biological treatment of byproducts can be realized with different technologies and processes, referable to three following typologies:

- composting of quality, on selected biomasses, for the production of amendments of high agronomic value, according to the parameters defined by the Law 748/84 on the fertilizers;
- biostabilization of organic matrixes of low quality (as organic fractions from mechanical separation of the undiversified wastes residues, sludges with heavy metals, etc.);
- anaerobic digestion for production of biogas, both on high quality organic matrixes for a successive agronomic utilization, and on lower quality or contaminated organic residuals. The byproduct (digestate) can be afterward used as fuel, aerobically stabilized, or arranged as sludge according to the Legislative Decree 99/1992 on the sludge usage in agriculture.

4. Composting

The composting is a controlled process, based on the control of the natural bioxidative process on organic substances (Figure 1) transformed by aerobes microorganisms naturally present in the environment and in the matrixes, in a stabilized and sanitized organic substance, defined compost (Casacchia et al., 2011; Insam & De Bertoldi, 2007).

The composting is classified by the law (Legislative Decree 22/97) as a recovery operation of byproducts, among the actions of "The recycle and/or retrieve of the organic substances not used as solvents"; being an interesting alternative in recycling large amounts of a wide range of the residues produced in Mediterranean areas, resulting in an organic fertilizer suitable to agricultural purposes, in improvement of o. m. content of soils, and reduction of fertilization costs.

The composting process is usually divided in two phases: Active Composting Time (ACT), and Curing.

During the ACT, there is an intense develop of degradative processes on to more easily fermentescible organic components. In this thermophylic phase the biomass reaches temperatures up to 65 °C, that also allow its hygienization. The ACT can develop, according to the characteristics of the matrix, in piles, with or without overturn and with or without forced airing, or in more complex systems (bioreactors), and varying from 21-28 days in the first case or 14-16 days in the second one.

During the curing phase, the degradative phenomena will be completed, and the synthesis of the humic-like substances occurred. This process, which has a duration of more than 45 days, need a smaller oxygen and drain of heat.

The main parameters of the composting process, that ensure an optimal development of microorganism able to transform the organic matrices, are synthetically reported in Table 4.

pH	Should be between 5.5 - 8, to assure the activity of useful microbes.
Humidity %	Range variable from 50-60% in the first weeks (ACT); to go down to 30-35% to the end of process, depending on the biomass characteristics and the processing typology.
Temperature	Range to develop microorganism: 40-45 °C (3 days over 55 °C for the Italian legislation, to kill pathogens)
C/N ratio	Nutrient balance: between 25 and 35 to avoid nitrogen losses for volatilization (< 25) or decrease speed of metabolic reactions (> 35)
Porosity	To ensure aerobic process; depending on particle size. The better air-filled pore space: 25-50%
Particle size	Surface air for microbial growth
O₂ concentration	Assured by forced airing and/or overturns. Optimal range for degradative process: 15-20%

Table 4. Controllable composting parameters.

During the composting, the organic substance evolves both quantitatively, reducing weight and volume, and qualitatively, with deep modifications of its chemical and structural characteristics. The parameters that defines the composted materials are:

- stability, that characterize a specific stage of decomposition or state of organic matter during composting, which is related to the type of residual organic compounds and the resultant biological activity in the material. The specific stage of organic matter represents the exhaustion of degradative processes due to the biological metabolic activity (low static and/or dynamic respiration index).
- maturity, is the degree of composting completeness and concerns improved qualities resulting from "ageing" or "curing" of a composted matter. Usually the level of maturity is detected by the phytotoxicity, verifiable with the homonym test.
- humification, or the endowment of humic-like molecules originated from reactions on the less fermentescible organic components. (Boulter et al., 2000; De Nobili & Petrussi, 1988; Franz et al., 2005; IPLA, 1992; Zucconi et al., 1981).

In order to the duration of the fermentative process, two typologies of compost can be obtained, based on their degree of maturation:

- the ready-to-use compost, of greater fertilizing efficiency, that is obtained in 3-4 months, at ACT end, that can be used for vegetable crops, gardens and in full field, on preseed and before the transplantation;
- the mature compost, more stabile and humified, that is obtained after 5-6 or more months of curing; usable as amendment in full field, or as substrate in the greenhouse nursery.

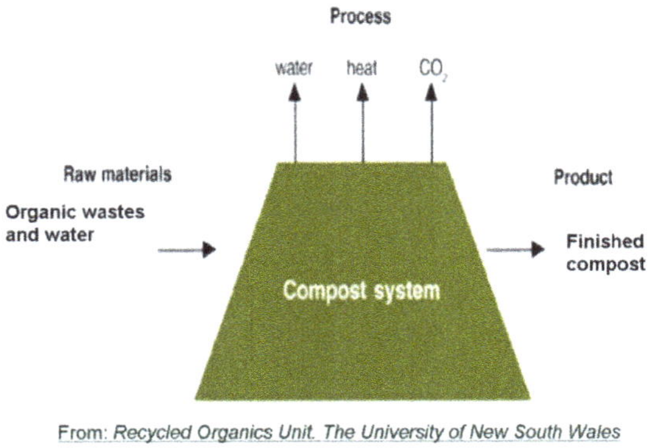

Figure 1. Composting process flow.

The composting of the agro industrial by-products and agricultural residual biomasses can be carried out according to two typologies:

- composting of crop residuals and animal manure by farmers; in this case the farmer that compost in situ produced biomasses, is not subject to any restriction, provided that the compost is used in the same firm;
- composting of extra farm biomasses; this kind of activity is not considered as normal agricultural practice, but solid wastes recycling. According to the prevailing interpretation, the extra-farm solid wastes included in the compostable materials list of the D.M. 05.02.1998, are subjected to the "simplified procedure", as from artts. 31 and 33 of the Legislative Decree 22/97 ("Ronchi" decree).

The on-farm composting does not require particular technologies, while the necessary machineries (carver, shovel loader, wagon mixer) are often already present in farm.

The biomasses to be composted, mixed with opportune percentages of appropriate structural and absorbent fractions (lignocelluloses bulking), are typically placed in pile on paved stage, preferably covered, and periodically overturned up to stabilization.

The shape and the dimensions of the pile will depend from the quantity and the quality of the material to be treated. The forms are of triangular, parabolic or trapezoidal section, of 1.5 - 1.8 m height, while width and length depending from the quantity of material or from the available space, and also in base to the machinery used for the overturns (hauled or self moving over-turner, shovel loader).

During the composting, must be constantly monitored temperature and moisture of the pile, taking care of maintain the porosity, to allow the aeration and thermic dispersion of biomass.

According to the moisture and granulometry of biomass, the overturns will have to take place to intervals of 2-3 days, in case of high moisture or low percentage of structural fraction (bulking); while greater intervals (up to 10-15 days) can be applied on more porous or coarse matrixes, and in the curing phase of compost.

The composting practices contemplate several operative options, in relationship to the different typologies of biomasses to be treated, and to the different managerial and environmental situations: from the simplest systems of aerobic stabilization on natural conditions, up to the industrial plants with the complete control of the process conditions (Alfano et al., 2008; Amirante & Montel, 1999; Baeta-Hall et al., 2005; Calvet et al., 1985; De Bertoldi et al., 1983; Diaz et al., 2007; Madejon et al., 1998; Montemurro et al., 2009; Tomati et al., 1996; Veronesi & Zampighi, 1997).

Although the OP is characterized from inadequate physical characteristics, in particular the OP2, that makes difficult the aerobic degradation, this by-product can be easily composted by using complementary residues such as pruning residues or cereal straw as bulking agents; also mixed with other agricultural wastes as animal manures or horticultural residues. During the process, the potentially dangerous organic substances in the olive wastes are degraded, allowing the safe and useful use of the obtained compost in agricultural fields; and valorizing these byproducts as a resource.

4.1. Composting normative

The composting sector is disciplined by the two following normatives: the wastes normative, that defines the conditions of production of the compost from wastes or selected fractions, and the fertilizers one that establish the marketing conditions of the compost.

- Wastes Normative: Legislative Decree n. 22, 05.02.1997 with modifications, and Technical Normative of realization of the Legislative Decree 22/97, and the D.M. 05.02.1998: Individuation of non-dangerous wastes submitted to the simplified retrieving procedures to the senses of the artts. 31 and 33 of the Legislative Decree 22/97.
- Fertilizers Normative: Law 748/84 (Annex 1C) and modifications and integrations.

The Legislative Decree 22/97, which follows the CEE/91/156 wastes directive, classifies the composting among the recovery operations of wastes (Annex "C", R3: Recycle/recovery of the organic substances not used as solvents, included the composting operations and other biological transformations).

With the Technical Normative, the D. M. 05.02.1998, have been defined the technical aspects tied up to the compost production, even if limited to the activities of recovery in simplified regime, with a reference to the Law 748/84 "Norms for the fertilizers discipline", that in the enclosure 1C, as modified by the D.M. 27.03.1998, defines the marketing characteristics for the compost, indicating the following four types of final products:

- Simple not composted vegetable amendment: not fermented product composed of barks and/or other vegetable residues as pomace, chaffs, peels, with exclusion of algae and other marine plants (it is a product that, for definition, cannot considered as raw material for composts);

-   Green composted amendment: obtained through a controlled process of transformation and stabilization of organic wastes constituted by urban green residues, crops residuals, other residues of vegetable origin, with exclusion of algae and other marine plants;
-   Mixed composted amendment: obtained through a controlled process of transformation and stabilization of organic residues that can be constituted by the organic selected fraction of the urban wastes, from animal slurry, included the animal sewages, from agro industrial residues, and from the wood and textile untreated throughput as well as from the matrixes of the Green composted Amendment;
-   Peaty composted amendment: obtained by mixture of peat with Green composted amendment or Mixed composted amendment.

For the realization and the management of the composting installation can be followed:

-   the simplified procedure (as artts. 31 and 33 of the Legislative Decree 22/97) when the installation, the first and the final products are entirely conforming to the prescriptions of the D.M. 05.02.1998, particularly for the selection of the organic materials. Furthermore, the characteristics of the final products should be conforming to the requisite in the enclosure 1C to the L. 748/84 (as modified by the D.M. 27.03.98). This procedure lay down a preventive communication of start of the recovery activity together with a technical explanatory document, in the respect of the discipline contained in the DPR 203/88, for the releases in atmosphere, and of that reported in the Legislative Decree 152/99, for landfill.
-   the ordinary procedure (as artts. 27 and 28 of the Legislative Decree 22/97), that require a specific authorization, released from the local administrations.

4.2. Composting procedures

The compost should to have an adequate degree of stabilization (maturity); and qualified amendments and nutritional properties.

The quality of the compost is function of the following factors:

-   the composition of the mixture (raw materials), as percentage among different components;
-   the management of the composting process, particularly the duration, the moisture and the oxygenation of mixture;
-   the final conditioning (refinement, addition of other products).

The composting of selected organic substances has to be considered as a recovery system for the production of high quality organic amendments to use for the agricultural purpose able to replace, even though partially, the manure in the extensive agriculture, the chemical fertilizers in the intensive crops and the peat in nursery.

The ACT phase, in relationship to the characteristics of the treated mixture, can happen in piles on open air, or confined in more complex systems (bioreactors) with forced airing, with or without overturns. During this phase must be checked the temperature, moisture, pH and oxygen. The duration of this phase should be enough to guarantee the attainment of an adequate Respiration Index (I.R.) of biomass, to be test with the IPLA (Institute of Plants, Wood and Environment) method (IPLA, 1992). This duration corresponds to about 21 days of treatment, in the case of stabilization in overturned piles and about 14-16 days in the case of bioreactors.

The curing phase is always carried out in piles and require a lower amount of oxygen and overturn in comparison to the ACT phase.

The biomasses to be composted must be previously grinded and mixed to optimize the characteristics of the matrix (C/N ratio, porosity, moisture, pH). A good starting matrix for the production of a Quality Composted Amendment (ACQ), has to have a moisture among 45 and 65%, C/N among 20 and 30, with at least a 30% in weight of lignocelluloses material. The ACQ can be commercialized to the senses of the Law 748/84, also in mix with other fertilizers, according to the same law prescriptions, as reported in Table 5. This material could be used in the agricultural activities or similar (maintenance of the public or private green, biofilters, etc.).

Parameter	Green Composted Amendment	Mixed Composted Amendment	Peaty Composted Amendment
pH	6.0-8.5	6.0-8.5	-
Humidity (%)	<50	<50	-
TOC (% d.m.)	>30	>25	>30
Organic Nitrogen (%)	>80	>80	>80
C/N ratio	<50	<25	<50
Humic and fulvic acids (% d.m.)	>2.5	>7	>7
Peat (%)	-	-	>50
Cd (mg kg^{-1})	<1.5	<1.5	<1.5
Cu (mg kg^{-1})	<150	<150	<150
Hg (mg kg^{-1})	<1.5	<1.5	<1.5
Ni (mg kg^{-1})	<50	<50	<50
Pb (mg kg^{-1})	<140	<140	<140
Zn (mg kg^{-1})	<500	<500	<500
Cr VI (mg kg^{-1})	<0.5	<0.5	<0.5
Inert (>10 mm) (%)	none	none	none
Salmonelle (on 25g sample)	none	none	none
Enterobacteriacee (CFU g^{-1})	< 100	< 100	< 100
Fecal Streptococcus (MPN g^{-1})	< 1000	< 1000	< 1000
Nematodes (on 50g sample)	none	none	none
Trematodes (on 50g sample)	none	none	none
Cestodes (on 50g sample)	none	none	none

Table 5. Limits of acceptability for the Quality Composted Amendment (ACQ), conforming to Annex 1C of the Law 748/84, as modified by the D.M. of 27.03.1998.

The composting process can also carried out with low-tech approaches and extreme operational simplification, excluding the forced airing and effecting the biomass overturns using machineries already in endowment to the farm (choppers, shovel loader, wagon mixers).

4.3. Technological systems: classification and description

The bioxidative processes are carried out in different operative systems, classified as (Amirante & Montel, 1999; Clodoveo et al., 2000; Goldstein, 1980; Willson et al., 1980):

-   intensive or extensive systems, according to both the technological complexity and the energetic needs;
-   closed or open systems, according to be confined or not by the open air;
-   static or dynamic systems, according to whether the mass has moved;
-   aired or not aired systems, in presence or absence of forced airing.

The intensive systems are destined to high fermentability biomasses. They are generally closed, dynamic and aired, and works through a first active phase (ACT), with control of process parameters, and a second phase (curing) with lower-level of technological complexity, similar to the extensive systems. The processing times vary between 25-30 and 120 days (in average 90 days). The energetic needs are around 40-60 kwh ton^{-1} for the machinery power (overturners, ventilation systems, grinders, sieves). The need areas are around 0.7-1.5 m^2 ton^{-1} of annual biomass to process.

The extensive systems are dedicate to biomasses of low fermentability and are open, or also confined, static and not aired (aired for diffusion and natural convection), and do not differentiate the operative phases, but they adopt an only lower technological step. Typically they consist in macropiles open air, without or very time deferred overturns. The process times varying around 6 months up to over 1 year. The energetic needs are low (10-20 kwh ton^{-1}), for the grinding, the optional overturn with generic machines (mechanical shovels) and the sifting. They need surfaces around 1.5 - 2 m^2 ton^{-1} of annual biomass to process.

4.3.1. Systems "closed" or "open"

In the closed systems the process is carried out in confined spaces (bioreactors) or in covered areas (sheds), for biomasses that require greater control of the fermentative parameters (high moisture, stinking as well as selected for high quality compost). Conversely, the open systems are destined to biomasses of low fermentability, high percentage (i.e. 70%) of lignocelluloses bulking, or for curing phases of biomasses already submitted to ACT.

4.3.2. Systems "static" and "dynamic"

These systems differ for the periodic or continuous biomass moving (dynamic) or for immobility (static). The base idea of the static systems is not to disturb the growth and the action of the fungi and of the microbial population, maintaining a microecological environment. This situation is important to create in the biomass good growth conditions with the purpose to facilitate the stabilization processes. This systems can work with matrixes well structured and homogeneous, with at least 40 - 50% in weight of lignocelluloses bulking, to avoid the compaction of the mass, and moisture less than 65%.

In the dynamic systems can be processed very moisture matrixes (> 65%) and with less than 30% in weight of lignocelluloses bulking.

4.3.3. Systems "aired" and "not aired"

The forced airing of the biomass is an important factor of optimization of the process particularly for the treatment of low consistence and higher fermentability biomasses. For the optimization of the forced airing it is necessary to calculate and monitoring:

-   the specific air flow, generally expressed in Normal cubic meters for hour and of weight biomass unit (NCM h^{-1} ton^{-1}). Generally, the airing needs for the drain are higher also of a dimensional order, in comparison to those of oxygen in stechiometric demand for the mineralization. The airing systems are therefore sized and used for the control of the temperature, while the biomass oxygenation is obtained as derived effect. This sizing involves nevertheless an excessive evaporation and drying of the biomass, with early interruption of biostabilizative processes. For this reason, in airing systems are necessary appropriate systems of control and remoisten.
-   the proportion between times of airing activation and turning off. The intermittent ventilation of the biomass, besides the energetic saving, has the purpose to allow during the periods of turning off the equalization of moisture and temperature in the different zones of the biomass, while the airing in continuous can behave more sensitive stratifications of them. The proportion between times of working and turning off is established in order to the better thermal level needed for the biomass, being the optimization of the temperatures another of the essential parameters to the efficiency of composting process. To the same flow of air and of "heating power" of the biomass (dependent from moisture and fermentability), greater periods of working airing proportionally involve lesser thermometric levels of the biomass (Higgins, 1982).

The different processing systems, classified according to the installations or the operational methodology, can be associate to one or more of following categories. Typically, among the open systems are included: Aired static pile; Overturned pile; Short pile. While, among the closed systems they are: Biocontainer; Dynamic trenches; Dynamic basins; Silos.

4.3.4. Aired static pile

The aired static pile constitute a system of relative technological simplicity. It was developed in America as "Beltsville system ", with the purpose to have a simple system of the wastes biomasses bioconversion to the farmers for agronomic reuse. In his different variations, it shapes as an open, static and aired system (Willson et al., 1980). Typically it foresees the disposition of the biomass in pile, with forced airing in aspiration below the pile and dispatch of the exhausted air to a compost-made biofilter. On the pile surface a layer of mature compost is placed, with not permeable and filtering actions. As static system, it requires of a high percentage of structural lignocelluloses bulking, and of a relatively low moisture of the starting matrix. It also requires a appropriate homogenization pretreatments of biomass. In some different systems the forced airing works by inflation, while for the pile coverage are adopted cloths or semi-permeable membranes, to limit the losses of moisture.

4.3.5. Overturned piles

In this system the biomass are disposed in piles of great dimensions (3-4 m of height; width up to over 20 m). The piles are typically open, not aired and overturned with ample intervals (weeks or months), with mechanical shovels or specific overturning machines in the case of continuous systems. This system is adopted for matrixes to tall prevalence of lignocelluloses material (green residues) or in the curing of biomasses after the ACT. The piles can be managed in batch, or in continuous, with progressive translation of the biomass during the overturns, toward the unloading section. This second solution eliminates the unused spaces between piles, while is forcing to adopt a fixed overturns periodicity, that would not correspond to the real needs of process.

4.3.6. Short pile

The short pile differ from the piles for their smaller dimensions (max 2.5 m height) and can be or not endowed with the forced airing. The overturn is generally frequent (from few days up to daily intervals) and are made with specific overturners. The short piles are planned as "closed system" for the active phase of high fermentable matrixes; or as "open" where the stinking impacts do not constitute trouble, or for low fermentable material, or for the curing of end-fermented biomasses.

4.3.7. Biocontainer and biocell

Biocontainer and biocell are horizontal closed reactors, typically static and airing. The biomass is placed in beds of the maximum height of around 3 meters. The biocontainer are made with metallic or concrete, insulate, and have unitary volumes of the order of many about ten cubic meters. These systems generally adopt the recycle of the air, and survey systems to manage the parameters of process (moisture, oxygen percentage in the inside atmosphere, temperature) and feed-back regulation of the air flows and the percentages of air recycle.

4.3.8. Dynamic trenches

Conversely to the biocontainer, in this system the reactor is divided in trenches served by one or more lines of forced airing, with modulation of the air flows in the different sections correspondents to the different times of process. The trenches race binary for the translation of self-moving overturners that also effect the progressive transfer of biomass (continuous system). The trenches are typically used for the management of the active phases of biomasses to elevated fermentability, in closed environments. As dynamic system, can effectively process also biomasses with high moisture content.

4.3.9. Dynamic basins

In this installation typology, typically closed, aired and dynamic, the composting biomass is placed in basins, and moved with special self-moving overturners. It is generally used for the active phase of higher fermentability biomasses or also for whole process (ACT and Curing, without discontinuity). This system result to be economically competitive for processing over 100 tons day^{-1} of biomasses.

4.3.10. Silos

The silos are vertical reactors, to one or multi step, closed and airing, with continuous or discontinuous loading and unloading of the biomass. The system can be static (for batch processes) or semi-dynamic (top loading of composting matrix and bottom unloading of stabilized biomass).

5. Composting experiences and results

Several experimental trials were carried out using different olive mill typology wastes (two phase mill fluid pomace; three phase mill pomace and waste water), mixed with others agricultural or agroindustrial byproducts, with the aim to evaluate both the composting process efficiency, and the amendment and nutritional efficiency of obtained composts on: i) maintain or improving soil properties and soil fertility; ii) growth and yield performance of herbaceous crops; iii) growth and yield performance of olive orchards (Albuquerque et al., 2007; Alfano et al, 2008; Baeta-Hall et al., 2005; Boulter et al., 2000; Calvet et al., 1985; Casacchia et al., 2012; Diacono & Montemurro, 2010; Hoitink et al., 1997; Montemurro et al., 2004, 2006, 2010, 2011b; Tejada et al., 2006, 2009; Toscano et al., 2009 b).

5.1. Results on olive orchards

In some experimental trials, the assessment of the composting process in natural conditions was studied on two different typology of matrices, respectively composed by olive mill wastes and other crop by-products as structural biomass: i) wet pomace by two phases mill with olive leaves and cereals straw; ii) pomace and waste water by three phases mill with carved pruning residuals. All materials were placed on a beaten-soil platform, and remixed to homogenize the matrices. The evolution of bioxidation was monitored controlling the temperature and humidity of matrices, and assuring oxygenation and thermal drain by weekly blending with mechanical shovel on tractor front loader. At the end of the fermentation period, both biomass microbial pools and chemical parameters were analyzed to evaluate the characteristics of obtained composts. Results of composting parameters trend and microbiological analysis, indicate a correct way of aerobic process; and the compost analyses confirmed that a "mixed composted amendments" was obtained (Law 748/1984).

To evaluate amendment and nutritional efficiency, the composts were spread at doses of approx. 150 kg tree^{-1} (60 tons ha^{-1}) on 6 x 4 scaled olive orchard, and buried with a light disk arrow tillage. At the following year's harvesting time, was compared the differences between treated and untreated soil characteristics, and yield responses of 15 treated vs. 15 untreated olive trees of a 15 years old "Nocellara messinese" cultivar for OP2 compost, and 20 years old "Leccino" cultivar for OP3 compost.

The amendment and nutritional efficiency of these kinds of composts was confirmed both by soil organic matter increment (+ 38.6%; + 40.6% for two phase and three phase compost, respectively), and trees productive responses at the following year harvesting time (+ 10.5% and + 15.1% oil yield increase on respective treated trees).

Even if the condition of fermentative process (open air, natural conditions) was not optimal, the obtained compost demonstrate that the aerobic fermentation of olive mill residuals can run in a correct way also in natural conditions, producing an hygienized and partially stabilized organic amendment, that can be better spread out at the optimal time, and without the negative effects related to the raw olive mill wastes supply. At the used doses, both composts increased soil fertility, improved water retention and the availability of nutrients in amended soils, and improved plant productivity. The showed nutritional effect of these composts would allow a reduction in the use of chemical fertilizers, in agreement with the energetic and economic sustainability principles in the use of renewable resources. In common situations where industrial installations are unfeasible, the natural composting process of olive mill wastes can therefore be a interesting alternative to the raw pomace and waste-water spreading on soils; and could represent a solution of sustainable disposal problem, allowing to increase soil organic matter contents and to reduce the desertification processes (Casacchia et al., 2012; Toscano et al., 2009*b*).

5.2. Results of olive mill waste water application on herbaceous crops

As mentioned before, the national law (Law 574/1996) allows the olive mill waste water (WW) spreading on soils with appropriate characteristics for agronomical use with the maximum amount of 80 m^3 ha^{-1} yr^{-1} for the centrifuge system and of 50 m^3 ha^{-1} yr^{-1} for the pressure method.

In a research carried out in the semi-arid environment of Southern Italy, the application of two WW rates without preliminary treatments was studied. To reach this objective, the effects on cereal and leguminous crops yield, quality, N uptake and on soil characteristics were recorded.

The results obtained indicate that the WW treatments (both doses) positively affected the yield of ryegrass, while a significant yield increase was found at the highest level of WW in proteic pea. The clover crop showed a species-specific sensitiveness, but the WW applications increased the protein content compared to the untreated plots. The WW rates also increased total organic content in the soil, in respect to the unfertilized control. At the end of this three-year experiment the values of soil total extracted carbon and humified organic carbon were higher compared to the initial ones. The values of soil available P and K of the control treatment found at the end of experiment were almost the same than those recorded at the beginning of the research confirming that the increases found in WW plots were due to the WW applications.

A two-year experiment was also carried out in controlled environment (lysimeters) to study the effects of applying untreated and treated WW as soil amendments on both rye-grass growth and soil characteristics. The results of this research indicated that the untreated and treated WW application increased growth parameters, indicating the possible use of WW as an amendment to rye-grass. A significant increase of total, extracted and humified organic carbon in soil, and humification parameters (degree and not humified organic carbon) were found, whereas no accumulations of heavy metals in the soil were measured at the end of the experiment. Furthermore, N content in WW was used by rye-grass for plant growth that increases N uptake and consequently the dry matter accumulation (Montemurro et al., 2007).

5.3. Results of pomace application on herbaceous crops

As already mentioned, the composting of olive pomaces could be recycled for agricultural purposes. In addition, a specific test could be performed to assess the phytotoxicity for both the raw pomace and the stabilized composts. The phytotoxicity is one of the most important criteria for evaluating the agronomical potential of organic materials as olive pomace, and can be measured by specific test. A method can be used to assess phytotoxicity of this residues, by combining the measurements of seeds germination and roots elongation of cress (*Lepidium sativum* L.). Different experiments indicate that the olive pomace composts were not phytotoxic. According to other researches, it can be suggested that repeated compost application might preserve the soil organic carbon content and supply macronutrients to crop. Finally, among different solutions, the addition to olive pomace of manure, as a nitrogen source, and pruning wastes, as bulking-agent, may generate organic amendments suitable for the organic cultivation management of some herbaceous crops.

Several experiments on herbaceous crops indicate that the use of organic pomace fertilizer, as a partial substitution of mineral fertilizers, reached the same yield of the highest mineral fertilizer treatment, ensured also an increase of soil total organic carbon and other soil properties. Furthermore, the application of these organic composted wastes also induce a lower nitrogen mineral soil level at the end of the experimental trials, indicating the possibility to reduce pollution risks. These findings were found in maize and barley in a same experimental research (Montemurro et al., 2006).

5.4. Results on greenhouse olive nursery

On the market are available different organic (peats, manures, urban wastes, barks, sawdusts) and inorganic (sand, pumice, clay, perlite, vermiculite) materials, usable in the preparation of olive nursery substrates. All these materials have some advantages, i.e. sterilization process not necessary, and commercial fertilizers can be added. Even though the peat is the more used organic material in the preparation of the substrates, other different material have been tested with the aim to replace it, include the olive industry byproducts.

In some experimental trials has been valued the compatibility and efficiency of composts, obtained from different olive mill wastes, in the composition of growth substrates for olive nursery, with the aim of mostly valorizing these biomasses as partial or total substitution of peat. Variable proportions of peat and/or two different olive waste composts deriving from the continuous two and three phases extraction systems, were added to a basis of river sand, used as control, for a total of six treatments for each of the three cultivars under observation (Carolea, Nocellara messinese and Tondina). The self-rooted plantlets were kept for 28 months in greenhouse, making periodic measurements of linear growth (cm) and number of internodes. At the end of trial, dry weights of leaves, branches and roots were detected. The results showed a greater nutritional efficiency of the substrates containing composts derived from the three phases olive mill wastes, with highly significant differences both in the linear growths and in the number of internodes; while the compost derived from two phases olive mill wastes do not showed particular benefits, giving results similar to peat, and to sand (control), presumably due to the higher C/N ratio of this compound (Santilli et al., 2012; Toscano et al., 2009*a*).

5.5. General findings

In the experimental conditions of southern Italy, the findings of the studies on repeated applications of WW indicate that it could be possible both to sustain herbaceous crops performance and support soil fertility, valorizing these wastes as organic amendment and reducing the risks of soil degradation, giving also a useful practice to reuse purposes as EU provides.

Also the composting experiences of olive mill and olive orchard wastes carried out in natural conditions, demonstrate as relatively easy feasible and noticeable effective the production of stabilized organic amendment of elevated agronomic value to very contained costs. These materials have their natural utilization to maintain soil organic fertility, as well as substrate for olive nursery, in perspective as substitute of peat. According to the modern principles of the energetic and economic sustainability of agro industrial processes, the olive mill and olive orchard wastes composting has been confirmed as a valid alternative to the raw waste water and pomace shedding on the soil. In particular, a higher efficiency was found in stabilizing the organic matter, which works better in the restoration and in long-term maintenance of soil fertility. Furthermore must be considered that the nutritional effect of compost can allow to reduce the use of chemical fertilizers in fertilization plans, mostly raising the value of this kind of agro industrial by-products management. So could be a better solution also to fit EU prescriptions concerning wastes reuse.

Author details

Pietro Toscano
Agricultural Research Council, Olive Growing and Oil Industry Research Centre, Rende (CS), Italy

Francesco Montemurro
Agricultural Research Council, Research Unit for the Study of Cropping Systems, Metaponto (MT), Italy

Acknowledgement

Financial support for this study was provided by the Italian Ministry of Agriculture, Food and Forestry Policy through the project GERMOLI "Salvaguardia e valorizzazione del GERMoplasma OLIvicolo delle collezioni del CRA-OLI".

6. References

Alburquerque, J.A., Gonzalvez, J., Garcia, D. & Cegarra, J. (2004). Agrochemical characterisation of "alperujo," a solid by-product of the two-phase centrifugation method for olive oil extraction. *Bioresource Technology* 91, pp. 195–200.

Alburquerque, J.A., Gonzalvez, J., Garcia, D. & Cegarra, J. (2007). Effects of a compost made from the solid by-product (alperujo) of the two phase centrifugation system for olive oil extraction and cotton gin waste on growth and nutrient content of ryegrass (Lolium perenne L.). *Bioresource Technology*, 98, pp. 940–945.

Alfano G., Belli C., Lustrato G. & Ranalli G. (2008). Pile composting of two-phase centrifuged olive husk residues: technical solutions and quality of cured compost. *Bioresource Technology*, 99, pp. 4694-4701.

Alianiello F., Trinchera A., Dell'orco S., Pinzari F. & Benedetti A. (1998). Somministrazione al terreno delle acque di vegetazione da frantoio: effetti sulla sostanza organica e sull'attività microbiologica del suolo. *Agricoltura Ricerca*, 173, pp. 65-74.

Amirante P. & Di Rienzo G.C. (1993). Tecnologie e impianti disponibili per la depurazione delle acque di vegetazione. *Collana di Agricoltura e Agroindustria" Tecnologie e Impianti per il Trattamento dei Reflui dei Frantoi Oleari"*, Ed. Conte, Lecce, pp. 24-57.

Amirante P. & Montel G.L. (1999). Tecnologie ed impianti per il compostaggio intra ed inter aziendali. *In Il compostaggio in ambito agricolo. Promozione ed impiego del compost in agricoltura biologica. I quaderni di Biopuglia*, 5, pp. 136-141. IAMB, Bari.

Andrich G., Balzini S., Zinnai A., Silvestri S. & Fiorentini R. (1992). Effect of olive oil wastewater irrigation on olive plant products. *Agricoltura Mediterranea*, 122, pp. 97-100.

Angelidaki I., Karakashev D., Batstone D.J., Plugge C.M. & Stams A.J. (2011). Biomethanation and its potential. *Methods Enzimol.*, 494, pp. 327-351.

A.R.S.S.A. (Agenzia Regionale per lo Sviluppo e per i Servizi in Agricoltura della Calabria), (2001). Utilizzazione agronomica delle acque reflue olearie. *Pgt. POM Mis. 2 Cod. A 10: Sistemi e metodi per la valorizzazione a fini agricoli dei residui delle industrie agroalimentari nel meridione d'Italia*. Ed. Zefiro, Cosenza.

Baeta-Hall L., Sàagua M.C., Bartolomeu M.L., Anselmo A.M. & Rosa, M.F. (2005). Biodegradation of olive husks in composting aerated piles. *Bioresource Technology*, 96, pp. 69–78.

Ballesteros I., Oliva J. M., Negro M. J., Manzanares P. & Ballesteros M. (2002). Ethanol Production from Olive Oil Extraction Residue Pretreated with Hot Water. *Applied Biochemistry and Biotechnology*, 98-100, pp. 717-732.

Benitez J., Beltran-Heredia J., Torregrosa J., Acero J.L. & Cercas V. (1997). Aerobic degradation of olive mill wastewaters. *Applied Microbiology and Biotechnology*, 47, pp. 185-188.

Ben Rouina B., Taamallah H. & Ammar E. (1999). Vegetation water used as a fertilizer on young olive plants. *Acta Horticolturae*, 474, pp. 353–355.

Bing U., Cini E., Cioni A. & Laurendi V. (1994). Smaltimento-recupero delle sanse di oliva provenienti da un "due fasi" mediante distribuzione in campo. *L'Informatore Agrario*, 47, pp. 75-78.

Bonanomi G., Giorgi V., Del Sorbo G., Neri D. & Scala F. (2006). Olive mill residues affect saprophytic growth and disease incidence of foliar and soilborne plant fungal pathogens. *Agricolture, Ecosystems and Environment*, 115, pp. 194-200.

Bonari E., Macchia M., Angelini L. G. & Ceccarini L. (1993). The Waste Water from Olive Oil Extraction: their influence on the germinative characteristics of some cultivated and weed species. *Agricoltura Mediterranea*, 123, pp. 273-280.

Boulter J.I., Boland G.J. & Trevors J.T. (2000). Compost: A study of the development process and end-product potential for suppression of turf grass disease. *World Journal of Microbiology and Biotechnology*, 16, pp. 115-134.

Briccoli Bati C. & Lombardo N. (1990). Effects of olive oil waste water irrigation on young olive plants. *Acta Horticolturae*, 286, pp. 489–491.

Brunetti G., Plaza C. & Senesi N. (2005). Olive Pomace Amendment in Mediterranean Conditions: Effect on Soil and Humic Acid Properties and Wheat (*Triticum turgidum* L.) Yield. *J. Agric. Food Chem.*, 53 (17), pp. 6730–6737.

Calvet C., Pagés M. & Estaún V. (1985). Composting of Olive Marc. *Acta Horticolturae, (ISHS)*, 172, pp. 255-262.

Capasso R., Cristinzio G., Evidente A., & Scognamiglio F. (1992). Isolation, spectroscopy and selective phytotoxic effects of polyphenols from vegetable waste waters. *Phytochemistry*, 31, pp. 4125-4128.

Casacchia T., Toscano P., Sofo A., & Perri E. (2011). Assessment of microbial pools by an innovative microbiological technique during the co-composting of olive-mill by-products. *Agricultural Sciences*, 2, pp. 104-110.

Casacchia T., Sofo A., Zelasco S., Perri E. & Toscano P. (2012). In situ olive mill residual co-composting for soil organic fertility restoration and by-product sustainable reuse. *Italian Journal of Agronomy*, 7, pp. 35-38.

Catalano M., Gomes T., De Felice M. & De Leonardis T. (1985). Smaltimento delle acque di vegetazione dei frantoi oleari. *Inquinamento*, 2, pp. 87-90.

Clodoveo M.L., Leone A., Montel G.L. & Tamborrino A. (2000). Impianti semplificati per il compostaggio dei reflui e controllo dei parametri di processo. *Agricoltura Ricerca*, 187, pp. 135–142.

De Bertoldi M., Vallini G. & Pera A. (1983). The biology of composting: a review. *Waste Mangement & Research,* 1, pp. 157–176.

De Nobili M. & Petrussi F. (1988). Humification Index (HI) as evaluation of the stabilization degree during composting. *J. of Fermentation Technology*, 66 (5), pp. 577-583.

Decreto Legge 162 del 10/05/1995: Disposizioni in materia di riutilizzo dei residui derivanti da cicli di produzione o di consumo in un processo produttivo o in un processo di combustione, nonché in materia di smaltimento dei rifiuti.

D. Lgs. 99 del 27/01/1992: Attuazione della direttiva 86/278/CEE concernente la protezione dell'ambiente, in particolare del suolo, nell'utilizzazione dei fanghi di depurazione in agricoltura. *Suppl. ord. G.U., serie Generale*, n. 38 Feb.

D. Lgs. 22 del 5 Febbraio 1997 (Ronchi): Attuazione delle Direttive 91/156/CEE sui rifiuti, 91/689/CEE sui rifiuti pericolosi, e 94/62/CE sugli imballaggi e sui rifiuti da imballaggio. *Suppl. ord. G.U. serie Generale*, n. 38 Feb.

D. Lgs. 389 del 8 Novembre 1997: Modifiche ed integrazioni al d.lgs. 22 del 5/2/1997 (Legge Ronchi), in materia di rifiuti, rifiuti pericolosi, di imballaggi e di rifiuti da imballaggio. *G.U. serie Generale*, n. 261 Nov.

D.M. 5 Febbraio 1998 (Min. Ambiente): Individuazione dei rifiuti non pericolosi sottoposti alle procedure semplificate di recupero ai sensi degli artt. 31 e 33 del d. lgs. N. 22 del 5 febbraio 1997. *Suppl. ord. N. 72 - G.U., Serie Generale*, n. 88 Apr.

Della Monica M., Potenz D., Righetti E. & Volpicella M. (1978). Effetto inquinante delle acque reflue della lavorazione delle olive su terreno agrario. Nota 1: Evoluzione del pH, dei composti azotati e dei fosfati. *Inquinamento*, 10, pp. 81-87.

Della Monica M., Potenz D., Righetti E. & Volpicella M. (1979). Effetto inquinante delle acque reflue della lavorazione delle olive su terreno agrario. Nota 2: Evoluzione dei lipidi, dei polifenoli e delle sostanze organiche. *Inquinamento*, 1, pp. 27-30.

Diacono M. & Montemurro F. (2010). Long-term effects of organic amendments on soil fertility. A review. *Agronomy for Sustainable Development*, 30, pp. 401-422.

Diaz L.F., De Bertoldi M., Bidlingmaier W. & Stentiford E. (2007). Compost science and technology. *Elsevier Science. Waste Management Series*, Vol 8. ISBN: 9780080439600.

Di Giovacchino L. & Seghetti L. (1990). Lo smaltimento delle acque di vegetazione delle olive su terreno agrario destinato alla coltivazione di grano e mais. *L'Informatore Agrario*, 45, pp. 58-62.

Di Giovacchino L., Basti C., Costantini N., Ferrante M.L. & Surricchio G. (2001). Effects of olive vegetable water spreading on soil cultivated with maize and grapevine. *Agricoltura Mediterranea*, 131, pp. 33–41.

Di Giovacchino L., Basti C., Seghetti L., Costantini N., Surricchio G. & Ferrante M.L. (2004). Effects of olive pomace spreading on soil cultivated with maize and grapevine. *Agricoltura Mediterranea*, 134 (1), pp. 15-24.

FAOSTAT (2012). Food And Agriculture Organization Of The United Nations (http://faostat.fao.org/)

Filidei S., Masciandaro G. & Ceccanti B. (2003). Anaerobic digestion of olive oil mill effluents: evaluation of wastewater organic load and phytotoxicity reduction. *Water, Air and Soil Pollution*, 145, pp. 79–94.

Filippi C., Bedini S., Levi-Minzi R., Cardelli R. & Saviozzi A. (2002). Co-composting of olive oil mill by-products: chemical and microbiological evaluations. *Compost Science and Utilization* 10, pp. 63-71.

Franz L., Ceron A., Paradisi L., Germani F., Bergamin L. & Caravello G. (2005): Stabilità biologica: studio comparativo di due differenti metodiche per la determinazione dell'indice di respirazione nel compost. *Rifiuti Solidi*, 19 (1), pp. 22-26.

Galli E., Pasetti L., Fiorelli F. & Tomati U. (1997). Olive mill wastewater composting: microbiological aspects. *Waste Management & Research*, 15, pp. 323-330.

Garcia-Gomez A., Roig A. & Bernal M.P. (2003). Composting of the solid fraction of olive mill wastewater with olive leaves: organic matter degradation and biological activity. *Bioresource. Technology,* 86, pp. 59–64.

Goldstein J. (1980). An overview of composting installation. *Compost Science and Land Utilization,* 21 (4), pp. 28-32.

Higgins A.J. (1982). Ventilation for static pile composting. *Biocycle,* 23, pp. 36-41.

Hoitink H.A.J., Stone A.G. & Han D.Y. (1997). Suppression of plant disease by compost. *Hort. Science,* 32, pp. 184-187.

Insam H. & De Bertoldi M. (2007). Microbiology of the composting process. In *Compost Science and Technology. Waste management series 8,* Elsevier Ltd., pp. 26–45, ISBN-13: 978-0-08-043960-0.

IPLA (Istituto Piante Legno Ambiente) (1992). Metodi di analisi dei compost. *Collana Ambiente, 6.* Ed. Regione Piemonte, Assessorato Ambiente, Torino.

ISTAT (2010). Agricoltura e ambiente. ISTAT, Roma (http://www.istat.it/).

Kavdir Y. & Killi D. (2008). Influence of olive oil solid waste application on soil pH, electrical conductivity, soil nitrogen transformation, carbon content and aggregate stability. *Bioresource Tecnology,* 99, pp. 2326-2332.

Legge 10 Maggio 1976, n. 319 (Merli). Norme per la tutela delle acque dall'inquinamento. Gazzetta Ufficiale n. 141 maggio, int. e modif. dalla Legge 24 dicembre 1979, n. 650. G. U. n. 352 Dic.

Legge 19 Ottobre 1984, n. 748. Nuove norme per la disciplina dei fertilizzanti. *Suppl. ord. Gazzetta Ufficiale* - n. 305 Nov.

Legge 11 Novembre 1996, n. 574. Nuove norme in materia di utilizzazione agronomica delle acque di vegetazione e di scarichi di frantoi oleari. *Gazzetta Ufficiale, Serie generale* n. 265, Nov.

Legge 9 Dicembre 1998, n. 426: Nuovi interventi in campo ambientale. *G.U. Serie generale,* n. 291, Dic.

Levi-Minzi R., Saviozzi A., Riffaldi R. & Falzo L. (1992). Lo smaltimento in campo delle acque di vegetazione. Effetti sulle proprietà del terreno. *Olivae,* 40, pp. 20-25.

Lombardo N., Briccoli-Bati C., Marsilio V. & Di Giovacchino L. (1995). Comportamento vegeto-produttivo di un oliveto trattato con acque di vegetazione. *Atti Conv.: Tecniche, norme e qualità in olivicoltura.* Potenza 15-17 dic. 1993, pp. 93-107.

Madejon E., Galli E. & Tomati U. (1998). Composting of wastes produced by low water consuming olive mill technology. *Agrochimica,* 42, pp. 135-146.

Marsilio V., Di Giovacchino L., Solinas M., Lombardo N. & Briccoli Bati C. (1990). First Observations on the Disposal Effects of Olive Oil Mills Vegetation Water on Cultivated Soil. *Acta Horticulturae,* 286, pp. 493-496.

McNamara C.J., Anastasiou C.C., O'Flaherty V. & Mitchell R. (2008). Bioremediation of Olive mill wastewarter. Review. *Intl. Bioterioration & Biodegradation,* 61, pp. 127-134.

Molinari G. & Bonfà F. (2005). Utilizzazione di biomasse per alimentazione di gruppi cogenerativi con motori a vapore alternativi per utenze piccole e medie. *Conferenza Nazionale sulla politica energetica in Italia.* Bologna, 18-19 Aprile.

Montemurro F., Convertini G. & Ferri D. (2004). Mill wastewater and olive pomace compost as amendments for rye-grass. *Agronomie,* 24, pp. 481–486.

Montemurro F., Maiorana M., Convertini G. & Ferri D. (2006). Compost organic amendments in fodder crops: effects on yield, nitrogen utilization and soil characteristics. *Compost Science & Utilization,* 14 (2), pp. 114–123.

Montemurro F., Maiorana M., Ferri D. & Convertini G. (2007). Treated and untreated olive waste water application on rye-grass meadow: chemical soil properties and yielding responses. *Agrochimica,* 51 (2-3), pp. 148-159.

Montemurro F., Diacono M., Vitti C. & Debiase G. (2009). Biodegradation of olive husk mixed with other agricultural wastes. *Bioresource Technology,* 100, pp. 2969-2974.

Montemurro F., Charfeddine M., Maiorana M. & Convertini G. (2010). Compost use in agriculture: the fate of heavy metals in soil and fodder crop plants. *Compost Science and Utilization,* 18, pp. 47-54.

Montemurro F., Diacono M., Vitti C. & Ferri D. (2011*a*). Potential use of olive mill wastewater as amendment: crops yield and soil properties assessment. *Communications in Soil Science and Plant Analysis,* 42, pp. 2594-2603.

Montemurro F., Diacono M., Vitti C. & Ferri D. (2011*b*). Short-term agronomical effects of olive oil pomace composts on Pisum arvense L. and Trifolium subterraneum L. and impacts on soil properties. *Communications in Soil Science and Plant Analysis,* 42, pp. 2256-2264.

Pacifico A. (1989). Acque di vegetazione. *Agricoltura e Innovazione, "Dossier: Acque di vegetazione",* Notiziario ENEA Roma, 11, pp. 33-53.

Palliotti A. & Proietti P. (1992). Ulteriori indagini sull'influenza delle acque reflue di frantoi oleari sull'olivo. *L'Informatore Agrario,* 39, pp. 72-76.

Paredes, C., Roig, A., Bernal, M.P., Sàcnchez-Monedero, M.A. & Cegarra, J. (2000). Evolution of organic matter and nitrogen during co-composting of olive mill wastewater with solid organic wastes. *Biology and Fertility of Soils,* 32, pp. 222–227.

Paredes C., Bernal M.P., Cegarra J. & Roig A. (2002). Bio-degradation of olive mill wastewater sludge by its co-composting with agricultural wastes. *Bioresource Technology,* 85, pp. 1-8.

Proietti P., Cartechini A. & Tombesi A. (1988). Influenza delle acque reflue di frantoi oleari su olivi in vaso e in campo. *L'Informatore Agrario,* 45, pp. 87-91.

Ramos-Cormenzana A. (1986). Physical, Chemical, Microbiological and Biochemical Characteristic of Vegetation Water. *In Intl. Symposium on Olive by-products valorization, Seville.* FAO, Madrid, pp. 19-40.

Roig A., Cayuela M.L. & Sánchez-Monedero M.A. (2006). An overview on olive mill wastes and their valorization methods. *Waste Management,* 26 (9), pp. 960-969.

Rozzi A. & Malpei F. (1996). Treatment and disposal of olive mill effluents. *Intl. Biodeterioration & Biodegradation,* 38, pp. 135-144.

Santilli E., Toscano P., Casacchia T., Lombardo L. & Briccoli-Bati C. (2012). Utilizzo di compost di reflui oleari nella preparazione di substrati nel vivaismo olivicolo. *II Conv. Naz. Dell'Olivo e dell'Olio,* PG 21-23 Sett., (In Press).

Saviozzi A., Levi-Minzi R., Riffaldi R. & Lupetti A. (1991). Effetti dello spandimento di acque di vegetazione su terreno agrario. *Agrochimica,* 35, pp. 135-148.

Tamburino V., Zimbone S.M. & Quattrone P. (1999). Accumulo e smaltimento sul suolo delle acque di vegetazione. *Olivae,* 76, pp. 36-45.

Tejada, M. & Gonzales, J.L. (2003). Application of a by-product of the two-steps olive mill process on rice yield. *Agrochimica* 47, pp. 94–102.

Tejada, M. & Gonzales, J.L. (2004). Effects of application of a by-product of the two-steps olive oil mill process on maize yield. *Agronomy Journal*, 96, pp. 692–699.

Tejada M., Garcia C., Gonzalez J.L. & Hernandez M.T. (2006). Organic amendment based on fresh and composted beet vinasse: influence on soil properties and wheat yield. *Soil Sci. Soc. Am. J.*, 70, pp. 900–908.

Tejada M., Hernandez M.T. & Garcia C. (2009). Soil restoration using composted plant residues: Effects on soil properties, *Soil and Tillage Res.*, 102, pp. 109–117.

Tomati U. & Galli E. (1992). The fertilizing value of waste water from the olive processing industry. *In: Humus, its structure and role in agriculture and environment. (J. Kubat ed.)* Elsevier, Barking, pp. 117–126.

Tomati U., Galli E., Pasetti L. & Volterra E. (1995). Bioremediation of olive-mill wastewaters by composting. *Waste Management & Research*, 13, pp. 509-518.

Tomati U., Galli E., Fiorelli F. & Pasetti L. (1996). Fertilizers from composting of olive mill wastewaters. *Intl. Biodeterioration & Biodegradation*, 38, pp. 155-162.

Toscano P., Turco D. & Briccoli-Bati C. (2009a). Utilizzo di compost di reflui oleari nella preparazione dei substrati di coltura nel vivaismo olivicolo. *Conv. Int. e finale Progetto RIOM CRA-OLI* Rende (CS), 11-12 Giu, pp. 149-152.

Toscano P., Casacchia T. & Zaffina F. (2009b). The "in farm" olive mill residual composting for by-products sustainable reuse in the soils organic fertility restoration. *18th Symposium Of The International Scientific Centre Of Fertilizers - More Sustainability In Agriculture: New Fertilizers And Fertilization Management.* Rome, 8 – 12 Nov., pp. 116-121.

Vallini G., Pera A. & Morelli R. (2001). Il compostaggio delle acque di vegetazione dei frantoi oleari. *Informatore Agrario*, Suppl. 1, 50, pp. 22-26.

Veronesi G. & Zampighi C. (1997). Evoluzione degli impianti di compostaggio dei residui solidi organici. *VI Convegno Nazionale di Ing. Agraria, Università degli Studi*, Ancona, 11-12 settembre.

Vigo F., Uliana C. & Traverso M. (1990). Acque di vegetazione da frantoi di olive. Trattamenti utilizzabili per una riduzione del carico inquinante. *Riv. It. Sostanze grasse*, 67 (3), pp. 131-137.

Vitolo S., Petarca L. & Bresci B. (1999). Treatment of olive oil industry wastes. *Bioresource Technology*, 67, pp. 129-137.

Willson G. B., Parr J. F., Epstein E., Marsh P. B., Chaney R. I., Colacicco D., Burge W. D., Sikora L. J., Tester C. F. & Hornick S. (1980). Manual for composting sludge by the Beltsville aerated pile method. *USDA, EPA 600/8-80 002*, Cincinnati, Ohio, USA.

Zucconi F., Forte M., Monaco A. & De Bertoldi M. (1981). Biological evaluation of compost maturity. *BioCycle*, 22, pp. 27–29.

Chapter 9

Technological Aspects of Olive Oil Production

Maurizio Servili, Agnese Taticchi, Sonia Esposto,
Beatrice Sordini and Stefania Urbani

Additional information is available at the end of the chapter

1. Introduction

Virgin olive oil (VOO) is obtained exclusively by mechanical extraction from the olive fruit and can be consumed crude without any further physical-chemical treatments of refining. Its sensory and health properties are intimately linked to its chemical characteristics, in particular to several minor components, which are strongly affected by the operative conditions of oil processing and can thus be considered as analytical markers of the quality of oil processing.

VOO contains different classes of phenolic compounds, such as phenolic acids, phenolic alcohols, hydroxy-isochromans, flavonoids, secoiridoids and lignans. The phenolic acids together with phenyl-alcohols, hydroxy-isochromans and flavonoids are present in small amounts in VOO (Montedoro et al., 1992). Secoiridoids which, in combination with lignans, are the main hydrophilic phenols of VOO, include the dialdehydic form of decarboxymethyl elenolic acid linked to 3,4-DHPEA or p-HPEA (3,4-DHPEA-EDA or p-HPEA-EDA) an isomer of the oleuropein aglycon (3,4-DHPEA-EA) and the ligstroside aglycon (p-HPEA-EA). The main lignans found in VOO are (+)-1-acetoxypinoresinol and (+)-1-pinoresinol (Brenes et al., 2000; Owen et al., 2000). Both secoiridoids and lignans affect the quality of the sensory and health properties of VOO (Servili et al., 2004a), determining bitter, pungent sensations.

Many compounds, mainly carbonyl compounds, alcohols, esters and hydrocarbons, have been found in the volatile fraction of VOO.

The C_6 and C_5 compounds, especially C_6 linear unsaturated and saturated aldehydes alcohols and esters represent the most important fraction of volatile compounds found in high quality VOOs.

The C_6 and C_5 compounds, produced from polyunsaturated fatty acids by the enzymatic activities exerted by the lipoxygenase (LOX) pathway and their concentrations, depend on the level and the activity of each enzyme involved in this LOX pathway.

The pathway (Figure 1) begins with the production of 9- and 13-hydroperoxides of linoleic (LA) and linolenic (LnA) acids mediated by lipoxygenase (LOX). The subsequent cleavage of 13-hydroperoxides is catalysed by very specific hydroperoxide lyases (HPL) and leads to C_6 aldehydes, of which the unsaturated aldehydes can isomerize from *cis*-3 to the more stable *trans*-2 form. The mediation of alcohol dehydrogenase (ADH) reduces C_6 aldehydes to corresponding alcohols, which can produce esters due to the catalytic activity of alcohol acetyl transferases (AAT).

Furthermore, an additional branch of the LOX pathway (Figure 1) is active when the substrate is LnA. LOX would catalyse the formation of stabilized 1,3-pentene radicals, which can either dimerize leading to C_{10} hydrocarbons (known as pentene dimers), or can react with a hydroxy radical, producing C_5 alcohols. The latter can be enzymatically oxidated to corresponding C_5 carbonyl compounds. These compounds are responsible for the most important sensory notes of VOO flavour, such as the "Green" and floral notes (Angerosa et al., 2001; Angerosa et al., 2004; Servili et al., 2009a).

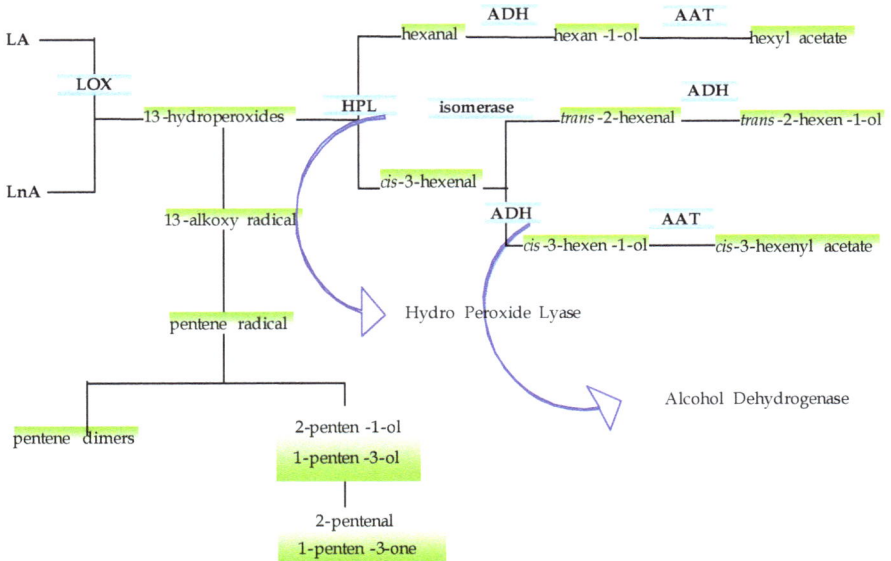

Figure 1. Lipoxygenase pathways involved in the production of C_6 and C_5 volatile compounds (Angerosa et al., 2004).

The new approach to VOO processing should include as its first objective the improvement of the quality of the sensory and health properties of oil. Since the presence of VOO hydrophilic phenols and volatile compounds is strictly related to the activities of various endogenous enzymes of olive fruit, their concentration in the oil is highly affected by the operative conditions of the mechanical oil extraction process. By taking into account the optimization of volatile and phenolic composition of VOO as the main goal of the new approach to the mechanical oil extraction process, the conditions during crushing and malaxation can be considered as the most critical points (Capella et al., 1997; Caponio et al., 1999; Servili et al., 2002; Servili et al., 2004a; Angerosa et al., 2004; Servili et al., 2009a).

2. The structure and composition of the olive fruit

The first factor for high quality production in the VOO industry is the structure of the raw material, thus the quality and sanitary status of the olive are very important.

The olive fruit is a drupe, the weight of which varies between 0.5g and 20g. The constituent parts of the fruit, which include the skin, pulp and stone, represent respectively 1.5-3.5%, 70-80% and 15-28% of the weight of the fresh drupe. The stone contains the seed with a weight ranging from 2% to 4% of the whole fruit (VV. AA., 2003).

It is difficult to define the average composition of the olive fruit, due to its remarkable biodiversity, which produces high compositional variability. Water (40-70%) and fat (6-25%) are the main constituents of the fresh fruit.

The fruit contains water-soluble compounds, including simple sugars, organic acids, nitrogenous substances, phenolic compounds and an insoluble fraction of colloidal nature. Colloids of the drupe include the components of the cell wall or the middle lamella, such as hemicelluloses, celluloses, pectins, enzymatic and structural proteins.

The most important reducing sugars found in olives are glucose, fructose and sucrose, whereas citric acid, malic acid and oxalic acid are the main organic acids of the olive drupe (VV. AA., 2003).

In the composition of the olive, the phenolic fraction, which includes the precursors of natural antioxidants present in VOO, is of major importance (Amiot et al., 1986; Servili et al., 1999a, 1999b, 1999c; Servili & Montedoro, 2002). Phenolic compounds present in very high quantities of drupe (from 0.5 to 2.5% fresh weight) include oleuropein and demethyloleuropein. These substances are found mainly in the peel and pulp, whereas the seed contains nüzhenide, not found in the pulp and which is not considered a precursor of the phenolic compounds of VOO (Servili et al., 1999b). Lignans were found both in the pulp and in the woody core, but the latter cannot be released in VOO during oil processing (Brenes et al., 2000; Garcia et al., 2001; Servili et al., 2007).

The peel and the pulp together contain more than 90% of the total phenolic concentration of the fruit, which varies significantly according to the cultivar and stage of ripening of the olives (VV. AA., 2003; Servili et al., 2004a). The oil fraction is present in the pulp (16.5-23.5% fresh weight) and in the seed (1-1 .5% fresh weight).

Oilseed cells typically contain cytoplasmic and vacuolar oil. The compartmentalization of the oil in the olive pulp cells is, in this context, unusual when compared with that of oilseeds. In fact, the pulp cells of a ripe olive oil contain almost exclusively vacuolar oil, whereas the amount of cytoplasmic oil in the oilseed cells is remarkable.

Vacuolar oil in the olive pulp is the result of the mechanical process of oil extraction, which explains why even the most remote civilizations in the Mediterranean area have been using this fruit as a natural source of dietary fat.

3. New technological approaches towards the quality of virgin olive oil

3.1. Olive fruit storage

Storage of the olives before the mechanical oil extraction process is a critical point, which can reduce the quality of VOO. The first adverse effects are seen in the decrease in oil phenols and in the reduction of volatile compounds responsible for the flavour of VOO (Angerosa et al., 1996; Kiritsakis, 1998a; Kiritsakis et al., 1998b; Angerosa et al., 1999; Servili et al., 2004a; Servili et al., 2009a). The activation during olive storage of endogenous polyphenoloxidase (PPO) and peroxidise (POD), which catalyze the degradation of the phenols, can explain the loss of phenolic compounds. The reduction of volatile compounds responsible for the oil flavour can be due to the inhibition of the LOX pathway.

Moulds, yeasts and bacterial contaminations, and their corresponding metabolisms, are the underlying cause of the off-flavour biogenesis in VOO. In several operative conditions involving long-term storage of the olive and high relative humidity, mould contamination increases the free acidity due to the production of fungal lipase, and simultaneously forms the characteristic sensory defect of "mould" (Angerosa et al., 1999, 2004). Several mould species can also produce mycotoxins. Sugar fermentation produces the formation of acetic acid and ethyl acetate, which are considered responsible for the "vinegar" off-flavour.

During storage, the fatty acid alkyl esters of ethanol and methanol develop. The formation of these compounds is used as a marker to recognize low quality virgin olive oil. In fact, these compounds, combined with other sensory and chemical parameters, are used to classify virgin olive oil according to the internationals standards set by the International Olive Council (I.O.C., 1996).

These observations lead to the conclusion that olive storage should be avoided. To maintain the quality of VOO, the olives should be processed within twenty-four hours after harvesting. A thin, 30-40 cm thick layer of olives should be stored in perforated boxes on pallets, in order to minimize the fermentation processes, which underlie not only the formation of sensory defects, but also water condensation on the surface of the olive skin, which can promote the attack by moulds. Perforated boxes and pallets are also the most suitable to transport the olives from the olive grove to the mill (Angerosa et al., 1996, 1999; Servili et al., 2004a; Servili et al., 2009a).

Another aspect, which affects the storage premises, is related to the handling of the pallets or boxes. This process must be done by avoiding the use of forklift trucks run by petrol engines, which produce polycyclic aromatic hydrocarbons. These compounds can contaminate the olives, and subsequently, the oil (Angerosa et al., 2004).

3.2. Leaf removal and washing

Leaf removal is always recommended, especially when harvesting is done mechanically. The presence of leaves during the mechanical oil extraction process does not add any positive characteristic to the oil but, on the contrary, can change its taste and aroma.

Olives are generally washed by continuous washing machines. (Di Giovacchino et al., 2002; Perez et al., 2003). Olive washing has a more or less significant effect on the quality of VOO, depending on the characteristics and sanitary state of the olives. Fresh olives, harvested at the correct degree of maturity and properly transported and stored, showed no direct effect in the quality of the oil due to washing.

The most critical aspect of the washing process concerns the purity of the water used during this stage. The water should be changed frequently during processing to prevent the use of washing water containing too many earthy particles, which can release the compounds responsible for the "Earthy" sensory off-flavour into the oil (Angerosa et al., 2004).

3.3. Olive crushing

The main hydrophilic phenols of VOO, such as secoiridoid aglycons, develop during crushing from the hydrolysis of oleuropein, demethyloleuropein and ligstroside, catalysed by the endogenous β-glucosidases (Servili et al., 2004a; Obied et al., 2008).The impact of crushing in the VOO phenolic and volatile compounds can be related to the differentiated distribution of the endogenous oxidoreductases and phenolic compounds in the constituent parts of the olive fruit (pulp, stone and seed). As reported in previous papers, the POD, in combination with the PPO, are the main endogenous oxidoreductases responsible for phenolic oxidation during processing. POD occurs in high amounts in the olive seed. The phenolic compounds, on the contrary, are largely concentrated in the pulp, whereas the stone and seed contain only small quantities of these substances (Servili et al., 2004a; 2007). As a result, the crushing methods, such as the olive stoning process or the use of mild seed crushers, which enable degradation of the seed tissues to be reduced by limiting the release of POD in the pastes, prevent the oxidation of hydrophilic phenols during malaxation, thus improving their concentration in the VOO (Figure 2) (Servili et al., 1999a, 2004a, 2007).

The operative conditions of crushing also affect the volatile composition of VOO (Table 1). As previously mentioned, almost all volatile compounds are responsible for the flavour of high quality VOOs when the olive pulp tissue is ruptured, thus the effectiveness of the crushing plays an important role in their production.

The traditional olive crusher used for many centuries was the stone crusher. The stone crusher consists of a basin formed by a plinth and a stainless steel edge with an opening for the unloading of olive paste at the end of milling. Two or four granite wheels rotate and revolve on a rough granite base at different distances from the centre of the tank. Rotation speed is normally 12- 15 rpm.

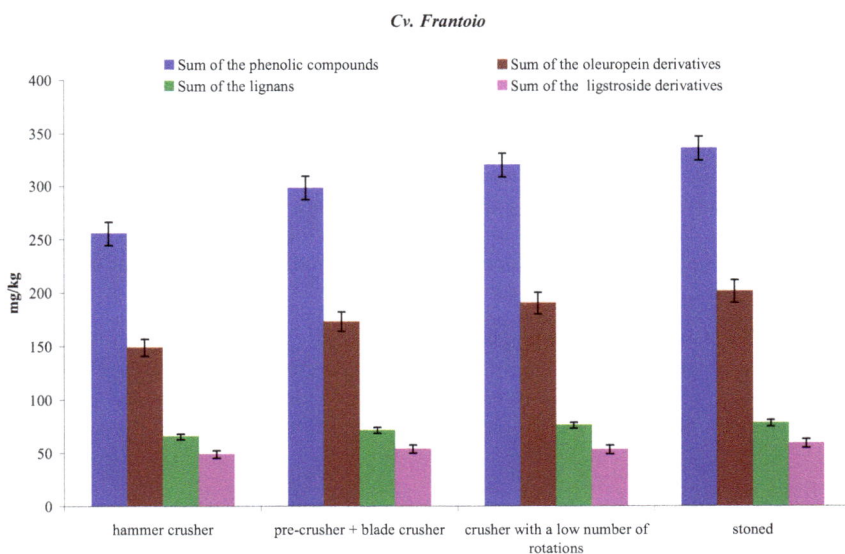

Figure 2. Effect of different crushing methods on the phenolic composition of VOO in Cv. *Frantoio* (mg/kg). The vertical lines are the mean value of three independent experiments ± standard deviation.

The popularity of the stone crusher extraction system using pressure gradually declined. In comparison with continuous crushers, this apparatus shows significant limitations in terms of olive oil quality. In particular, it reduces the phenolic concentration as the olive pastes are in long, extensive contact with the air during processing. Contact with the air stimulates PPO and POD, producing a high oxidation of phenolic compounds. Other weaknesses of the stone crusher are its low working capacity, the high hourly machine footprint, and its low ability to release the chlorophyll found in the olive skin, responsible for the green colour of EVOO, into the oil. This aspect is particularly relevant when the stone crusher is combined with a solid-liquid centrifugal separation. The crushing operation in oil extraction by centrifugation is generally replaced by the use of continuous crushers.

Continuous crushers include the hammer crushers, which were historically the first to be introduced as an alternative to stone crushers. These machines have some benefits that are attributable to their high working capacity, low footprint, and low installation costs compared to the stone crushers. At the same time, the hammer crushers show some disadvantages, such as the strong emulsifier effect produced on olive paste during crushing, a considerable increase in paste temperature and the high degradation of the seed tissues which, as mentioned earlier, can encourage phenolic oxidation (Angerosa & Di Giacinto., 1995; Servili et al., 1999a; Servili & Esposto 2004).

The new approach to olive crushing is based on the differentiated crushing of the constituent parts of the fruit, such as the skin, pulp and seed. In other words, the degradation process of the olive tissues should be strong for the skin and pulp, in order to facilitate the release of oil and pigments, whereas impact on the seed should be limited. This reduces the transfer of POD found in the seed to the olive paste pulp, which can increase the oxidation of phenolic compounds during malaxation. The new generation of mild seed crushers, which include the blade crusher, teeth crusher and double stoker crusher, reduce the phenolic degradation and simultaneously improve the concentration of volatile compounds, especially of hexanal, trans-2-hexenal and C_6 esters, with a consequent positive increase of the intensity of "cut grass" and "floral" sensory notes (Table 1) (Angerosa et al., 2004).

	hammer crusher			pre-crusher + blade crusher			crusher with a low number of rotations			stoned		
Aldehydes												
Pentanal	236.5	±	4	273.4	±	2.1	17.9	±	1	66.5	±	6.7
Hexanal	280	±	2.9	511.4	±	35.7	553.7	±	0.3	579.6	±	5.3
2-Hexenal (E)	43600.6	±	327	44718.9	±	208	39811.6	±	587	52228.1	±	521
2,4-Hexadienal (E,E)	19.4	±	0.1	42	±	3.5	341.6	±	14.4	88.9	±	5.4
Alcohols												
1-Pentanol	167	±	5.2	94.5	±	4.7	23.3	±	0.7	62.6	±	1.7
2-Penten-1-ol (E)	166	±	11.3	91.4	±	5.1	52.4	±	3.5	104.0	±	2.1
1-Penten-3-ol	960.3	±	53.2	899	±	43.3	522	±	49.2	300.0	±	4.6
1-Hexanol	1788	±	57	2152	±	74	512	±	41	1501.0	±	18.4
3-Hexen-1-ol (Z)	88.4	±	22.2	103.6	±	10.1	49.2	±	2.3	77.0	±	8.0
3-Hexen-1-ol (E)	22.2	±	0.2	20.2	±	0.1	9.9	±	0.2	20.4	±	2.3

Table 1. Effect of different crushing methods on the volatile composition of VOO in Cv. *Frantoio* (μg/Kg). Results are the mean value of three independent experiments ± standard deviation.

Several researches have shown that olive stoning during the mechanical extraction process of VOO increases the phenolic concentration in VOO (Figure 2) (Angerosa et al., 1999; Mulinacci et al., 2005; Lavelli & Bondesan, 2005; Amirante et al., 2006) and, at the same time, modifies the composition of volatile compounds produced by the LOX pathway, increasing the concentration of those volatile substances correlated to the "green" sensory notes (Table 1) (Servili et al., 2007). These results are particularly important, because they would appear to demonstrate that the enzymes involved in the LOX pathway have a different activity in the pulp and in the seed of the olive (Table 1) (Servili et al., 2007).

The stoning process, moreover, produces a significant reduction in industrial oil yields. This problem is caused by stone elimination which, when present in the pastes, produces an important draining activity, which increases the efficiency of the separation of the oil from the olive paste during solid-liquid extraction.

It should be noted that the use of enzymatic preparations with depolymerizing activities, degrading the colloidal structure of fruit might partly solve the problem of low yields due to the extraction of pastes for stoning. So far, however, the European Union (EU regulation) does not currently allow the addition of enzymatic preparations.

3.4. Malaxation process

The mixing and heating (25-35 °C) of the olive pastes during malaxation causes the breakdown of water-oil emulsion, allowing oil droplets to form larger droplets, which separate easily from the aqueous phase during the solid-liquid and liquid-liquid separation processes.

The operative conditions applied during malaxation of the olive pastes largely affect VOO quality (Servili et al., 1994; Montedoro et al., 2002; Servili et al., 2004a; Inarejos-Garcia et al., 2009). As in the case of the crushing process, malaxation can also produce significant modifications in the minor components of VOO with particular emphasis on volatile and phenolic compounds. The technological parameters of temperature and oxygen management show the highest impact on the volatile and phenolic composition of VOO.

The problems concerning temperature management during malaxation have been widely studied for over twenty years and a substantial negative relationship between the processing temperature and the quality of the VOO has been shown (Garcia et al., 2001; Di Giovacchino et al., 2002; Servili et al., 2003a; Servili et al., 2004a; Kalua et al., 2006). However some aspects of the relationship between the operative conditions of malaxation and oil quality must be better defined.

As previously reported, the most sensitive quality markers linked to the effect of processing temperature are the phenols and volatile compounds with their sensory impact. The literature on phenolic compounds clearly shows that the phenolic concentration of VOO could be more or less drastically reduced in relation to the increase in the mixing temperature. In particular, the derivatives of oleuropein, demethyloleuropein and ligstroside are highly affected by the processing temperature, whereas lignans are less affected (Servili et al., 2004a).

The optimal temperature of activity for PPO and POD may explain the loss of phenolic compounds in the oil depending on the processing temperature. These enzymes catalyze the oxidative degradation of phenolic compounds during the mixing process and show an optimal temperature of activity at approximately 50° C and 45° C respectively for PPO and POD.

As a result, the oxidation of phenols by PPO and POD within the range from 20° C to 35° C would be progressively higher, depending on the operating temperature. This explains the widely published data about the differences in phenolic concentration between oils obtained at different temperatures of between 20° and 40° C (Sánchez & Harwood, 2002; Angerosa et al., 2001).

These results are, in any case, obtained by performing malaxation with the pastes under continuous contact with air, as shown in the traditional mixer (Servili et al., 1998, 2003a).

However, when the process is performed in the new-generation malaxer, known as a "covered malaxer", which can control contact of the olive pastes with oxygen during mixing, the results obtained in terms of relationships between phenol concentrations in VOO and the processing temperature are completely different.

During processing, the olive pastes release CO_2 and the dissolved O_2 is simultaneously consumed by the oxidoreductase activities. As a result, the reduction of the O_2 content obtained in the covered malaxer inhibits the PPO and POD activities, improving the concentration of hydrophilic phenols in the olive pastes and in the corresponding VOO (Figure 3) (Servili et al., 2008).

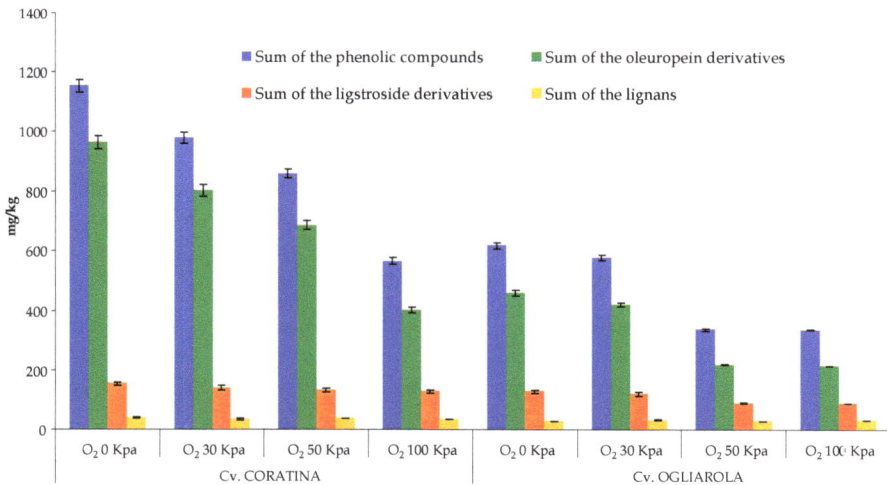

Figure 3. Phenolic composition (mg/Kg) of VOOs obtained after malaxation in different initial atmospheric compositions (Servili et al., 2008).
0 kpa saturated with N_2; 30 kpa corresponding to the air composition. The vertical lines are the mean values of three independent experiments, standard deviation is reported in brackets.

As a result, the oxidative reactions occurring in the pastes during malaxation can explain the relationships between VOO phenolic concentration and malaxation temperatures (Servili et al., 2004a; 2009a; 2009b). The O_2 dissolved in the pastes during malaxation, activate POD and PPO, which oxidize phenolic compounds according to the temperature and consequently reduce their concentration in VOOs obtained by pastes malaxed at high temperatures. The traditional malaxer, which contains a high amount of O_2 dissolved in the paste during the process due to contact with the air, represents a classical example of the aforementioned relationship between high temperatures and VOO phenolic loss. Low amounts of O_2, on the contrary, inhibit the oxidative reactions of phenols during malaxation and, in this case, their concentration in the VOO increases according to the temperatures, because of a major release of phenols into the oil (Servili et al., 2008; 2009a; 2009b).

Thus, O_2 control during malaxation can be considered a new technological parameter which, in combination with the traditional ones (time and temperature of the process), can be used to optimize the VOO phenolic and volatile concentrations (Servili et al., 2004a, 2008, 2009a). In this regard, the time of exposure of the olive pastes to air contact (TEOPAC) was studied as a process parameter to regulate the O_2 availability in the paste and, as a result, the amount of phenols in the VOO (Servili et al., 2003a, 2003b).

Furthermore, the natural increase of an inert gas of this respiration catabolite, such as CO_2, released during malaxation after the destruction of the olive cell, may be combined with the use of nitrogen or argon to reduce the O_2 contact with the olive pastes during malaxation (Parenti et al. 2006a, 2006b; Servili et al., 2008).

The latter have to be carefully controlled according to the time of malaxation. Moreover, special attention must be paid to the traditional malaxers which work continuously in an air saturated atmosphere. In fact, in this case, the longer the time at the same temperature conditions, the greater the loss of phenolic compounds in the oil (Servili et al., 1994; Di Giovacchino et al., 2002; Servili et al., 2004a). On the other hand, there is not a direct link in the confined malaxer between the time of malaxation and the loss of phenolic compounds. However, malaxation times of over 35-40 minutes do not involve an extraction yield increase and, therefore, even though a loss in the quality of the oil is not observed, longer periods of malaxation are negative for a correct plant management.

Time and temperature of malaxation also affect the volatile profile and, therefore, the sensory characteristics of the resulting EVOOs (Angerosa et al. 2004; Servili et al. 2009a, 2009b). A large part of the volatile compounds, which explain the flavour of VOO, is due to the activity of the enzymes involved in the lipoxygenase pathway (Figure 1).

This group of enzymes promotes the formation of aldehydes, alcohols at C_5 and C_6 saturated and unsaturated and esters. The main effect of malaxation time is the increase of C_6 and C_5 carbonyl compounds, especially of trans-2-hexenal, which represent an important contribution to the flavour of olive oil due to their low odour threshold, whereas high temperatures of malaxation promote a fall of esters and cis-3-hexen-1-ol and an accumulation of hexan-1-ol and trans-2-hexen-1-ol, both considered by some authors as producing a not entirely agreeable odour (Angerosa et al., 2004; Servili et al., 2009a).

The enzymes involved in the LOX pathway such as lipoxygenase, hydroperoxidelyase, alcoholdehydrogenase and alcohol acyltransferase show an optimal temperature between 15 and 25 °C, whereas their activity decreases after 30 °C. Therefore, the malaxing process carried out at temperatures of over 35 °C can produce a reduction in the volatile compounds generation during malaxation.

From the aforementioned considerations, it is, therefore, clear that the mixing temperature control at levels lower than the 28-30 °C represents an advisable stage in the extraction process in order to get high quality VOO (Angerosa et al., 2001; Angerosa et al., 2004; Servili et al., 2009a). We must also take into account that even temperatures of paste below 22 °C in the new generation of confined malaxer lead to a decrease in the solubilisation of phenolic compounds and chlorophylls.

Considering the relationships between the volatile composition and O_2 concentration of VOO during malaxation, the results reported in literature indicate that the O_2 concentration in the pastes seems to have no effect on LPO activity during the malaxation (Table 2) (Servili et al., 2008).

	Initial O_2 partial pressure in the malaxer chamber headspace (kPa)							
	0^a		30^b		50		100	
	OGLIAROLA cv.							
ALDEHYDES								
2-Pentenal $(E)^z$	291.5	(31.8)ab	343.0	(31.1)a	247.5	(11.7)b	269.5	(13.5)b
Hexanal	939.5	(9.2)a	1546.0	(200.8)b	1011.5	(27.6)a	1499.5	(16.3)b
2-Hexenal (E)	43645.0	(912.2)a	39130.0	(1054.7)b	37315.0	(233.3)b	38170.0	(1258.7)b
ALCOHOLS								
1-Pentanol	28.5	(2.1)a	128.0	(6.8)b	122.5	(3.5)b	158.0	(1.4)c
2-Penten-1-ol (E)	55.5	(3.5)a	63.0	(4.6)a	50.5	(9.2)ab	38.5	(6.4)b
1-Penten-3-ol	567.0	(17)a	871.0	(4.7)b	690.0	(1.4)c	809.5	(3.5)d
1-Hexanol	8357.0	(102.6)a	9699.0	(106.1)b	11660.0	(99)c	13675.0	(63.6)d
3-Hexen-1-ol (E)	35.0	(1.2)a	41.0	(3.5)a	47.5	(2.1)b	61.5	(2.1)c
3-Hexen-1-ol (Z)	286.5	(4.9)a	434.0	(20.6)b	341.0	(11.3)c	400.5	(7.3)d
2-Hexen-1-ol (E)	7662.5	(75.7)a	8616.0	(87.9)b	9355.0	(353.6)c	9780.0	(60.8)c
	CORATINA cv.							
ALDEHYDES								
2-Pentenal (E)	548.5	(16.3)ab	509.7	(5.8)b	636.7	(17.9)c	613.0	(51.2)ac
Hexanal	1187.0	(9.9)a	1624.3	(30)bc	1532.1	(27.3)b	1744.0	(12.2)c
2-Hexenal (E)	51565.0	(827.3)a	52900.0	(565.7)ab	54340.5	(355.7)b	53920.0	(332.1)b
ALCOHOLS								
1-Pentanol	40.0	(5.7)a	54.3	(5)b	39.4	(5)a	48.0	(3.2)ab
2-Penten-1-ol (E)	87.5	(0.7)a	67.0	(0.2)b	105.8	(5.7)c	105.0	(8.3)c
1-Penten-3-ol	890.0	(2.8)a	820.0	(1.2)b	1093.5	(33.7)c	1185.0	(91.2)c
1-Hexanol	2326.0	(49.5)a	3694.2	(2)b	1788.0	(57.2)c	2170.0	(123.1)a
3-Hexen-1-ol (E)	25.5	(0.7)ab	31.6	(3.8)a	20.0	(1.9)b	21.0	(1.9)b
3-Hexen-1-ol (Z)	561.0	(4.2)a	513.6	(9.6)b	486.3	(11.1)b	498.0	(31.2)b
2-Hexen-1-ol (E)	3654.5	(30.4)a	5905.0	(321)b	3350.1	(80.5)a	4185.0	(35.6)c

Table 2. Volatile composition (µg/kg) of EVOOs obtained after malaxation in different initial atmospheric compositions (Servili et al. 2008).
[a] saturated with N_2; b corresponding to the air composition. [z] Data are the mean values of three independent experiments, standard deviation is reported in brackets. Values in each row with different letters (a-d) are significantly different from one another at p < 0.01.

In conclusion, it can be considered that, as regards the process variables adopted during malaxation, the temperature should be set within the range of 24-27 °C, both for the traditional and for the new malaxer (i.e., the confined malaxer), while process times greater than 35-40 minutes may result in a large loss in terms of product quality (in particular in the traditional malaxer) without producing significant positive effects on the oil extraction yield.

Special attention must be paid to the control and regulation of oxygen percentage in contact with the olive paste during malaxation to optimize the phenolic content as well as the flavour of VOO. This new operating parameter can be used to act on the phenolic concentration of the pastes and, therefore, of the oils, excluding negative collateral effects on the volatile compound content of the product (Servili et al., 2003a; Migliorini et al., 2006; Servili et al., 2008). In fact, the traditional Italian *cvs* differ with respect to their content of phenolic substances and, because of this difference, the phenolic concentration in oil must be optimized; this optimization can be obtained by regulating their oxidative degradation level during malaxation. Thus, malaxation should be carried out without oxygen for the *cvs* with low phenolic concentration, whereas malaxation should be carried out with controlled supplementation of oxygen for those *cvs* characterized by higher phenolic concentrations (Figure 3, Table 2).

Moreover, it must be pointed out that no additional gases, such as nitrogen or argon, are required inside the head space in the confined malaxer to avoid the presence of oxygen. In fact, if the malaxer is filled with crushed paste during the process, the olive tissues of pastes naturally release carbon dioxide (CO_2) (Weichmann, 1987), whereas the limited amount of oxygen they adsorb during the crushing process will be consumed rapidly by endogenous enzyme activities. As a consequence, the malaxer head space will be naturally saturated by an inert gas, such as carbon dioxide (Servili et al., 2003a; Parenti et al., 2006a, 2006b; Servili et al., 2008).

3.4.1. Use of the technological co-adjuvants in malaxation

According to the UE 2568/1991 and 1989/2003 standards, co-adjuvants can be added during malaxation to break down emulsions and at the same time to guarantee a high oil extraction yield. In particular, the most frequently employed co-adjuvants are micronized talc and in some countries, although not in Europe, enzyme preparations are used.

Such enzyme preparations act on the structural colloids of the cells of not only the pulp, but also the skin and assist the activity of the endogenous enzymes (pectinases, cellulases and hemicellulases) resulting in an increase in the olive oil yield and in its quality (Vierhuis et al., 2001). In particular, it has been demonstrated that enzyme preparations tend to increase the phenolic content in VOO (Table 3). However, these effects are found only when working with olives characterized by an early ripening (Ranalli & Serraiocco, 1996; Ranalli & De Mattia, 1997; Vierhuis et al., 2001; Montedoro et al., 2002). In fact, the use of enzymatic preparations in olives when they have reached an advanced stage of ripening does not lead to any qualitative or quantitative benefit. This is due to the fact that the structural colloids in mature olives, on which the aforementioned enzymes should act, have already been degraded by the endogenous enzymatic activity during the ripening of the fruit (Heredia et al., 1993). Therefore, the addition of enzymatic preparations would be useless in this case.

However, the EU regulations 2568/1991 and 1989/2003 do not allow the addition of enzymatic preparations, whereas they authorise the use of micronized talc as co-adjuvant. The use of talc as a co-adjuvant has been proved important to increase oil extraction yield without any interference to the quality of VOO. The amount of talc used ranges from 0.7 to 1.5% of the weight of the olives being milled after the first 10 minutes of the process. In fact, its addition to difficult pastes improves the paste structure and reduces emulsions. This product acts on the olive pastes by increasing the drainage effect and, therefore, improving the efficiency of solid-liquid separation during centrifugation of the crushed pastes (Servili et al., 2004b). The micronized talc can be added to the pastes during malaxation (0.7-1.5% of the paste weight) after the first 10 minutes of the trial. Several studies carried out on Italian cultivars show that the use of the micronized talc does not involve any negative effects on oil quality and, in some cases, its use leads to a meaningful increase in the extraction yield (Servili et al., 2004b).

	crushed paste			malaxed paste			malaxed paste + enzyme preparations		
3,4-DHPEA[a]	2.7	±	0.3[a]	0.7	±	0.1[b]	1.09	±	0.1[c]
p-HPEA	2.3	±	0.4[a]	1.2	±	0.1[b]	1.02	±	0.1[b]
3,4-DHPEA-EDA	515.0	±	23[ab]	317.0	±	16[c]	439	±	16[d]
p-HPEA-EDA	24.8	±	1.9[a]	25.8	±	1.4[ab]	29.4	±	0.8[c]
Lignans	32.5	±	1.4[a]	24.2	±	0.8[b]	28.5	±	0.9[c]
3,4-DHPEA-EA	357.0	±	13[a]	177.0	±	8.0[b]	218	±	8.0[c]

Table 3. Phenolic composition of virgin olive oil (mg/kg) with and without enzymatic treatment during malaxation (Vierhuis et al., 2001).
[a] The phenolic content is the mean value of three independent experiments (standard deviation). Values in each row bearing the same superscripts are not significantly different from one another ($P < 0.05$).

3.5. Olive oil extraction systems

Different extraction technologies, such as pressure and centrifugation and selective filtration (i.e. "surface tension" or "percolation") enabling the separation of oily must from the olive paste can be used (Boskou, 1996; Di Giovacchino et al., 1994, 1995).

3.5.1. Pressure extraction system

Pressing is one of the oldest methods of oil extraction and has evolved considerably over the centuries. In olive oil mills equipped with this system the press separation of the oil from the paste is currently carried out using open hydraulic presses, whereas close cage presses have almost disappeared not only due to high purchase prices, but also to their maintenance costs. The previously malaxed paste is subsequently stratified on stacked filter mats, each covered with approximately 0.5 inches (1.25cm) of paste and interposed with metal disks. This operation is carried out mechanically thanks to a dispenser, which takes the paste from the malaxer and stores it on the nylon and/or polypropylene filter mats. Both types of filter mats have a central hole to allow the expressed oil and water (olive juice) to exit in both directions. From a theoretical point of view, this system guarantees intrinsic oil quality. However, its use presents a few problems, mainly due not only to its low working capacity per hour, in which case the storage of olives lengthens, but also to the proper use of the filter mats and to the types of materials used to build the equipment. The critical aspects of the process regarding the use of the press, which impacts on the quality of the oil, are concerned with both the proper management of the filter mats and the use of construction materials made of stainless steel. As regards the filter mats, it is important to point out that they can represent a source of contamination, due to fact that they may introduce fermentation and an oxidation defect into the oil, causing sensorial defects (Angerosa et al., 2004). This effect can arise both from the contamination with oils obtained from poor batches of olives and from fermentation processes of the vegetation water and pomace fragments, which remain in the filter mats, when they are kept in storage during the different processing stages. The latter problem occurs particularly when the oil harvest is interrupted by bad climatic conditions and it is impossible to work continuously. In order to minimize the risk of defects developing in the VOO, it would be desirable: i) to work in a continuous cycle; ii) to change the stacked filter mats frequently during the process and to clean them periodically using a pressure washer; iii) to store the aforementioned filter mats at a low temperature (0 °C-5 °C) to avoid fermentative processes during breaks in the oil processing.

As regards the materials used to construct the press, all the metallic parts which come into contact with the product must be made of, or at least covered with stainless steel to avoid the transfer of metals, especially those metals which can speed up the oil oxidation, to the oil during the extraction process.

3.5.2. Extraction by centrifugation

The majority of VOO is currently extracted by centrifugation in Mediterranean countries. The idea of exploiting direct centrifugation of the malaxed paste to extract the oil dates back to the late nineteenth century, when the use of the first decanter applied to food industries was widespread. In the olive oil sector, this idea determined technological innovations in the VOO mechanical extraction process, which were opposed to the traditional press. The first operating patents, including the patent by Corteggiani, date back to 1956, followed by new companies producing olive oil machines in the early sixties. This machine, called a decanter, consists of a drum containing a cylindrical and a conical part with a horizontal axis, inside which an additional cylinder worm is placed, which acts as a screw conveyor. The differential speed of the latter is slower than that of the outer drum in order to discharge the solid part. In recent years, this extraction system has evolved considerably in order to reduce the amount of water used during the process. In fact, the decanters can be classified as follows:

1. Traditional three-phases decanters, featuring water addition ranging from 0.5 to 1 m³/ton.
2. Two-phases decanters, which can operate without the addition of water and do not produce vegetation water as a by-product of the extraction oil process.
3. New three-phases decanters, working at low water consumption ranging from 0.2 to 0.3 m³/ton.

The traditional three-phases decanters, which allow the oil to be separated both from the vegetation water and from the pomace, feature a humidity level of between 50% and 55% and dilute the pastes produced to reduce their viscosity. In doing so, they facilitate separation of the oil-vegetation water with a dilution ratio ranging from 1:0.5 to 1:1 (from 50 to 100 l of water for every 100 Kg of paste to be decanted). In addition to the enormous amounts of vegetation water which have to be drained, this implies a decrease in the oil quality, principally due to the washing away of the phenolic compounds of the product, with massive decreases in this important antioxidant fraction (Ranalli & Angerosa, 1996; Servili et al., 1999c, 1999d; Stefanoudakii et al., 1999; Di Giovacchino et al., 2001; Servili & Esposto, 2004).

The evolution of this technology has lead to the production of two and three-phases decanters with low water consumption. By using these new systems, the extracted oils feature a higher phenolic concentration than those extracted by means of the traditional centrifugation process, because the loss of these hydrophilic phenolic compounds in the vegetation water is reduced (Table 4) (Servili et al., 1999d). In this context, it is important to point out how these new extraction systems, which do not take into account the addition of water, enable high quality VOOs to be obtained. The focus of this problem consists in linking together the low processing temperatures with a reduced use of water dilution in the pastes: by using centrifugation systems, these two process variables should allow high

Phenolic compounds	Coratina cv.		Ogliarola cv.	
	two phases	three phases	two phases	three phases
3,4 DHPEA	0.87 ± 0.02	0.58 ± 0.08[b]	0.66 ± 0.11[a]	0.50 ± 0.11[a]
p-HPEA	3.74 ± 0.07[a]	2.34 ± 0.08[b]	3.30 ± 0.10[a]	4.22 ± 0.10[b]
Vanillic acid	0.41 ± 0.01[a]	0.19 ± 0.01[b]	0.26 ± 0.01[a]	0.14± 0.05[b]
Caffeic acid	0.16 ± 0.01[a]	0.12 ± 0.02[b]	0.09 ± 0.01[a]	0.21 ± 0.03[b]
3,4 DHPEA-EDA	522.2 ± 13.5[a]	427.2 ± 13.8[b]	30.09 ±1.03[a]	18.53 ± 0.68[b]
p-HPEA-EDA	78.16 ± 0.52[a]	67.26 ± 2.55[b]	20.99 ± 0.82[a]	22.40 ± 0.33[a]
Lignans	38.41 ± 0.10[a]	35.62 ± 1.11[b]	48.00 ± 3.40[a]	46.72 ± 5.78[a]
3,4 DHPEA-EA	351.7 ± 11.0[a]	244.9 ± 13.6[b]	68.01 ± 6.00[a]	52.04 ± 3.11[b]
Total polyphenols	673 ± 4[a]	585 ± 7[b]	304 ± 5[a]	263 ± 4[b]
Induction period [h]	17.8 ± 0.1[a]	15.5 ± 0.2[b]	5.2 ± 0.1[a]	4.6 ± C.1[b]

Table 4. Effect of water reduction during centrifugation on the phenolic composition of virgin olive oil (mg/kg) (Servili et al., 2002).
Data are the mean values of three independent experiments ± standard deviation. Values in each row, with cvs with different letters, are significantly different from one another (p < 0,01).

quality oils and machines featuring high yields to be obtained (Ranalli & Angerosa, 1996; Servili et al., 1999d; Stefanoudakii et al., 1999; Di Giovacchino et al., 2001).

As regards the by-products obtained by centrifugation, it is important to recall here that, whereas the aforementioned problem concerning the use of three-phases traditional systems is represented by the enormous amounts of vegetation water to be drained (0.7-1.2 m³/ton), the main problem, when using the two-phases systems, is not only a decrease in oil quality, but also the high humidity level of the pomaces (50% for the traditional three-phases decanters, 55-60% for the two-phases). This last aspect implies two disadvantages: i) where to store by-products of the extraction process; ii) how to transport to the pomace oil factory and their subsequent use for residual oil recovery by solvent extraction.

The pomaces produced by two-phases system are generally employed for the production of compost or for spreading to improve agriculture soil, following a process similar to that used for the vegetation water of olives. In this context, three-phases decanters with low water consumption can represent an adequate alternative with respect to the two-phases extraction system, because they allow a quality of oil to be obtained which is comparable to that of oils obtained using the two-phases system, and pomaces with a reduced humidity content, similar to those produced by the traditional three-phases systems. On the contrary, they produce a certain amount of vegetation waters, which imply water draining procedures which comply with the regulations of the law.

3.6. Separation of the oil from vegetation water

The liquid coming from the extraction system is called "oily must", and consists of oil and vegetation water, which is separated by using vertical centrifuges. The oily must also contains solid particles and mucilage (0.5 to 1.0%). These substances are suspended if they are very small, whereas they are easily separated from the liquid if they are of an appreciable size (seed fragments and/or epicarp of olive fruit fragments). In particular, separation from the liquid is carried out by using a sieve (1~2 mm mesh) placed at the top of the olive oil storage tanks.

Disk stack centrifuges, suitable for separating solid impurities with a specific weight ranging from 1.050 to 1.150, are used to separate the oil from the oily must. The basic principles of centrifugation are well known. If a vigorous rotational motion is applied to the oily must (oil, vegetation water and impurities), the lighter part (i.e. the oil) is collected close to the axis of rotation, while the heavier part (i.e. the water) is collected further away. Finally, the impurities are collected even further away.

Moreover, spillways enable the oil as well as the vegetation waters and impurities to be recovered. A fraction of the impurities is deposited on the rotating drum, which must be periodically cleaned, even though self-cleaning decanters are widely used. In fact, these decanters work in a continuous cycle, providing a periodical and automatic discharge of the sediment, which could compromise the centrifugation process. The decanters most frequently employed in the oil mills consist of a series of perforated, truncated cone-shaped disks, mounted on the hollow shaft-mounted drum in order to leave a free space of approximately 1 mm between the disks.

Centrifugal force forces the oily must poured in from the top through the hollow drum shaft upwards and it is divided into three different layers of oil, vegetation water and impurities, according to their specific weight. The drum diameter of decanters used in the oil mills ranges from 400 to 700 mm, with a rotational velocity ranging from 5000 to 12000 rpm. The working capacity of these machines in terms of litres of oily must poured in per hour is very high and it varies between 500 and 2000 l/h. The work carried out by the vertical centrifuges is qualitatively satisfactory, even though there is often a loss of oil in the vegetation waters. This loss cannot exceed 500 g/ton of processed olives if centrifugation process is to be considered adequate.

3.7. Olive oil storage

During storage, the phenolic composition of EVOO is modified by the endogenous enzymatic activities contained in the cloudy phase. These enzymes may reduce the "pungent" and "bitter" sensory notes, the intensity of which is strictly linked to the content of aglycon secoiridoids, and, at the same time, can produce olfactory and taste defects. Oil filtration partially removes the water and enzymes from EVOOs, and enables the EVOO phenolic content to stabilize during its storage. The filtration process of EVOO is a procedure carried out in two steps: first, the suspended solids are removed, and second, the elimination of humidity gives the oil a brilliant aspect. Normally, organic or inorganic materials are used in conjunction with a variety of filtration equipment to enhance or enable the separation of suspended solids and water-oil. The type of such equipment, often called filter aids, depends on the final objective (Montedoro et al., 2005).

The olive oil profile changes during its storage, due to the simultaneous, drastic reduction in compounds from the LOX pathway and to the neo-formation of volatile compounds, responsible for some common defects referred to as "rancid", "cucumber" and "muddy sediment" (Morales & Aparicio, 1997; Angerosa et al., 2004; Servili et al,. 2009a).

This runs parallel to the increase in saturated aldehydes nonanal, and above all hexanal in the oxidation process, but it cannot be considered a useful marker of oxidation, since it is also present in the aroma of high quality EVOOs (Angerosa et al., 2004; Servili et al., 2009a).

Furthermore, the presence of sediment as a result of the decantation of unfiltered olive oil during its storage can determine, under suitable temperature conditions, the production of unpleasant compounds responsible for the typical "muddy sediment" defect due to the fermentation which produces compounds, probably of the butyric type (Angerosa et al., 2004; Servili et al., 2009a).

Author details

Maurizio Servili, Agnese Taticchi, Sonia Esposto, Beatrice Sordini and Stefania Urbani
University of Perugia, Faculty of Agriculture, Perugia, Italy

4. References

Amiot, M.J., Fleuriet, A., Machiex, J.J. (1986). Importance and evolution of phenolic compounds in olive during growth and maturation. *J. Agric. Food Chem.* (34): 823-826.

Amirante, P., Clodoveo, M., Dugo, L., Leone, G., Tamborrino, A. (2006). Advance technology in virgin olive oil production from traditional and de-stoned pastes: Influence of the introduction of a heat exchanger on oil quality. *Food Chem.* (98): 797-805.

Angerosa, F.& Di Giacinto, L. (1995). Quality characteristics of virgin olive oil in relation to crushing method. Note II. *Riv. Ital. Sostanze Grasse.* (72): 1-4.

Angerosa, F., Basti, C., Vito, R., Lanza, B. (1999). Effect of fruit stone removal on the production of virgin olive oil volatile compounds. *Food Chem.* (67):295-299.

Angerosa, F., Lanza, B., Marsilio, V. (1996). Biogenesis of fusty defect in virgin olive oils. *Grasas Aceites.* (47):142-150.

Angerosa, F., Mostallino, R., Basti, C., Vito, R. (2001). Influence of malaxation temperature and time on the quality of virgin olive oils. *Food Chem.* (72): 19-28.

Angerosa, F., Servili, M., Selvaggini, R., Taticchi, A., Esposto, S., Montedoro, GF. (2004). Volatile compounds in virgin olive oil: occurrence and their relationship with the quality. *J. Chromatogr. A.* (1054): 17-31.

Boskou, D. (1996). Olive oil composition. *Olive oil chemistry and technology.* Boskou, D. (Ed), AOC Press, Champaign, Illinois, USA. pp. 52-83.

Brenes, M., García, A., García, P., Aarrido, G. (2000). Rapid and complete extraction of phenols from olive oil and determination by means of a coulometric electrode array system. *J. Agric. Food Chem.* (48): 5178-5183.

Capella, P., Fedeli, E., Bonaga, G., Lerker, G. (1997). Manuale degli oli e dei grassi. *Tecniche Nuove,* Ed. (Milano).

Caponio, F., Alloggio, V. & Gomes, T. (1999). Phenolic compounds of virgin olive oil: Influence of paste preparation techniques. *Food Chem.* (64): 203-209.

Di Giovacchino, L. & Serraiocco, A. (1995). Influenza dei sistemi di lavorazione delle olive sulla composizione dello spazio di testa degli oli. *Riv. It. Sost. Grasse.* (72): 443-450.

Di Giovacchino, L., Costantini, N., Serraiocco, A., Surricchio, G., Basti, C. (2001). Natural antioxidants and volatile compounds of virgin olive oils obtained by two or three-phases centrifugal decanters. *Eur. J. Lipid Sci. Technol.* (103) :279-285.

Di Giovacchino, L., Sestili, S., Di Vincenzo, D. (2002). Influence of olive processing on virgin olive oil quality. *Eur. J. Lipid Sci. Technol.* (104): 587-601.

Di Giovacchino, L., Solinas, M., Miccoli M. (1994). Effect of extraction systems on the quality of virgin olive oil. *J. Am. Oil Chem. Soc.* (71):1189-1193.

Esposto, S., Montedoro, GF., Selvaggini, R., Riccò, A., Taticchi, A., Urbani, S. & Servili, M. (2008). Monitoring of virgin olive oil volatile compounds evolution during olive malaxation by an array of metal oxide sensors. *Food Chem.* (113): 345-350.

European Economy Community. (1991). Commission Regulation (EEC) No. 2568/91. on the characteristics of olive oil and olive-residue oil and on relevant method of analysis.

European Economy Community. (2003). November 6, Regulation 1989/03 amerding Regulation (EEC) No 2568/91 on the characteristics of olive oil and olive-pomace oil and on the relevant methods of analysis modifies the CEE n. 2568/91 on olive oils and pomace olive oils characteristics and relative analysis methods. *Official Journal L.* 295/57 13/11/2003.

Garcia, J.M., Yousfi, K., Mateos, R., Olmo, M., Cert, A. (2001). Reduction of oil bitterness by heating of olive (Olea europaea) fruits. *J. Agric. Food Chem.* (49): 4231-4235.

Heredia, A., Guillén, R., Jiménez, A., Bolaños, J. F. (1993). Activity of glycosidases during development and ripening of olive fruit. *Lebensm Unters Forsch.* (196): 147-151.

Inarejos-Garcia, A. M., Gomez-Rico, A., Desamparados Salvador, M., Fregapane, G. (2009). Influence of malaxation conditions on virgin olive oil yield, overall quality and composition. *Eur. Food Res. Technol.* (228) (4): 671-677.

I.O.C. International Olive Council. (1996). COI/T.20/Document 15/ Rev. 1. Organoleptic assessment of olive oil. Resolution RES-3/75-IV/96, 20 November; Madrid, Spain.

Kalua, C.M., Bedgood, D.R.Jr., Bishop, A.G., Prenzler, P.D. (2006). Changes in volatile and phenolic compounds with malaxation time and temperature during virgin olive oil production. *J. Agric. Food Chem.*(54): 7641-7651.

Kiritsakis, A. K., Nanos, G.D., Polymenopoulos, Z., Thomai, T., Sfakiotakis, E.M., (1998b). Effect of fruit storage conditions on olive oil quality. *J. Am. Oil Chem. Soc.* (75):721-724.

Kiritsakis, A.K. (1998a). Flavor components of olive oil. A review. *J. Am. Oil Chem. Soc* (75): 673-681.

Lavelli, V. & Bondesan, L. (2005). Secoiridoids, tocopherols, and antioxidant activity of monovarietal extra virgin olive oils extracted from destoned fruits. *J. Agric. Food Chem.* (53): 1102-1107.

Migliorini, M., Mugelli, M., Cherubini, C., Viti, P., Zanoni, B. (2006). Influence of O_2 on the quality of virgin olive oil during malaxation. *J. Sci. Food Agric.* (86): 2140-2146.

Montedoro, G.F., Servili; M., Baldioli, M., Miniati, E. (1992). Simple and hydrolyzable phenolic compounds in virgin olive oil. 1. Their extraction, and quantitative and semiquantitative evaluation by HPLC. *J. Agric. Food Chem.* (40): 1571- 1576.

Montedoro, GF., Selvaggini, R., Begliomini, A. L., Baldioli, M., Esposto, S., Servili, M. (2005). Questa filtrazione s'ha da fare. *Olivo & olio.* (5): 32-40.

Montedoro, GF., Servili, M., Baldioli, M. (2002). The use of biotechnology means during oil mechanical extraction process: relationship with sensory and nutritional parameters of virgin olive oil quality. *Acta Horticult.* (586): 557-560.

Morales, M.T. & Aparicio, R. (1997). Changes in the volatile composition olive oil during oxidation: Flavors and off-flavors. *J. Agric. Food Chem.* (45): 2666-2673.

Mulinacci, N., Giaccherini, C., Innocenti, M., Romani, A., Vincieri, F. F., Marotta, F., Mattei A. (2005). Analysis of extra virgin olive oils from stoned olives. *J. Sci. Food Agric.* (85):662- 670.

Obied, H. K., Prenzler, P. D., Ryan, D., Servili, M., Taticchi, A., Esposto, S., & Robards, K. (2008). Biosynthesis and biotransformations of phenol-conjugated oleosidic secoiridoids from *Olea europaea* L. *Natural Product Reports* (25): 1167–1179.

Owen, R.W., Giacosa, A., Hull, W.E., Haubner, R., Wurtele, G., Spiegelhalder, B., Bartsch, H. (2000). Olive-oil consumption and health: the possible role of antioxidants. *Food Chem. Toxicol.* (38): 647-659.

Parenti, A., Spugnoli, P., Masella, P., Calamai, L. (2006a). Carbon dioxide emission from olive oil pastes during the transformation process: technological spin offs. *Eur. Food Res. Technol.* (222): 521-526.

Parenti, A., Spugnoli, P., Masella, P., Calamai, L., Pantani, O.L. (2006b). Improving olive oil quality using CO2 evolved from olive pastes during processing. *Eur. J. Lipid Sci. Technol.* (108): 904-912.

Perez, A.G., Luaces, P., Rios, J.J., Garcia, J.M., Sanz, C. (2003). Modification of volatile compound profile of virgin olive oil due to hot-water treatment of olive fruit. *J Agric Food Chem.* (51): 6544-6549.

Ranalli, A & Angerosa, F. (1996). Integral centrifuges for olive oil extraction. The qualitative characteristics of products. *J. Am. Oil Chem. Soc.* (73): 417-422.

Ranalli, A. & Serraiocco, A. (1996). Quantitative and qualitative effects of a pectolytic enzyme in olive oil production. *Grasas Aceites.* (47): 227-236.

Ranalli, A. & De Mattia, G. (1997). Characterization of olive oil production with a new enzyme processing aid. *J. Am. Oil Chem. Soc.* (74): 1105-1113.

Sánchez, J. & Harwood, L. (2002). Biosynthesis of triacylglycerols and volatiles in olives. *Eur. J. Lipid Sci. Technol.* (104): 564-573.

Servili, M. & Esposto, S. (2004). Tecnologia e qualità degli oli vergini di oliva., L'estrazione dell'olio di oliva: aggiornamento sulle conoscenze biochimiche-tecnologiche e impiantistiche in relazione alla qualità dell'olio e allo smaltimento dei rifiuti. *Atti del Corso di aggiornamento tenuto a Spoleto, 28-31 ottobre 2004. Spoleto, Accademia Nazionale dell'Olivo e dell'Olio, Spoleto*, pp. 41-79.

Servili, M. & Montedoro, GF. (2002). Contribution of phenolic compounds to virgin olive oil quality. *Eur. J. Lip. Sci. Technol.* (104): 602-613.

Servili, M., Baldioli, M., Montedoro, GF., (1994). Phenolic composition of virgin olive oil in relationship to some chemical and physical aspects of malaxation. *Acta Hortic.* (1): 331-336.

Servili, M., Baldioli, M., Selvaggini, R., Macchioni, A., Montedoro, GF. (1999b). Phenolic compounds of olive fruit: one- and two-dimensional Nuclear Magnetic Resonance characterization of nüzhenide and its distribution in the constitutive parts of fruit. *J. Agric. Food Chem.* (47): 12-18.

Servili, M., Baldioli, M., Selvaggini, R., Mariotti, F., Federici, E., Montedoro, GF. (1998). Effect of malaxation under N2 flush on phenolic and volatile compounds of virgin olive oil. *13 the International symposium on plant lipids*, Seville, July, pp. 307-310.

Servili, M., Baldioli, M., Selvaggini, R., Miniati, E., Macchioni, A., Montedoro, GF. (1999a). HPLC evaluation of phenols in olive fruit, virgin olive oil, vegetation water and pomace and 1D- and 2D-NMR characterization. *J.Am. Oil Chem. Soc.* (76): 873-882.

Servili, M., De Stefano, G., Piacquadio, P., Di Giovacchino, L., Sciancalepore, V. (1999d). Effect of extraction systems on the phenolic composition of virgin olive oils. *Eur.Journal Lip. Sci. Technol.* (101): 328-332.

Servili, M., Esposto, S., Fabiani, R., Urbani, S., Taticchi, A., Mariucci, F., Selvaggini, R., Montedoro, GF. (2009b). Phenolic compounds in olive oil: antioxidant, health and organoleptic activities according to their chemical structure. *Inflammopharmacology.* (17): 1–9.

Servili, M., Esposto, S., Selvaggini, R., Taticchi, A., Urbani, S., Montedoro, GF. (2004b). Talco micronizzato. Primi risultati. *Olivo e Olio.* (10): 20-24.

Servili, M., Esposto, S., Taticchi, A., Urbani, S., Di Maio, I., Sordini, B., Selvaggini, R., Montedoro, GF., Angerosa, F. (2009a). Volatile compounds of virgin olive oil: Their importance in the sensory quality. *Advances in Olive Resources.* Liliane Berti and Jacques Maury Eds. pp. 45-77.

Servili, M., Mariotti, F., Baldioli, M., Montedoro, GF. (1999c). Phenolic composition of olive fruit and virgin olive oil: distribution in the constitutive parts of fruit and evolution during the oil mechanical extraction process. *Acta Hort.* (474): 609-613.

Servili, M., Piacquadio, P., De Stefano, G., Taticchi, A., Sciancalepore, V. (2002). Influence of a new crushing technique on the composition of the volatile compounds and related sensory quality of virgin olive oil. *Eur. J. Lip. Sci. Technol.* (104): 483-489.

Servili, M., Selvaggini, R., Esposto, S., Taticchi, A., Montedoro, GF., Morozzi, G. (2004a). Health and sensory properties of virgin olive oil hydrophilic phenols: agronomic and technological aspects of production that affect their occurrence in the oil. *J. Chromatogr. A.* (1054): 113-127.

Servili, M., Selvaggini, R., Taticchi, A., Esposto, S., Montedoro, GF. (2003a). Volatile compounds and phenolic composition of virgin olive oil: optimization of temperature and time of exposure of olive pastes to air contact during the mechanical extraction process. *J. Agric. Food Chem.* 27(51): 7980-7988.

Servili, M., Selvaggini, R., Taticchi, A., Esposto, S., Montedoro, GF. (2003b). Air exposure time of olive pastes during the extraction process and phenolic and volatile composition of virgin olive oil. *J. Am. Oil Chem. Soc.* 7 (80) 685-695.

Servili, M., Taticchi, A., Esposto, S., Urbani, S., Selvaggini, R., Montedoro, GF. (2007). Effect of olive stoning on the volatile and phenolic composition of virgin olive oil. *J. Agric. Food Chem.* (55): 7028-7035.

Servili, M., Taticchi, A., Esposto, S., Urbani, S., Selvaggini, R., Montedoro, GF. (2008). Influence of the decrease in oxygen during malaxation of olive paste on the composition of volatiles and phenolic compounds in virgin olive oil. *J. Agric. Food Chem.* (56):10048–10055.

Stefanoudakii, E., Koutsaftakis, A., Kotsifaki, F., Angerosa, F., Di Girolamo, M. (1999). Quality characteristics of olive oils dual-phases, three-phases decanters and laboratory mill. *Acta Hort.* (474): 705-708.

Vierhuis, E., Servili, M., Baldioli, M., Schols, H.A., Voragen, A.G.J., Montedoro, GF. (2001). Study of the effect of enzyme treatment during the mechanical extraction of olive oil on phenolic compounds and polysaccharides. *J. Agric. Food Chem.* 3(49): 1218-1223.

VV.AA. (2003). *Olea*. Trattato di olivicoltura a cura di Fiorino P., Edagricole Ed. (Bologna).

Weichmann, J. (1987). *In Postharvest physiology of vegetables*. Marcel Dekker, Inc. New York. pp 13.

Chapter 10

Packaging and Storage of Olive Oil

Amalia Piscopo and Marco Poiana

Additional information is available at the end of the chapter

1. Introduction

Storage is a very important step of any food, including olive oil. In fact, olive oil shelf life can be influenced by different factors, from olive quality to processing technologies, however, the selection of proper storage conditions, including packaging, can be of great importance. The Mediterranean diet is recommended as food model for the prevention of various chronic-degenerative pathologies (cardiac diseases, cancer etc.) and olive oil is surely the cornerstone of this type of nutrition with fruits, vegetables, legumes and fish. Its peculiar nutritional characteristics depend on the presence of antioxidant components and monounsaturated fatty acids that are predominant on unsaturated ones, with positive results in the increase of HDL (High-Density Lipoproteins) and reduction of LDL (Low-Density Lipoproteins) oxidation.

The consumer expresses its judgment on olive oil quality considering only some sensory characteristics, such as the more or less pungent taste, fruity and mild flavour and within this context a wide range of preferences can be found, because the sensory quality may match specific dishes, cultural aspects or simple dietary habits. Incorrect storage practices influence the sensory quality of the oil, as rancidity and off-flavours may develop. The lipid oxidation, in fact, is one of the factors of olive oil quality deterioration. The rate of oxidation depends on the availability of oxygen, the presence of light and the temperature. In absence of light, the auto-oxidation follows a free radical mechanism with the formation of hydroperoxides. These labile compounds further decompose to produce a complex mixture of volatile compounds such as aldehydes, ketones, hydrocarbons, alcohols, and esters responsible for the already mentioned deterioration of olive oil flavour known as "oxidative rancidity". Many phenolic compounds of the olive oil contribute to its resistance to oxidative rancidity. After light exposition, photo-oxidation occurs through the action of natural photosensitizers (i.e. chlorophyll), which react with triplet oxygen to form the excited state singlet oxygen. It then gives a free radical from unsaturated fatty acids, leading to the production of hydroperoxides and eventually to carbonyl compounds, which result in the development of undesirable off flavours in oils.

Packaging can directly influence olive oil quality by protecting the product from both oxygen and light. The shelf life of the oils exposed to intense artificial light and diffused daylight is shorter than that of oils kept in the dark. Moreover, the storage temperature, the use of nitrogen atmosphere and the reduction of the oxygen in the headspace volume can appreciably control quality changes during storage time.

Materials which have been used for olive oil packaging include glass, metals (tin-coated steel) and more recently plastics and plastics coated paperboard. Among plastics, polyethylene terephthalate (PET) has captured a large portion of the olive oil retail market due to its many advantages including clarity, chemical inertness, low oxygen permeability, and excellent mechanical properties. Incorporation of pigments and/or UV blocking agents or oxygen scavengers may improve plastics properties with regard to quality retention of olive oil. Besides PET, polyethylene (PE) in the form of LDPE-coated paperboard/aluminium foil laminates, i.e. brick-type cartons and bag-in-box pouches and polypropylene (PP) are being used today for the packaging of vegetable oils including olive oil. In this chapter, factors affecting olive oil quality during storage will be discussed and various packaging solutions will be approached.

The aim is to give evidence the best practices to adequately manage this important unit operation, in order to improve/maintain the quality characteristics of the olive oil.

2. Nutritional value of olive oil

Mediterranean food tradition is sustained by three basic essentials: wheat, olives and grapes. Nevertheless, olive oil is the central element inherent to this diet, and its importance is due to the increasing consumption around the world, because of its nutritional and sensory properties. European Union (EU) is the leading producer of olive oil and within the EU, the Mediterranean members are the biggest producers, in fact Mediterranean area accounts for 95% of world production and 85% of world consumption of olive oil (IOOC, 2007).

Several clinical data have shown that consumption of olive oil can provide heart health benefits, such as favourable effects on cholesterol regulation and LDL cholesterol oxidation, and that it exerts anti-inflammatory, antithrombotic and antihypertensive effects (Lairon, 2007; Perez-Jimenez et al., 2007).

Virgin olive oil is a genuine fruit juice obtained from olive drupes (*Olea europaea* L.), using exclusively mechanical procedures, without further treatments or chemical additions. The saponifiable fraction is more representative (about 98%) than the unsaponifiable one and it comprises principally triacylglycerols, esters formed by glycerol and fatty acids, mainly unsaturated acids whose the major is oleic acid (about 65-80%). The olive oil contains also a relatively reduced level of polyunsaturated essential fatty acids (PUFA), linoleic and linolenic acids (C18:2 ω-6, C18:3 ω-3). Such composition gives good resistance to chemical and biological oxidation, in contrast with other edible oils in which polyunsaturated fatty acids prevail on monounsaturated ones (Rastrelli et al., 2002). Thus, the importance of virgin olive oil is related to its high levels of monounsaturated fatty acids but also to the presence of minor components including aliphatic and triterpenic alcohols, sterols, hydrocarbons, volatile compounds, and several antioxidants. In fact, the unsaponifiable fraction which covers a small percentage (0,5-3%) plays a significant role on human health.

Among the antioxidant compounds, squalene, a triterpenoid hydrocarbon and precursor of sterol biosynthesis, occurs in olive oil with a high concentration (up to 1%). Other active constituents in the olive oil are represented by tocopherols α-, β -, γ -, δ -: the concentration of α–tocopherol (Vitamin E), traditionally considered as the major antioxidant, covers almost 88% of the total tocopherols. The tocopherol content of olive oil depends not only on the presence of these compounds in olive fruit but also on several other factors, involved in the transportation, storage and olive fruit processing. Normally in animal fluids and tissues, vitamin E works in synergy with coenzyme Q (CoQ) to protect cells and tissues against lipoperoxidation, and some authors detected CoQ_9 and CoQ_{10} in olive oil (Pregnolato et al., 1994; Psomiadou & Tsimidou, 1998). According to Viola (1997), the ratio of vitamin E to polyunsaturated fatty acids in olive oils is better than in other edible oils.

Phenolic compounds make an important contribute to the nutritional properties, sensory characteristics and the shelf life of olive oil, because they improve the resistance to the autoxidation. Olive fruit contains simple and complex phenolic compounds: those derived from the hydrolysis of oleuropein contribute to the intensity of the bitterness of virgin olive oil, and especially hydroxytyrosol, tyrosol, caffeic, coumaric and p-hydroxybenzoic acids influence the sensory characteristics of olive oil (Kiritsakis, 1998). The concentration and composition of phenolic compounds in virgin olive oil is strongly affected by many agronomical and technological factors, such as olive cultivar (Tura et al., 2007), place of cultivation (Vinha et al., 2005), climate, degree of maturation (Kalua et al., 2005; Sicari et al., 2009; Giuffrè et al., 2010), crop season (Gómez-Alonso et al., 2002), irrigation (Tovar et al., 2001) and production process (Cinquanta et al., 1997; Ranalli et al., 2001).

Regarding other compounds in the olive oil, its colour is mainly related to the presence of chlorophyll and pheophytin a (Psomiadou & Tsimidou, 2001).

Carotenoids are also responsible for the colour of olive oil: the major are β-carotene and lutein in virgin olive oil (Gandul-Rojas & Mínguez-Mosquera, 1996). The presence of these constituents depends on several factors, such as cultivar, soil and climate, and fruit maturation as well as applied conditions during olive fruit processing (Gallardo-Guerrero et al., 2002).

3. Deterioration reaction in olive oil

The olive oil quality is strongly related to the physiological conditions of the fruit from which it is extracted. Important chemical changes occur inside the drupe during ripening. They are related to the synthesis of organic substances, especially triglycerides, and to other enzymatic activities that may affect virgin olive oil quality. From olive oil extraction to storage operations the most common variations, recognized generally with the term "rancidity", are divided into hydrolytic and oxidative rancidity.

The hydrolytic rancidity is a change due to the presence of water in the drupe and the catalytic action of an enzyme, the lipase, often derived by microorganisms. The reaction consists of a triglyceride hydrolysis to give glycerol and fatty acids, which result in an increase of free acidity. The release of mono-, diglycerides and finally fatty acids is obtained by the enzymatic hydrolysis.

Besides lipases, also peroxidises and lipoxygenases are numbered among the involved enzymes and they are responsible to the more or less selective formation of hydroperoxides, which will damage the native antioxidants and the polyphenoloxidases that reduce the polyphenol content, in particular on the olive paste. The enzyme activity becomes slower due to the inhibition effect of the polyphenol oxidation products. This variation regards all olive oil process, in particular the olive oil extraction systems.

The oxidative rancidity or autoxidation is due to the reaction between the oxygen and unsaturated fatty acids, free and esteryfied. It follows the characteristic trend of radicalic reactions with an induction, a propagation and a termination phase. The induction period is characterized by production of free radicals by unsaturated fatty acids or lipid peroxides (so called hydroperoxides), which constitute the primary autoxidation products. The direct attack of atmospheric oxygen on the unsaturated fatty acid chain is unfavoured from the thermodynamic point of view, because the activation energy of the reaction is high (145-270 kJ mol^{-1}). The direct attack could be attained by singlet oxygen which can be formed by a photochemical reaction. After the first formation, hydroperoxides are degraded to give a chain-reaction in the propagation phase: it is the principal oxidation step. Oil-quality changes related to the production of oxidized by-products that alter the sensory and nutritional characteristics of the oil include the production of carbonyl compounds, a decrease of the α-tocopherol concentration, and the generation of off-flavour compounds. These secondary products responsible of rancid odour and flavour are represented by saturated and unsaturated aldehydes, ketones, volatile alchools, hydrocarbons, cyclic oxygenated compounds, etc. (Frankel, 2005; Morales et al., 1997). In the end of the autoxidation, reactions lead mainly to the formation of polymers (dymers oxygenated and not oxygenated according as involved reagents).

When vegetable oils are exposed to light, photo-oxidation occurs through the action of natural photosensitizers (i.e. chlorophyll), which react with triplet oxygen to form the excited state singlet oxygen. Singlet oxygen then forms free radicals from unsaturated fatty acids leading to the production of hydroperoxides and eventually to carbonyl compounds resulting to the development of undesirable off flavours in oils (Skibsted, 2000).

The reaction responsible of oxidative rancidity is so promoted by light, heat, metals traces (Fe, Cu, Co, Ni, Mn). Substrates of these reactions are principally unsaturated free fatty acids which in general oxidize faster respect to triglycerides and phospholipids. The oxidation velocity is affected mainly by the unsaturation degree. The saturated fatty acids oxidize at a temperature higher than 60° C, whereas the polyunsaturated fatty acids oxidize at lower temperatures.

Other unsaturated substrates can be submitted to similar oxidation reactions, among which some hydrocarbons in the oils, in particular squalene, vitamin A, carotenoids and vitamin E (α-tocopherol). Oils and vegetable fats are more or less rich in α-tocopherol that has a natural antioxidant activity delaying the lipid oxidation. Tocopherols are known to act as antioxidants by donating a hydrogen atom to chain-propagating peroxyl radicals (Kamal-Eldin & Appelqvist, 1996). The oxidation of vitamins A and E and carotenoids can be also due to the peroxides action formed by unsaturated fatty acids in the secondary oxidation. This process leads to loss of vitamin activity and colour, while the essential fatty acids oxidation involves a decrease of nutritional value (Sciancalepore, 1998).

In autoxidation, virgin olive oil stability is correlated with the polar phenol content: a linear relationship between the phenolic content and oxidative stability of extra virgin olive oil has been in fact noted (Di Giovacchino et al., 1994). The main active phenols are the o-diphenol hydroxytyrosol and its oleosidic forms (Tsimidou, 1998, Aparicio et al., 1999) but relation between total phenol content and oil stability is not always validated. The olive oil stability to oxidation decreases in fact during the storage time, but a proportional trend to the phenols decrease is not observed. It is possible that an equilibrium attains (or tends to achieve) between polyphenols and their oxidation products, already present or formed successively to its protection versus fatty acids. This equilibrium opposes to the normal antioxidant activity of polyphenols still intact which, in these conditions, do not exert the same antioxidant activity (Lercker, 2004).

Apart from contributing to colour, carotenoids protect the oil from photo-oxidation by quenching singlet oxygen and acting as light filters (Fakourelis et al., 1987).

Chlorophyll pigments are also well-known to act as photosensitizers during light exposure. They are able to catalyze photoxidation with a velocity higher (up to 30000 times more). Chlorophylls exert an antioxidant activity dependent on the derivative present, the lipids substrate and storage temperature (Endo et al., 1985; Gutierrez-Rosales et al., 1992).

The squalene hydrocarbon has a slight antioxidant activity which is concentration dependent (Psomiadou & Tsimidou, 1999)

3.1. Deterioration agents in olive oil: Oxygen, light and temperature

Among physical agents, the oxygen plays a fundamental role in the oil alteration: in contact with air it loses many qualitative characteristics as colour, flavour, odour and vitamins.

As index of oxidative deterioration, the peroxide formation in olive oil stored in closed tins is in fact generally insufficient to lead to development of the typical rancid odour, because of the limited amount of oxygen in the headspace. The lipid oxidation substrates, as previously observed, are the unsaturated molecules, while the unsaturation degree influences the oxidation velocity. The high level of natural antioxidants, associated with an excellent fatty acid composition, confers the olive oil high stability against oxidation, apart from the recognized nutritional value.

The adoption of proper operations is also important in the preliminary phase of olive oil storage, as for example the laminar filtration by using of cartridges closed in steel cylinders that reduce the air exposition, or the filtration under nitrogen, in controlled atmosphere.

The oxygen plays an important role on the entire and complex interactions in the olive oil. Also antioxidant molecules are strongly influenced by oxygen presence. A study of changes in total polar phenols, α-tocopherols, β-carotene, lutein, chlorophyll and squalene contents during autoxidation indicated that most interactions may be suppressed if oxygen availability is limited (Psomiadou & Tsimidou, 2002).

Another principal harmful agent of olive oil is light, both direct and diffused and in general the UV area in the light spectrum at high energy, from 290 to 400 nm is the most charged. The electromagnetic radiations promote some chemical and biochemical reactions responsible of qualitative degradation as the lipid oxidation in the presence of air. In the absence of air, however, direct sunlight causes a decrease in peroxide and Kreis values of the oil. In particular, in a study on several olive oil packaging typologies, Kiritsakis (1984) revealed that the oil oxidation proceeded slowly in darkness, faster in diffused light and even faster in direct sunlight.

Moreover several natural substances have the property to absorb much the incident light and, for it, to originate alterations and/or undesirable variations. Pigments have this characteristics and colour degradation is in fact an unavoidable and well known event: many vitamins (as A and B2) are photosensitive.

The temperature controls many of reactions catalyzed by enzymes. For each increase or diminution of it a variation of biological food activity is shown. At low temperatures the kinetic energy of reagent molecules decreases, the drop of mobility and collisions necessary for the formation of enzyme-substrate complex and their products occurs. Some studies (Gambacorta et al., 2004; Kanavouras & Coutelieris, 2006; Pristouri et al., 2010) demonstrated that high storage temperatures modify the olive oil quality, but less than the light exposition.

4. Storage of olive oil

According to olive variety, extra virgin olive oil has different sensorial attributes, e.g. fragrance, flavour, colour and nutrient composition parameters. Thus, its quality can be considered from diverse points of view: normative, commercial, nutritional, therapeutic and sensorial. An excellent olive oil is produced with good olive growing, harvesting, storage and processing practices. This condition considers the preservability an olive oil qualitative characteristic.

Some quality parameters of extra virgin olive oil (acidity, peroxide value and UV absorbance) can vary depending on time and storage method, reaching undesirable values at the end of the shelf life. Acidity increases with time, both in oils stored in the dark and exposed to light (Gómez-Alonso et al., 2007). Nowadays, consumers are imposing an increasing demand for a higher quality of extra virgin olive oil during the shelf life period. This is due to the expectation for a safe food and also for a reduction to a minimum of undesired changes in sensorial quality (Hrncirik & Fritsche, 2005).

The olive oil storage could be shorter as possible: its assumption could not exceed the production year: the oil gains the excellence about 3-8 months after production, after that its quality starts decreasing. This behaviour depends on the olive cultivar and on the alteration of fatty fraction.

During storage the oil tends to loss the typical pigmentation and the aromatic characteristics of bitter and spicy, showing too many transparency and brilliance. Immediately after oil extraction, the flavour appears too strong and also characterized by an unpleasant aftertaste. If the olive oil is stored properly, as in well-sealed packages, it can reach the second storage year maintaining its sensory properties.

A good quality olive oil with a low acidity degree preserves its opacity for a long time. This is a characteristic of some oils, as those extracted from first harvested olives not yet completely ripe, which are very fruity and with low free acidity. The consumption of these oils, also named "new oils" is recommended up to a few weeks after bottling. For many consumers the cloudy olive oil represents a certainty of authenticity and it is also preferred in particular when the oil is purchased directly at the oil mill. It is characterized by a certain water amount (about 2-4 g/kg) and residual solids as pectins, hemicelluloses, cellulose and mucilages in form of emulsion containing active or latent enzymatic activities as lipoxygenase, polyphenoloxidase and, principally, some esterases able to hydrolyze oleuropein, di-methyl-oleuropein and ligstroside aglycon derivatives responsible of bitter and sour flavour in virgin olive oils. The association between the natural and the healthy/nutritional aspect does not correspond to a scientific assessment because the presence of cloud induces the phenols degradation, which results in sensory changes.

Moreover fermentative processes develop on olive polysaccharides with transfer of compounds which confer sludge and warmed defects to the extracted oil. Thus, the filtration operation appears necessary, indeed is advisable before bottling and also before storage with the scope to maintain a major shelf life, avoiding settling and frequent decanting operations, which in turn induce a higher oxidation. The industrial filtration can be conducted by several systems, which differ for used material: hydrophilic cotton or cellulose at high porosity are better than fossil flour which reduces the antioxidant property because acts directly on o-diphenols, decreasing the amount in the olive oil and so its shelf life (Ricci, 2007).

The oil storage is favoured by the antioxidants content that prevents the rancidity, but their activity initially slows and then stops with successive formation of free radicals. This action can be enhanced by improper storage practices: for example, when storage temperature is not controlled, or when the oil is held in contact with the direct light and or at high temperatures at consumer's house, or when an improper sealing is applied after the first container opening. Concerning this last aspect, the use of screw caps is more advisable than metal pourers which expose the oil to the oxidative agents.

As the extra virgin olive oil extraction is conducted with controlled temperatures lower than 28 °C, as the extra virgin olive oil storage requests the same attention. It has to be done with control of temperature, which may range from 10 °C to 18 °C: the correct storage temperature is 14-15 °C. The high temperatures increase the chemical variation velocity and the major oil fluidity. This last effect promotes the oxygen permeation. When storage temperature decreases at 8-9 °C, white deposits can appear in the oil, due to triglyceride crystallization. The higher is the content of saturated glycerides, the higher is the freeze process. The step from the oil crystallization to the solidification by lower temperatures (3-4 °C) slightly increases the oil stability to the oxidation and gives substantial modifications in the sensory profile. If the clear colour maintains beside the transfer of solidified oils to warmer places, it possible that a margarinization process involved or mucilage could be present.

Finally, it is important that storing and packaging facilities respect hygiene and healthiness standards. Aromatic polycyclic hydrocarbons and benzene present in the atmosphere are absorbed by the oils contained in not hermetically sealed packages (Ricci, 2007).

5. Packaging of olive oil

After olive pressing, oils are transferred in tanks and after pouring they are put into storage containers where they attain their sensory characteristics. Containers are of different materials and capacity ; they are made mainly with AISI 304 inox steel which is also used for the containers intended for olive oil clarification.

The packaging in the selling container is the last unit operation: it is very important for product stability, maintaining high quality level and also value added, if carried out properly.

Reduction of oxygen in the packaging headspace and light exposure are key factors in lowering lipid oxidation and off-flavour development, thus keeping quality of olive oil.

5.1. Olive oil packaging method

The Regulation 1019/2002/CE (Official Journal European Union, 2002) sanctioned the prohibition of the consumer sale of the on tap olive oils. This effect promotes the consumer protection in product quality and the valorisation of farm identity. In this sense olive farms can provide packaging or ask for it to external companies. As for the other foods, the oil storage and packaging mill area must be maintained clean, ventilated, illuminated and all the unit operations should follow the hygienic-health rules according to Regulation 852/2004/CE (Official Journal European Union, 2004). As previously discussed, olive oil must be stored at constant temperature, about 14-15 °C, protected from light and air. When oils are stored in big containers (inox steel is the best material), these have to be preferably maintained quite full or better with the headspace filled with nitrogen..

For smaller quantities, the bottling with depression and the pre-bottling with inert gas are nowadays diffused by using bottlers apparatus , which saturate the headspace with nitrogen in bottle filling operation. In particular, bottles are hooked up by a special machine, which invert them and blow inside liquid nitrogen, which becomes gas when comes in contact with the environment, dilutes and moves out the oxygen present in the free volume cf the bottle.

Then, bottle straighten up and proceeds toward the filling operation during which nitrogen stays inside because is heavier than air. As the bottle fills up and the oil level goes up, the nitrogen is carried outside avoiding anyway the oxygen penetration. The velocity of successive capping guarantees that only presence of nitrogen between cap and oil (Soressi, 2009).

Using of inert gases, as argon and nitrogen, can solve many problems and provide an optimal product storage in several production steps, as pressing, kneading and, most of all, storage and bottling. A study demonstrated in fact, that extra virgin olive oils collected from the previous harvesting season and stored under nitrogen atmosphere could be packaged in glass bottles without appreciable quality changes, as compared with seasonal extra virgin olive oil packaged in similar bottles (Guil-Guerrero & Urda-Romacho, 2009). Nitrogen is the most used gas to protect oil from the air contact, able to remove oxygen by container headspace and pre-formed volatile contaminants. Bottling lines are very versatile for packaging typologies and easily cleanable. A past research on Italian plant typologies revealed that the prevalent packaging line is the semi-automatic (50%), followed by the automatic (27%) and manual (23%) ones. The first and the third show a low specialization process level and a limited bottling capacity.

Bottle capping can be carried out by means of cork or more frequently by metal caps. A good quality cork has not to be porous and present fungal contaminants, however it is not suggested as may cause early deterioration, due to its high oxygen permeability. Metal caps are provided of a screw and a plastic soft liner that permit a perfect airtight acting as dripper. Also a thermally retractable capsule can be present on cap to seal the packaging and improve the product from the esthetical point of view. Just these last are the most used ones (72%) respect to ones with a simple screw (15%) and cork (11%) (Ricci, 2007).

5.2. Olive oil packaging materials

In Italy and in the world the majority of consumers purchases bottled olive oil. Nowadays attention of many actors is focused on olive oil packaging, in particular on container design that often communicates the natural concept and territory link associated to the contained olive oil. Due to the changed life-style, oil container manifests also a functional property.

The nature of the packaging material has a notable influence on oil quality (Gutierrez et al., 1988). In general, a packaging is chosen depending on several criteria as product stability, environmental conditions to which food will be exposed during storage and distribution, and the product nature. In particular, the basic factors that may alter the quality of packed oils are:

- Dissolved oxygen in the oil, that is the oxygen that remains in the container free space after it is sealed and the oxygen diffused through the walls
- Light, which passes through containers, activates the oxidation process
- Trace metals, in particular copper and iron
- Autocatalytic oxidation
- Temperature during storage
- Humidity during storage
- Dissolution of some substances from the container to the oil (Richardson, 1976).

The principal used materials for olive oil packaging are represented by glass, steel, tinplate and, in Italy to a lesser degree, plastic and polycoupled ones.

5.2.1. Glass

Glass represents an ancient material used in container structure: first glass containers derived by rocks date back to 3000 BC. It is formed principally by silica obtained by sand, flint or quartz. Silica is fused at very high temperatures (about 1720 °C) to form silicated glass. In most cases, silica is mixed with variable proportions to several raw materials: e.g. sodium and potassium carbonates (which act as stabilizer and protect the glass from the water solubilisation), lead (which confers transparency and lightness) and aluminium (which increases its hardness and lastingness). Glass is neither in a solid nor in a liquid state but glassy. Its molecules are arranged in disarray but with a sufficient cohesion to give mechanical stiffness. The structural basic unit is represented by the silica-oxygen tetrahedron, in which a silica atom is surrounded by four oxygen atoms to form a tetrahedron. Big groups tend to dispose with disorder (amorphous structure) giving the glass fragile with the tendency to rupture if submitted to an excessive tension. This induces effects on its thermal resistance that is function of container type, packaging faces and exposed surface to tension (internal if warmed and external if cooled) (Robertson, 2009).

In Italy glass is the king container for the olive oil packaging. Glass bottles can have different forms and capacities and their colour can vary largely from white to green and more dark tints. In Figure 1 several size of glass bottles are presented to demonstrate the large offer in the Italian olive oil market also of aromatized oils with spices or hot pepper.

From an Italian research about the most diffused containers, 64%of olive farms use the green glass whereas only the 10% of the total chooses the transparent one (Ricci, 2007). The best containers for the olive oil are made in opaque and dark glass: it is advisable to prefer very dark ones because little light can pass through. Since the consumer appreciates also the light glass because it is transparent and so shows the oil colour, then it is advisable to supply the bottle with a paper case which can protect the product by light. Guil-Guerriero and Urdo Romacho (2009) reported a shelf life study carried out on oils produced from *Picual*, *Hojiblanca* and *Arbequina* cultivars packaged in dark and transparent glass bottles. These oils showed an increase in some parameters; the variation of peroxide value was significant in extra virgin olive oils stored in transparent glass. Similar results were observed by Del Caro et al. (2006) and Vacca et al. (2006) on *Bosana* extra virgin olive oils. Several studies conducted on olive oil shelf life attested the glass as the best material for the storage (Pristouri et al., 2010), especially when oil was stored in the dark (Kanavouras et al., 2004), and for its acidity and peroxide value, with respect to other packages (Rababah et al., 2011). Finally, in a comparative study among glass, High Density Polyethylene (HDPE) and PET, results clearly indicated the glass was the best in the following ranking Glass > HDPE > PET (Ben Tekaya et al., 2007).

Figure 1. Several types of commercial glass bottles for olive oil.

5.2.2. Steel and tinplate

First uses of this material date back to 1700 in domestic environment. Steel is an iron/carbon alloy: the higher the carbon percentage, the higher the rupture limit.

The tinplate is a sheet of soft steel, which is more workable because of its lower carbon content, coated on both faces by tin oxides layers and on face destined to the contact with food also by organic synthetic lacquers. Tinplate container is mainly of rectangular shape; its side seam is welded and protected with a food-approved special varnish: the welding method assures a secure side seam, avoiding the dissolution of lead into the product (Tsimis & Karakasides, 2002).

The inox steel is an alloy containing 11% of chrome which reacts with oxygen originating an autopassivation condition due to the chrome oxides presence (Piergiovanni & Limbo, 2010). It is the best material for the preservation of big volumes of olive oil. Compared to glass, it has the same water resistant properties (so protects the product from oxygen, humidity and microrganisms), and major advantages, e.g. handless, cleanness, shock resistance and total light protection.

A research of Grover (1982) demonstrated that the quality of oils decreases if packed in re-used tinplate containers, whereas new containers did not alter oils during the one year storage period. A more recent study confirmed that the storage of oils in stainless and dark glass appears more adequate, with respect to other packaging materials, as clear PET and clear glass (Dabbou et al., 2011).

Tin plate is used for the olive oil retail both in bigger volume size (3-5 L) and in smaller bottles (1 L). A recent study of Rababah et al. (2011) on olive oil samples stored for 60 months at light in bottles of 2.5 L denoted higher sedimentation amount in tinplate than in the other containers and a strong reduction of sensory attributes, probably due to the reaction of other components in olive oil, e.g. phenol compounds, with tinplate that negatively affected the descriptive attributes.

5.2.3. Plastic materials

Plastic materials are partially or totally synthetic organic substances. Their principal components are represented by polymers constituted by carbon, hydrogen and, in some cases, oxygen, nitrogen, chloride, silica and sulphur. They have a relatively recent history but in the time gained a large number of applications for their versatility, low production cost, low weight, their good performance and their 'thermoplastic' nature which guarantees the recyclability and consequent low environmental impact.

Plastic containers are formed by blow-moulding (extrusion blow-moulding or injection blow-moulding). One of their disadvantages is the wall permeability to gases and vapours. If transparent, they also transmit light. Migration of small molecular weight substances (e.g. monomers, oligomers and additives) from plastic to food can also occur, thus affecting the quality of food (Tsimis & Karakasides, 2002). Oil contained in bottles with high air permeability (PE, PP, etc.) should be sold within 4 weeks, in contrast to polyvinyl chloride (PVC) bottles, which can hold olive oil for 3 months without appreciable quality loss.

PVC is a popular packaging material for edible oils in many countries, mainly due to its transparency, adaptability to all types of closure, total compatibility with existing packaging lines, and suitable for personalized design features (Kanavouras et al., 2004). PVC was used in food packaging without any doubts in 1970s. Nowadays for the dissolution of Vinyl Chloride Monomers (VCM) in oil during the storage time and for issues such as the environmental protection, the ample supply, plastic shaping, and its mechanical properties, PET has been supplanting PVC in the edible-oil market.

PET bottles are produced by bioriented extrusion and their advantages are: low water and oxygen permeability, high hardness and stiffness. PET is also more resistant to oil and fats, than the other plastic materials and for these properties it fits well for olive oil packaging. The disadvantages of this material are the processing condition (to control the humidity content), costs and the migration of acetaldehyde from bottle to food (Dipalma, 1986). As previously affirmed, plastic materials are, however, porous and thus permit the penetration of humidity and gases. Pristouri et al. (2010) demonstrated that between PET and PP, PET provided a better protection to olive oil than the other one due to its significantly lower oxygen transmission rate. Moreover Kiritsakis and Dugan (1984) reported that peroxide values were higher for olive oil packaged in plastic containers as compared to those packaged in glass bottles in the dark. Kanavouras et al. (2006) revealed instead that plastic containers had a particularly stronger protective role when oil was stored in the light, with respect to transparent glass. Besides in Italy the glass bottle is the most used, in other countries the olive oil packaging in plastic containers is developing. Contemporary trends in olive oil packaging include dark coloured glass bottles and PET bottles which have incorporated oxygen scavengers. These are used in order to prolong the shelf life of foods whose degradation kinetic depends on the partial pressure of oxygen inside the container. In particular, Del Nobile et al. (2003) confirmed that oxidation kinetics slower than that found with glass bottles can be obtained by bottling the olive oil into materials containing an oxygen scavenger. However, the slowest decay kinetics were obtained by bottling the oil in PET containers and by reducing the oxygen concentration prior to bottling to 10% of the equilibrium value. Also Gambacorta et al. (2004) studied the effect of several percentages of oxygen scavengers and of barrier resin included in PET on olive oil quality, and indicated that the materials having higher oxygen barrier properties can slow down the quality decay kinetics of extra virgin olive oil.

5.2.4. Policoupled material

Policoupled materials are mainly represented by a triple coupled material formed by Low Density Polyethylene (LDPE)/paper/LDPE   and a thin aluminum foil between further polyethylene layers as barrier respectively to gas, light and liquids. It is a parallelepiped with rectangular section, the most known version of which is the Tetra Brik®, also named Tetra Prisma.

Tetra Prisma allows an ideal storage for distributors and consumers. For the packaging of vegetable oils including olive oil is being used today in some Mediterranean countries, e.g. Spain and Greece, in the form of brick-type cartons and bag-in-box pouches. It is a very revolutionary packaging because guarantees a total and prolonged protection of olive oil chemical and physical characteristics up to two years. This material permits also to update the packaging aspect, with a more coloured, snappy and modern graphic design. Besides preserving the quality, Tetra Prisma allows the producer a major differentiation in packaging, logistic efficiency, lower probabilities of rupture during transfer and, not least, a 100% recyclable package. From the consumers point of view, the pack is indestructible, easy to take and handle. The realized cap contributes to protect from oxygen and it is for a prolonged use, with the aim to avoid the drip, the product spill after opening and to facilitate the pouring, regulating the flow and cutting the drip. In Italy, the olive oil packaged in Tetra Prisma was proposed in 2005 with opposite considerations (Soressi, 2007). Studies have been conducted also on this material type to evaluate its performance as container for oil. It is efficient in maintaining antioxidants in stored product for several months, due to its efficient barrier to light and oxygen, unlike the transparent glass and plastic materials, despite these last show a low gas permeability (Mendez & Falque, 2007).

5.3. Olive oil packaging shape and size

Glass is the preferred packaging material in Italy for olive oil, and the most used bottle shape is the *marasca* type. It is chosen for three aspects: small price, the best shelf management and the functionality. The most diffused bottle volume is 0.75 L, followed by 0.5 L; the use of other volumes is very marginal. A research developed in 2005 on 48 producers of biological oil in the South and the Centre of Italy indicated that the bottle largely used is that of 0,75 L (about 85% of the total), followed by 0,50 L (about 58%) and 0,1 L (about 17%) ones (Paffarini, 2007). Low capacity packaging is gaining market approval above all for exported products and for those directed to a medium-high consumer segment. Packaging in 0,1 L bottles is also used for very valuable oil to evidence the product value. Producers propose on market sophisticated packaging created properly to contain one or more bottles, as those for tasting of 6 mL volume. For the other materials (steel, tinplate and plastic) the most used volumes have been previously cited.

It is noteworthy, the recent distribution of 10 mL single-dose sachets, constituted by 100% extra virgin olive oil or olive oil mixed with vinegar, that is illustrated in Figure 2 with tinplate and PET containers. This type of packaging developed after the coming into effect of Italian Law 81/2006 (Official Bulletin, 2006), which bans the use of of olive oil containers without label in restaurants .

Moreover containers that can be used without common rules of hygiene and safety are also prohibited. So, it is admitted to present bottles provided of label (although their filling up is also allowed) and single-dose sachets formed by appropriate machines, which guarantee more safety and traceability of their material and because they are unable to be reused in fraudulent practice. The use of cruets is so banned because the origin of contained oil is impossible to be easily traced.

With the increase of container typologies and the research of more refined packaging in materials and graphic the supplementary cost merges into the final price that is, however, a product quality indicator (Paffarini, 2007).

Figure 2. PET, tinplate and single-dose sachets as olive oil packaging

5.4. Olive oil labelling

The olive oil labelling has two principal functions: safety for consumer and aesthetic quality. It has to follow legal requirements and so present several technical informations, some of these facultative, other obligatory. For the graphic point of view, it results as presentation of the olive oil in market shelves.

When consumer purchases olive oil, he is also more interested on symbolic contents reported in label, as the original production territory and extraction process, with respect to the sensory and chemical-physical quality of product. The oil packaging in fact satisfies curiosity about origin and characteristics by means of label, back label or attached paper, especially buying particular containers in the organized distribution. Thus, packaging elements have gained a relevant role to communicate not only product characteristics but also its history, playing on emotions that are much linked to biological oils, for example, of which label can evocate proper quality of biological agriculture.

The new Regulation 182/2009/CE (Official Journal European Union, 2009) established that labelling must obligatory indicate the extra-virgin and virgin olive oil origin. This rule does not regard a protected geographical indication (PGI) or a protected designation of origin (PDO) qualified olive oils because they are submitted to another specific normative, that is the Regulation 510/2006/CE (Official Journal European Union, 2006).

Three typologies were defined by Regulation 182/2009/CE to indicate olive oil origin:

1. Oil obtained in the same member State where olives are harvested: in this case origin can be recalled indicating State name followed by phrases as "Produced in", "Obtained in ....", and also "100% produced in ...". The member State name can be substituted by a Communitarian reference. Then using of "produced in Italy" is allowed only if olives are produced and milled in Italy for the 100%
2. Oil obtained in a member State by olives derived from other member States: in this case the origin indication can be referred as " (Extra) virgin olive oil obtained in .....by olives harvested in ...". The member State name can be substituted by a Communitarian reference If it is necessary indication of more States, they are cited in decreasing order in relation to the brought quantity
3. Mix of Communitarian and/or not Communitarian oils: in this case origin indication mode is one of the following:
a. "Mix of Communitarian olive oils", or a reference to the Community
b. "Mix of not Communitarian olive oils", or a reference to the not Communitarian origin
c. "Mix of Communitarian and not Communitarian olive oils", or a reference to Communitarian and not Communitarian origin.

Regarding the indication about sensory characteristics, the Regulation 182/2009/CE sanctions that they must be reported only in labelling of extra-virgin and virgin olive oils, besides confirm of the voluntariness. Moreover, adjectives related to a positive attribute for the oil (e.g. "intense", "medium", "light", linked to fruity, green, ripe, etc.) are admitted exclusively if derived by an objective evaluation based on the International Olive Oil

Council method described in XII attachment of Reg. 2568/91/CE (Official Journal European Union, 1991) and revisions. The Regulation 182/2009/CE is a sort of Italian victory that can let acknowledge in a Communitarian regulation the possibility to indicate the production area of olive oils.

The question of mixing of olive oil with other vegetable oils is also well defined in the new Regulation that establishes that member States can ban in the territory the production of mix for internal consumption. Instead, they cannot prohibit on the territory the sale of mixes produced by other States and ban the mix production for the commercialization in another member State or the exportation. This regulation stopped an age-old argument in which Italian farms were penalized in particular in not Communitarian markets where did not compete at par with other member States for offer of oil mixtures. This product type is in fact requested particularly in USA and Asiatic markets.

In olive oil packaging labelling is distinguished in:

- Principal, normally rectangular, big, obligatory, that refers all terms requested by regulations
- Back label, not ever present, smaller and collocated generally on the back of the bottle.

It is obligatory to indicate on the label:

- Commercial and sale denomination of oil
- Oil category according to Reg. 1019/02/CE art. 3 (Official Journal European Union, 2002). For extra virgin olive oil: "Olive oil of superior category obtained directly by olives and only by mechanical extraction"
- Production or packaging plant location
- Packaging batch
- Product volume (e.g. 0,75 L followed by symbol)
- Preferred consume date (as consume preferably up to the end….month and year)
- Storage conditions (keep in a dry environment away from light and heat)
- Recommendation "after using not disperse into environment"
- Other facultative indications as:
  - "First cold-pressing" and "cold-pressed" are reserved to virgin oils obtained by extractions not above 27° C. To guarantee the truth of their declaration, olive farm owners could indicate the installed techniques for each stock of processed olives. The term "extracted" could be substituted by "obtained" or "produced"
  - Acidity could be reported only if accompanied by peroxides value, waxes content and spectrophotometrical values (K232, K270, Delta K). Values can be followed by the phrase "maximum values at packaging"
  - Polyphenol content only for olive oil that contains almost 5 mg of hydroxytirosol and its derivatives per 20 g of olive oil. This indication must be accompanied by information to the consumer that the beneficial effect is obtained after a daily consumption of 20 g of olive oil. This last parameter has been fixed by a recent Regulation (Official Journal European Union, 2012) on allowed healthy information on food products.

Anyone who reports acidity and other parameters in labelling must organize the necessary evidence to prove the conformity of product to indicated limits.

6. Conclusions

Olive oil storage and packaging are final steps of the production process and are as important as the other ones. Deterioration agents can decrease the quality of this important food also during these unit operations, so a correct control and monitoring of some indicators can be useful to the olive oil shelf life. Storage environment and its characteristics (temperature most of all) contribute to it.

Packaging typologies also influence the stored olive oil properties with different results depending on materials. During the time various containers were adopted and evaluated with evident feedback for qualitative aspects (some material have been in fact banned for healthy causes). Moreover, the traditional uses can give to different preferences (e.g. in Italy is very difficult to substitute the glass apart from its ascertained value as packaging material for olive oil).

Future developments are surely expected because of many attentions on this subject also by scientific community and probably some materials will be improved and other container types will be proposed in a more and more advanced food market.

Author details

Amalia Piscopo and Marco Poiana
Dipartimento di Biotecnologie per il Monitoraggio,
Agroalimentare e Ambientale (Bio.M.A.A.), Mediterranean University of Reggio Calabria, Italy

7. References

Aparicio, R., Roda, L., Albi, M.A. & Gutierrez, F. (1999). Effect of various compounds on virgin olive oil stability measured by Rancimat, *Journal of Agricultural and Food Chemistry* Vol. 47: 4150-4155.

Ben Tekaya, I., Ben Tekaya Ben Amor, I.,.Belgaied, S., El Atrache, A. & Hassouna M. (2007) Étude du conditionnement de l'huile d'olive dans les emballages en plastique. *Science des Aliments* Vol. 27 (No.3): 214-233.

Cinquanta, L., Esti, M., & La Notte, E. (1997) Evolution of phenolic compounds in virgin olive oil during storage, *Journal of American Oil Chemists's Society* Vol. 74: 1259.

Dabbou, S., Gharbi, I., Dabbou, S., Brahmi, F., Nakbi, A. & Hammami, M. (2011). Impact of packaging material and storage time on olive oil quality, *African Journal of Biotechnology* Vol. 10 (No. 74): 16937-16947.

Del Caro, A., Vacca, V., Poiana, M., Fenu, P., & Piga, A. (2006). Influence of technology, storage and exposure on components of extra virgin olive oil (Bosana cv) from whole and de-stoned fruits, *Food Chemistry* Vol. 98 (No. 2) 311-316.

Del Nobile, M. A., Bove, S., La Notte, E., & Sacchi, R. (2003). Influence of packaging geometry and material properties on the oxidation kinetics of bottled virgin olive oil, *Journal of Food Engineering* Vol. 57: 189–197.

Di Giovacchino, L., Solinas, M., & Miccoli, M. (1994) Effect of extraction systems on the quality of virgin olive oil. *Journal of American Oil Chemists's Society* Vol. 71: 1189-1194.

Dipalma, G. (1986). Tra PET e PVC confronto all'americana, *Poliplasti* Vol. 383: 38-44.

Endo, Y., Usuki, R. & Kaneda, T. (1985). Antioxidant effects of chlorophylls and pheophytin on the autoxidation of oils in the dark. II. The mechanism of antioxidative action of chlorophylls, *Journal of American Oil Chemists's Society* Vol. 62: 866-871.

Fakourelis, N., Lee, E.C. & Min, D. B. (1987). Effects of Chlorophyll and β-Carotene on the Oxidation Stability of Olive Oil, *Journal of Food Science* Vol. 52: 234-235

Frankel, E.N. (2005). *Lipid oxidation* (2nd ed.), Bridgewater: Barnes P.J. & Associates, The Oily Press.

Gallardo-Guerrero, L., Roca, M., & Mínguez-Mosquera, M.I. (2002), Distribution of Chlorophylls and Carotenoids in Ripening Olives and Between Oil and Alperujo When Processed Using a Two-Phase Extraction System. *Journal of American Oil Chemists's Society* Vol. 79: 105–109.

Gambacorta, G., Del Nobile, M.A., Tamagnone, P., Leopardi, M., Faccia, M. & La Notte, E. (2004). Shelf life of extra virgin olive oil stored in packages with different oxygen barrier properties, *Italian Journal of Food Science* vol. 16, 417-428.

Gandul-Rojas, B. & Mínguez-Mosquera, M. I. (1996). Chlorophyll and carotenoid composition in virgin olive oils from various Spanish olive varieties, *Journal of the Science of Food and Agriculture* Vol. 72: 31-39.

Giuffrè, A.M., Piscopo, A., Sicari, V. & Poiana, M. (2010). The effects of harvesting on phenolic compounds and fatty acids content in virgin olive oil (cv Roggianella), *La Rivista Italiana delle Sostanze Grasse*, Vol. 87 (No. 1): 14-23.

Gómez-Alonso, S., Salvador, M. D., & Fregapane, G. (2002). Phenolic compounds profile of Cornicabra virgin olive oil, *Journal of Agricultural and Food Chemistry* Vol. 50: 6812–6817.

Gómez-Alonso, S., Mancebo-Campos, V., Salvador, M.D. & Fregapane, G. (2007). Evolution of major and minor components and oxidation indexes of virgin olive oil during 21 months storage at room temperature, *Food Chemistry* Vol. 100: 36-42.

Grover, M.R. (1982). Studies on shelf-life of vegetable oils packed in tin containers, *Journal of Food Science and Technology* Vol. 19: 268-270.

Guil-Guerrero, J.L. & Urda- Romacho, J. (2009). Quality of extra virgin olive oil affected by several packaging variables, *Grasas y Aceites* Vol. 60 (No. 2): 125-133.

Gutierrez, F.R., Herrera, C.G., & Gutierrez, G.Q. (1988). Estudio de la Cinética de Evolutión de los Indices de Calidad del Aceite de Oliva Virgen Durante su Conservatión en Envases Comerciales, *Grasas y Aceites* Vol. 39: 245–253.

Gutierrez-Rosales, F., Garrido Fernández, J., Gallardo-Guerrero, L., Gandul-Rojas, B. & Mínguez-Mosquera, M.I. (1992). Action of chlorophylls on the stability of virgin olive oil, *Journal of American Oil Chemists's Society* Vol. 69: 866-871.

Hrncirik, K. & Fritsche, S. (2005). Relation between the Endogenous Antioxidant System and the quality of Extra Virgin Olive Oil under accelerated storage conditions, *Journal of Agricultural and Food Chemistry* Vol. 53: 2103-2110.

IOOC – International Olive Oil Council (2007). Sensory analysis of olive oil – Method – Organoleptic assessment of virgin olive oil. COI/T.20/Doc. No. 15/2nd Review. Madrid, September.

Kalua, C. M., Allen, M. S., Bedgood, D. R., Bishop, A. G., & Prenzler, P. D. (2005). Discrimination of olive oils and fruits into cultivars and maturity stages on phenolic and volatile compounds, *Journal of Agricultural and Food Chemistry* Vol. 53: 8054–8062.

Kamal-Eldin, A. & Appelqvist, L. A. (1996). The chemistry and antioxidant properties of tocopherols and tocotrienols, *Lipids* Vol. 31: 671-701

Kanavouras, A., Hernandez-Münoz, P., Coutelieris, F. & Selke, S. (2004). Oxidation-Derived Flavor Compounds as Quality Indicators for Packaged Olive Oil, *Journal of American Oil Chemists's Society* Vol. 81, 251–257.

Kanavouras, A., & Coutelieris, F. (2006). Shelf-life predictions for packaged olive oil based on simulations, Food Chemistry Vol. 96 (No. 1), 48-55.

Kiritsakis, A. (1998). *Olive oil– Second Edition, From the tree to the table*. Food and Nutrition Press, Inc. Trumbull, Connecticut, 006611, USA.

Kiritsakis, A. K., & Dugan, L. R. (1984). Effect of selected storage conditions and packaging materials on olive oil quality, *Journal of the American Oil Chemists' Society* Vol. 61: 1868–1870.

Lairon, D. (2007). Intervention studies on Mediterranean diet and cardiovascular risk, *Molecular Nutrition & Food Research* Vol. 51: 1209-1214.

Lercker, G. (2004). Aspetti tecnologici e caratteristiche degli oli di oliva, *Atti della "Giornata in commemorazione del Prof. Lotti"*, Pisa, 8 novembre 2002, 75-99.

Mendez, A. & Falque, E. (2007). Effect of storage time and container type on the quality of extra-virgin olive oil, *Food Control* Vol.18 (No.5): 521-529.

Morales, M.T., Rios, J. J., & Aparicio, R. (1997). Changes in volatile composition of virgin olive oil during oxidation: flavors and off-flavors, *Journal of Agricultural and Food Chemistry* Vol. 45, 2666-2673.

Official Bulletin (2006). Legge 11 marzo 2006, n. 81 "Conversione in legge, con modificazioni, del decreto-legge 10 gennaio 2006, n. 2, recante interventi urgenti per i settori dell'agricoltura, dell'agroindustria, della pesca, nonche' in materia di fiscalità d'impresa".

Official Journal European Union (1991). Commission regulation (EEC) No 2568/91 of 11 July 1991 on the characteristics of olive oil and olive-residue oil and on the relevant methods of analysis.

Official Journal European Union (2002). Commission regulation (EC) No 1019/2002 of 13 June 2002 on marketing standards for olive oil.

Official Journal European Union (2004). Regulation (EC) No 852/2004 of 29 April 2004 of the European parliament and of the council of 29 April 2004 on the hygiene of foodstuffs.

Official Journal European Union (2006). Council regulation (EC) No 510/2006 of 20 March 2006 on the protection of geographical indications and designations of origin for agricultural products and foodstuffs.

Official Journal European Union (2009). Commission regulation (EC) No 182/2009 of 6 March 2009 amending Regulation (EC) No 1019/2002 on marketing standards for olive oil.

Official Journal European Union (2012). Commission regulation (EU) No 432/2012 of 16 May 2012 establishing a list of permitted health claims made on foods, other than those referring to the reduction of disease risk and to children's development and health.

Paffarini C. (2007) Così il packaging aiuta a catturare il consumatore, *Olivo e Olio* Vol. 3: 20-22.

Perez-Jimenez, F., Ruano, J., Perez-Martinez, P., Lopez-Segura, F. & Lopez-Miranda, J. (2007). The influence of olive oil on human health: not a question of fat alone, *Molecular Nutrition & Food Research* Vol. 51: 1199-1208.

Piergiovanni, L. & Limbo, S. (2010). *Food Packaging. Materiali, tecnologie e qualità degli alimenti*, Springer.

Pregnolato, P., Maranesi, M., Mordenti, T., Turchetto, E., Barzanti, V. & Grossi, G. (1994). Coenzimes Q10 and Q9 in some edile oils. *La Rivista Italiana delle Sostanze Grasse*, Vol. 71: 503-505.

Pristouri, G., Badeka, A., & Kontominas, M. G. (2010). Effect of packaging material headspace, oxygen and light transmission, temperature and storage time on quality characteristics of extra virgin olive oil, *Food Control* Vol. 21: 412–418

Psomiadou, E. & Tsimidou, M. (1998). Simultaneous HPLC determination of tocopherols, carotenoids, and chlorophylls for monitorino their effect on virgin olive oil oxidation. *Journal of Agricultural and Food Chemistry* Vol. 46: 5132-5138

Psomiadou E. & Tsimidou M (1999). On the role of squalene in olive oil stability. *Journal of Agricultural and Food Chemistry* Vol. 47: 4025-4032.

Psomiadou, E. & Tsimidou, M. (2001) Pigments in greek virgin olive oil: occurrence and levels *Journal of the Science of Food and Agriculture* Vol. 41: 640-647.

Psomiadou E. & Tsimidou M (2002). Stability of virgin olive oil.1. Autoxidation studies, *Journal of Agricultural and Food Chemistry* Vol. 50: 716-721.

Rababah, T.M., Feng, H., Yang, W., Eriefej1, K. & Al-Omoush, M. (2011). Effects of type of packaging material on physicochemical and sensory properties of olive oil, *International Journal of Agricultural and Biological Engineering* Vol. 4 (No.4): 66-72.

Ranalli, A., Contento, S., Schiavone, C. & Simone, N. (2001). Malaxing temperature affects volatile and phenol composition as well as other analytical features of virgin olive oil, *European Journal of Lipid Science and Technology* Vol. 103: 228–238.

Rastrelli, L., Passi, S., Ippolito, F., Vacca, G. & De Simone, F. (2002). Rate of degradation of a-Tocopherol, squalene, phenolics and polyunsaturated fatty acids in olive oil during different storage conditions, *Journal of Agricultural and Food Chemistry* Vol. 50: 5566-5570.

Ricci, A. (2007). Serbevolezza, i segreti di un olio che dura di più, *Olivo e Olio* Vol. 10: 48-52, 56-7.

Richardson, K.C. (1976). Shelf life of packaging foods. *SIRO Fd Res. Q.* Vol. 36 (No. 1): 1-7.

Robertson, G.L. (2009). *Food packaging. Principles and Practice*, Taylor & Francis Group, LLC.

Sciancalepore, V. (1998) *Industrie agrarie*. Unione Tipografico-Editrice Torinese s.p.a. Torino

Sicari, V., Giuffrè, A.M., Piscopo, A. & Poiana, M. (2009). Effect of the "Ottobratica" variety ripening stage on the phenolic profile of the obtained olive oil, *La Rivista Italiana delle Sostanze Grasse*, Vol. LXXXII (No. 4): 215-221.

Skibsted, L.H. (2000). Light induced changes in dairy product, *Bulletin of the International Dairy Federation* Doc. No. 345, 4-9.

Soressi M. (2007). Brik, sullo scaffale arriva un outsider, *Olivo e Olio* Vol. 2: 8-9.

Soressi, M. (2009). Quando l'azoto previene l'ossidazione dell'olio, *Olivo e Olio* Vol. 4: 30.

Tovar, M. J., Motilva, M. J. & Romero, M. P. (2001). Changes in the phenolic composition of virgin olive oil from young trees (*Olea europaea* L cv. Arbequina) grown under linear irrigation strategies, *Journal of Agricultural and Food Chemistry* Vol. 49: 5502–5508.

Tsimidou, M. (1998). Polyphenols and quality of virgin olive oil in retrospect, *Italian Journal of Food Science* Vol. 10: 99-116 .

Tsimis, A. & Karakasides, N. G. (2002) How the choice of container affects olive oil quality- a review, *Packaging Technology and Science* Vol. 15: 147-154.

Tura, D., Gigliotti, C., Pedo, S., Failla, O., Bassi, D. & Serraiocco, A. (2007). Influence of cultivar and site of cultivation on levels of lipophilic and hydrophilic antioxidants in virgin olive oils (Olea Europea L) and correlations with oxidative stability, *Scientia Horticulturae* Vol. 112: 108–119.

Vacca, V., Del Caro, A., Poiana, M. & Piga, A. (2006) Effect of storage period and exposure conditions on the quality of Bosana extra-virgin olive oil. *Journal of Food Quality* Vol. 29: 139-150

Vinha, A. F., Ferreres, F., Silva, B. M., Valentao, P., Gonçalves, A., Pereira, J. A., et al. (2005). Phenolic profiles of Portuguese olive fruits (Olea europaea L): Influences of cultivar and geographical origin, *Food Chemistry* Vol. 89: 561–568.

Viola, P. (1997). Olive Oil and Health. *International Olive Oil Council*, Madrid Spain, pp. 26-27.

Chapter 11

Modern Methodologies to Assess the Olive Oil Quality

Giovanni Sindona and Domenico Taverna

Additional information is available at the end of the chapter

1. Introduction

1.1. Food quality and safety assessment

Food quality and safety assessments are the main issues that modern food industries are required to fulfill in their production chains, being these properties directly correlated to the claims of a holistic nutrition representing the fundamentals of a healthy balanced diet. In the same direction goes, very recently, the need of checking the origin of an aliment even this is aspect is apparently not directly correlated to the main issues above presented.

All the steps of olive production chain require, as for any other food, a careful evaluation of the identity and absolute amounts of micro and macro components at the "molecular level".

Protect Denomination of Origin (P.D.O.), Protected Geographical Identification (I.G.P) and other acronyms associated to olive oil quality, are often based on the apparent evaluation of macroscopic properties which hardly deals with a scientific approach to the problem, and whose official protocols of analysis lack often of the required specificity. A case study might be represented by the food safety incident worldwide known as *the melamine milk scandal*. Hundreds of babies fed with milk diluted with water and fraudulently treated with melamine to artificially increase the apparent protein content, developed kidney stones. Melamine is a plasticizer of molecular formula $C_3H_6N_6$ thus with a nitrogen content equal to 67% of its molecular weight. The international official protocols for protein assay are still based on the two hundred years old Kjeldahl method, whereby the nitrogen content is evaluated from the amount of the released ammonia. This was the fate of melamine too! Exactly in the same period researchers from the University of Shangai have published a mass spectrometric assay of melamine in milk based on the specificity of isotope dilution method by means of a contemporary, reliable and affordable approach.[1]

The assessment of quality and safety of all the steps of a given food chain needs to be based on the exploitation of high-tech analytical methodologies. Mass spectrometry represents a valuable tool widely applicable to the identification and assay of those quality markers present in olive oil.

1.2. The principle of mass spectrometry

Mass spectrometry (MS) is an analytical technique based on the principle of gas phase ion chemistry which allows the identification of analytes of any molecular weight and in any physical state. The availability of (i) multifaceted sample preparation, (ii) high throughput methods for on-line multistep mixture separation, (iii) multistep (MS/MS) ion analysis, (iv) high resolution acquisition and (v) specific isotopic analysis, confer to MS the unique feature of a methodology that can be directly applied to complex natural mixtures such as olive oil.

1.2.1. Multifaceted sample preparation

Any sample preparation protocol for MS evaluation of specific properties of foods requires versatile, reliable and affordable approaches which suit structure properties and physical state of the examined analytes. The sample preparation methods may include extensive chemical modification, solvent extraction and different concentration approaches as in the case of triglyceride (TGA) analysis from the alkyl esters of the corresponding fatty acid constituents. This approach classically transform low vapor pressure species to relatively volatile molecules amenable to gas-chromatographic (GC) separation followed by electron impact (EI) ionization and single quadrupole MS analysis. It should be mentioned, even this aspect goes beyond the scope of the present review, that the interest in the direct transformation of TGAs into alkyl fatty ester mixtures, including the methyl derivative (FAME) has found quite recently an enormous interest in the obtainment of biofuels under chemical and/or enzymatic catalysis.[2]

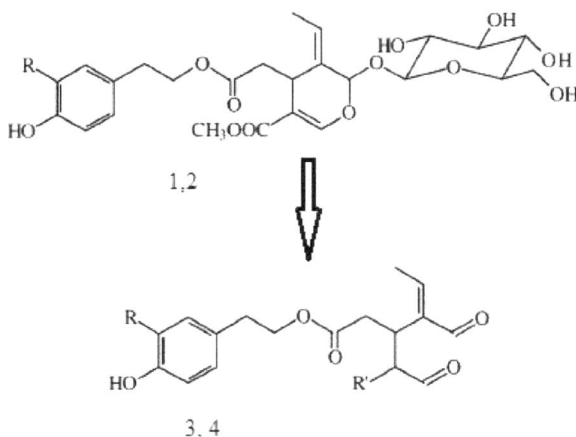

Scheme 1. Deglycosylation of secoiridoids 1,2 leading to the dealdehydes 3, 4.

When minor compounds or active principles present in olive oil need to be identified the sample preparation usually requires simple extraction of the analytes followed usually by a direct injection into a multistep mass spectrometer (MS/MS). A case study is represented by the anti-inflammatory dialdehydes formed in olive oil by enzymatic degradation of secoiridoids (scheme 1).[3]

A simple check on the existence of these molecules in olive oil can be performed from the crude methanol extracts by means of LC-MS/ MS (Figure 1).[4]

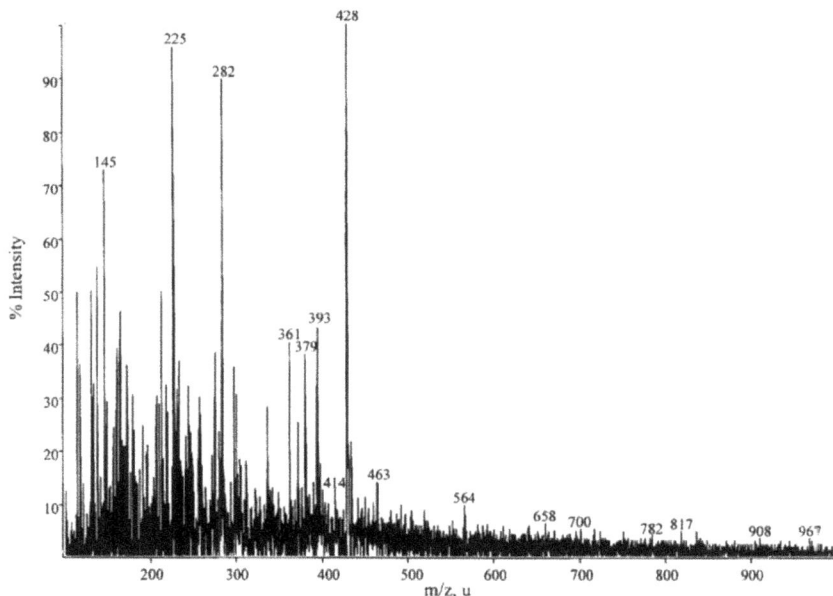

Figure 1. LC-MS chromatogram of a methanol extract.

The structure of many of the analytes present in the figure 1 was evaluated by further MS/MS experiments. The species at m/z 379, 414 and 428 were, than, associated to the protonated dialdehyde 4 (R= hydroxytyrosol, Scheme 1) and to the ammoniated species of the corresponding geminal diol and hemiacetal. The latter are expected reaction products of the very reactive system represented by the dialdehydes 3 and 4, when exposed to water and methanol interaction. The observed results could be considered as a remarkable case of *freezing* fast equilibrating species and detect them on-line, separately; the formation, however, of stable derivatives from a given analyte, present in a complex mixture such as olive oil, represents a *false positive* which can be eliminated by *selecting* a proper derivatization procedure as it will considered later.[5]

The recently introduced MS ambient ionization methods[6] are applicable to direct evaluation of complex food matrices such as olive oil. In this case too, especially in the assay of a specific component, in situ derivatization may still be needed.[7]

When the isotope dilution approach is considered either for quantitative investigation or for a reliable identification of analytes in complex natural mixtures[8], sample preparation requires the addition of suitable labelled internal standards which should not undergo intramolecular or intermolecular isotope scrambling.

1.2.2. High throughput methods for on-line multistep mixture separation

The exploitation of MS coupling with separation techniques was initially hampered by the experimental parameters of a conventional ion source operated at high vacuum. The availability of multistage mass spectrometers has favored direct injections protocols of even complex mixtures, nevertheless, many applications require the on-line separation of each analyte before being submitted to identification and assay. Modern protocols are based on the combination of GC, HPLC, capillary electrophoresis, and other separation techniques directly linked MS/MS instruments. Ionization techniques such as electrospray (ESI), atmospheric pressure chemical ionization (APCI), matrix-assisted laser desorption ionization (MALDI), and differently performing ambient ionization devices are nowadays mainly used in food analysis. The role of GC, as well as HPLC in high-resolution separation of complex mixtures is unquestionable. Similarly, mass spectrometry has attained an indisputable position in analytical chemistry as a highly structure-specific technique that can provide structural identity of a wide range of compounds. Both chromatography and mass spectrometry have, however, their limitations as stand-alone operations. The separation power of any chromatography system is, actually, finite since it will be nearly impossible to achieve complete separation of all components of a complex mixture. The identification of a compound from a chromatographic peak is less than reliable when it has to be indentified by comparing its retention time with that of a standard reference. Therefore, the commonly used separation techniques such as GC and LC cannot provide unequivocal identity of the analyte when used with conventional detection systems. A compound-specific detector is thus an essential adjunct to characterize unambiguously the components that elute from any separation system. In this respect, mass spectrometry offers the unique advantages of high molecular specificity, detection sensitivity, and dynamic range. Mass spectrometry, only, has the advantage to provide confirmatory evidence of an analyte because of its ability to distinguish closely related compounds on the basis of the molecular mass and structure-specific fragment ion information formed by gas phase processes. The confidence in identification of a target compound, however, diminishes when it is present in a mixture, such as olive oil. Because of the universal nature of mass spectrometry detection, the data obtained might also contain signal due to other components of the mixture. The coupling of a separation device with mass spectrometry thus benefits mutually. The result is a powerful two-dimensional analysis approach, where the high-resolution separation and the highly sensitive and structure-specific detection are both realized simultaneously. Some of the benefits that accrue when a separation technique and mass spectrometry are coupled can be summarized as follows:

• The capabilities of the techniques are enhanced synergistically; both instruments may, therefore be operated at subpar performance levels without compromising the data outcome.

- The high selectivity of mass spectrometry detection allows one to identify co-eluting components.
- The certainty of identification is enhanced further because, in addition to the structure-specific mass spectral data, the chromatographic retention time is also known.
- Multicomponent samples can be analyzed directly without prior laborious off-line separation steps, resulting in a minimal sample loss and saving of time.
- The sensitivity of analysis is improved because the sample enters the mass spectrometer in the form of a narrow focused band.
- Less sample is required than the amount required for off-line analysis by the two techniques separately.
- Because of the removal of interferences, the quality of mass spectral data is improved and any mutual signal suppression is minimized.

Various chromatographic techniques have been employed for separation, preparative isolation, purification and identification of such kind of compounds in food. At date, comprehensive two-dimensional separations are a definite trend in chromatography. Comprehensive two-dimensional gas chromatography (GC×GC) emerged some 15 years ago.[9] The latter technique allows for the separation of complex mixtures of volatile analytes GCxGC delivers more GC information in a shorter time than do other methods. GCxGC, at its roots, is a multidimensional separation technique in which the resolving power of two or more different columns is applied to some or all of the components in a sample. As the name implies, comprehensive separations apply all the available resolving power of both columns to all the peaks in a sample. In GCxGC each peak transits the first column, is trapped at the end of the first or the beginning of the second column and is released onto the second column (Figure 2).

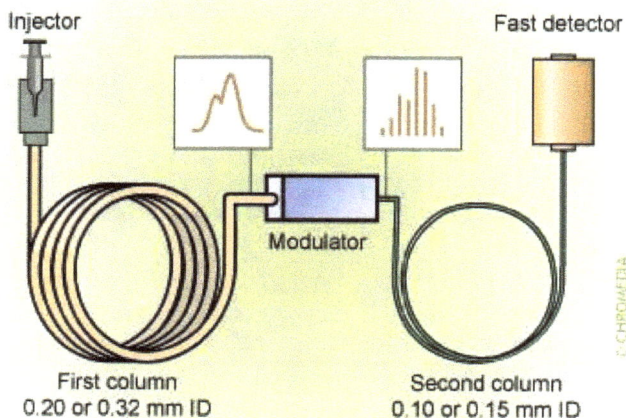

Figure 2. GCxGC device.

The major difference between GCxGC and heartcutting is that for GCxGC the trapping time and second column separation speeds are fast enough so that many heartcuts can be analysed in the second column during the course of a single first-column run. The heartcuts can be taken rapidly enough that individual peaks from the first column are sliced across several sequential secondary column runs, but going much faster than that has no advantage. Achieving this level of performance requires a high-speed separation for the secondary column and encourages ordinary speeds for the first column. Faster separations on the first column can drive the second-column speed requirement higher than what is practical. So, the main advantages of the technique are vastly-expanded separation space and the ability to resolve hundreds or thousands of peaks; improved sensitivity; and more obvious structure in retention patterns for peaks. The increase in sensitivity is due to the fact that most modulators compress and focus peaks into narrow bands, improving the signal, and at the same time separate coeluting materials from the analyte of interest, decreasing the noise.[10] Comprehensive Two-dimensional Gas Chromatography (GCxGC).[11] Comprehensive two-dimensional separation is achieved also by liquid chromatography; the roots of comprehensive two-dimensional liquid chromatography (LC×LC) are older. Two-dimensional thin-layer chromatography can be considered a form of LC×LC, dating the technique back at least some 50 years. The contemporary version of LC×LC was first demonstrated by Erni and Frei.[12] The common way to perform LC×LC involves a first-dimension separation that is relatively slow, with a typical analysis time of 1 h or longer. Fractions of the effluent from this first-dimension column are collected in a loop of a switching valve. When this valve is switched, the fraction is injected onto a second column, which provides a much faster separation, typically with an analysis time of a minute or less.

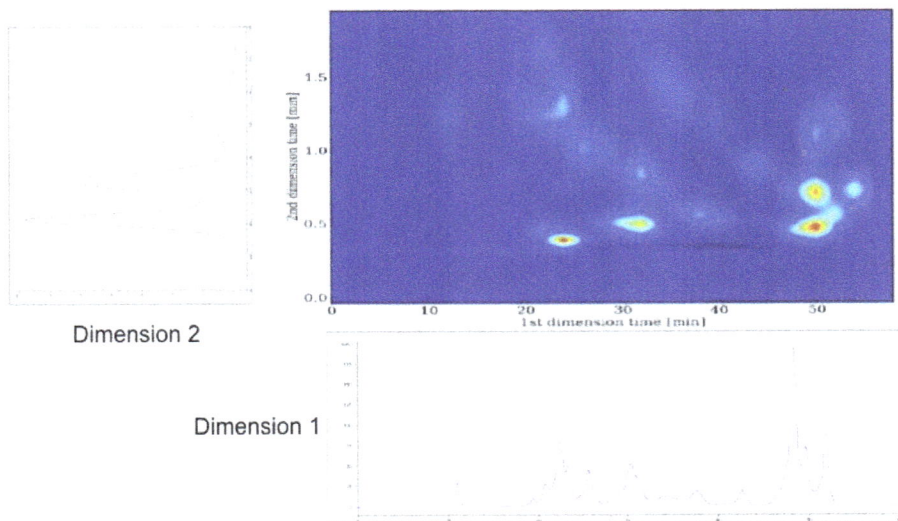

Dimension 2

Dimension 1

Figure 3. LCxLC, 2 dimensions chromatograms.

While one fraction is being analyzed on the second-dimension column, a new fraction is being collected in another loop connected to the same valve. The result of a comprehensive two-dimensional separation is a series of many (for example, 100) fast second-dimension chromatograms (Figure 3).

1.2.3. Multistep (MS/MS) ion analysis

The concept of *in-space tandem mass spectrometry* is illustrated in Figure 4. It involves two mass spectrometry devices; the first one (MS-1) performs the mass selection of a desired target ion from a stream of ions produced in the ion source. This mass-selected ion undergoes either unimolecular fragmentation (equation 1) or a chemical reaction in the intermediate region.

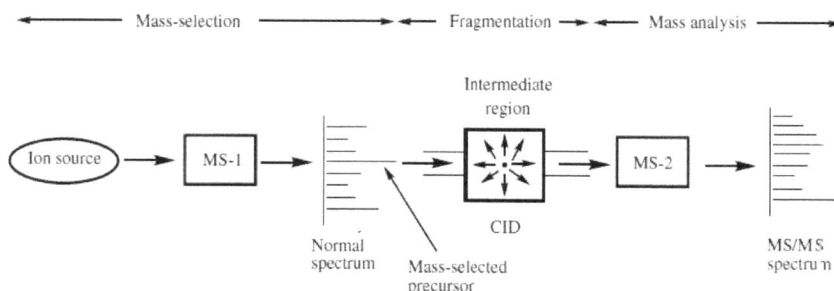

Figure 4. Basic principle of tandem mass spectrometry.

The second MS system (MS-2) performs the mass analysis of the product ions that are formed in the intermediate step.

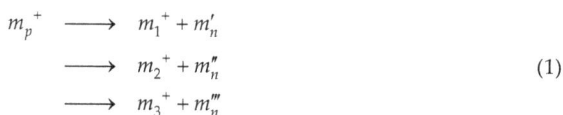

$$m_p^+ \longrightarrow m_1^+ + m_n'$$
$$\longrightarrow m_2^+ + m_n''$$
$$\longrightarrow m_3^+ + m_n''' \tag{1}$$

The first stage of MS/MS separates a mixture of ions into individual components thus resolving a mixture of compounds, whereas the second stage acts as an identification system for the mass-resolved ions. By convention, the mass-selected ion is called the *precursor ion* (m_p^+) and its fragments are called *product ions* (e.g., m_1^+, m_2^+, m_3^+, etc. in equation 1). The mn species in equation 1 are neutral losses. Because of the incontrovertible link between the precursor ion and all of its product ions, the molecular specificity of MS/MS is garanteed. This unique attribute of tandem mass spectrometry is a highly useful feature that plays a role in the unequivocal identification of a target compound in real-world samples. Tandem mass spectrometry is not restricted to two stages of mass analysis (i.e., MS/MS or MS^2); it is also possible to perform *multistage MS* (i.e., higherorder MS) experiments, abbreviated as MS^n. These experiments can determine the genealogical relation between a precursor and its ionic products. For example, MS^3 indicates three stages of tandem mass spectrometry, which involves mass selection of one of the products, say either m_1^+, m_2^+, or m_3^+, formed from the precursor m_p^+ of the first-stage MS (equation 1), and determination of the second-generation products of that mass-selected ion. Multistage MS experiments are performed mostly with ion-trapping instruments. A maximum of 12 MS/MS experiments has been envisioned with a quadrupole ion trap. Beam-type instruments can also be used for MS^n experiments but require as many discrete mass analyzers as there are number of stages in the experiment, making it difficult to perform more than four stages of MS/MS experiments. As an example, a three-sector magnetic field instrument can perform up to MS^3 experiments. Practical applications of tandem mass spectrometry require data to be acquired in the following four scan modes. A pictorial representation of these scans and their symbolism is given in Figure 5.

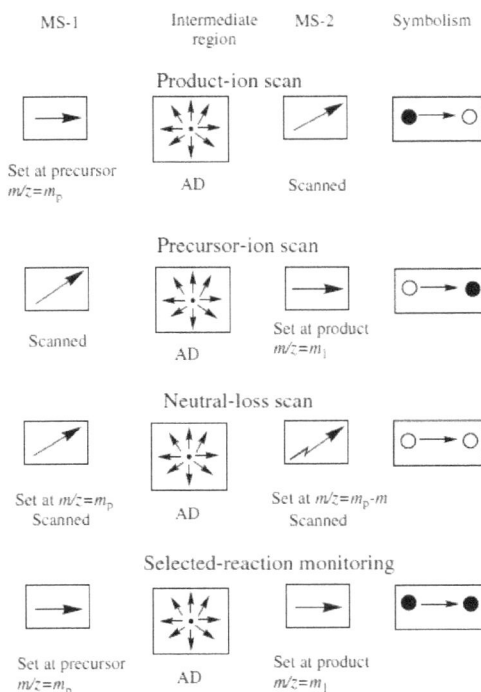

Figure 5. Representation of four scan modes of tandem mass spectrometry. AD refers to ion activation and dissociation, and the filled and open circles stand for fixed and scanning mass analyzers, respectively.

The *product-ion scan* (the old, now-unaccepted term, still used by some, is *daughter-ion scan*) is the most common mode of MS/MS operation. That spectrum is useful in the structure elucidation of a specified analyte. Information obtained in this scan is similar to that derived from a normal mass spectrum, except that the spectrum contains only those product ions that are formed exclusively from a mass-selected precursor ion. To acquire this spectrum, the first mass analyzer is set to transmit only the precursor ion chosen, and the second mass spectrometer is scanned over a required m/z range. Another popular MS/MS scan is the *precursor-ion scan* (the past, now-unaccepted term is *parent-ion scan*). It provides a spectrum of all precursor ions that might fragment to a common, diagnostic product ion The spectrum is obtained by adjusting the second mass spectrometer to transmit a chosen product ion (e.g., $m1$) and scanning the first mass analyzer over a certain m/z range to transmit only those precursor ions that fragment to yield the chosen product ion. This scan is useful for the identification of a closely related class of compounds in a mixture. A typical example is the detection of Sudan azo dyes within a food. A major product of fragmentation of azo dyes is the ion at m/z 156. The precursor-ion scan of this m/z value will detect the presence of azo dyes selectively in a complex mixture (see Figure 6).

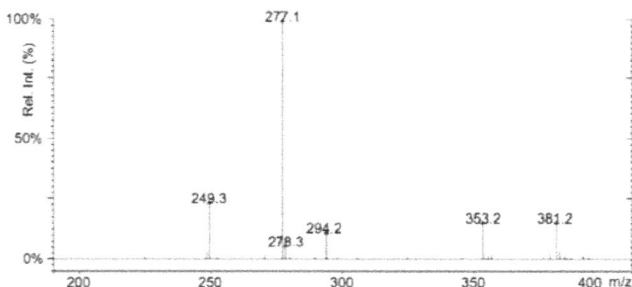

Figure 6. Parent ion scan MS/MS spectrum of the fragment m/z 156 present in the APCI spectra of 5 azodyes extracted from baked products and sampled by flow injection.

In a *constant-neutral-loss scan*, all precursors that undergo the loss of a specified common neutral are monitored. To obtain this information, both mass analyzers are scanned simultaneously, but with a mass offset that correlates with the mass of the specified neutral. Similar to the precursor ion scan, this technique is also useful in the selective identification of closely related class of compounds in a mixture. For example, the loss of 44 Da is a common reaction of carboxylic acids. Through the constant neutral loss scan, the identity of all carboxylic acids present in a complex mixture can be revealed. Similarly, by monitoring the 98-Da neutral loss, the presence of phosphopeptides can be detected in a complex mixture.[13] The fourth scan, *selected-reaction monitoring* (SRM), is useful in quantitative measurements of analytes present in complex mixtures. Conceptually, this scan mode is similar to the product-ionscan. However, instead of scanning the second mass spectrometer in abroad mass range, the two mass analyzers are adjusted to monitor one or more chosen precursor–product pairs of the analyte. This operation is identical to the selected-ion monitoring mode (SIM) of data acquisition. Monitoring more than one reaction is termed *multiple-reaction monitoring* (MRM). All four scan modes can be implemented with magnetic sector- and quadrupole-based true or hybrid tandem instruments. Time-of-flight (TOF) and tandem-in time devices are also suitable for product scan experiments, but they are unable to perform the other three scans. Further, within the multistep tandem mass spectrometry analyses have to be considered also Ion Mobility (IM). Those mass spectrometers are hybrid instruments that combine an IM separation system with conventional MS systems. An *ion mobility spectrometer* (IMS) can also serve as a stand-alone ion detection system.[14] A IMS uses gas-phase mobility rather than the m/q ratio as a criterion to separate ions.[15, 16] The mobility of ions is measured under the influence of an electrical field gradient and cross flow of a buffer gas, and depends on ion's collision cross section and net charge. A typical IMS consists of a reaction region and a much longer drift region; both regions contain a series of uniformly spaced electrodes that are connected via a series of high resistors to provide a uniform electric field strength. The two regions are separated by an electrical shutter. Buffer gas is also circulated in the drift tube. The ions are generated in the reaction chamber and are allowed to enter the drift region by opening the electrical shutter for a brief period. Under the influence of an electrical field, ions drift into the drift tube, where they are separated according to their size-to-charge ratio. The mobility of ions is a combined effect of ion acceleration by the electric field and retardation by collisions with the buffer gas. At the end of the drift region is placed an ion detector (e.g., a Faraday cup) for the detection of the separated ions. For more accurate mass analysis, an IMS is coupled to a quadrupole or TOF mass analyzer.[17, 18]

Similar to LC–MS systems, the IMS serves as a separation device and the quadrupole or TOF mass analyzer as a detection device, but has the added advantage that separation times are in milliseconds. ESI and MALDI ion sources have both been coupled to IM–MS instrument.[19, 20] Latter consists of a MALDI source, an ion mobility cell, a CID cell, and a TOF mass analyzer. Ions exit the drift tube when the axial field strength of the ion mobility cell is ramped up, and enter the source region of a –TOF instrument, where the ions are detected intact in the usual manner. Alternatively, ions can be fragmented in the CID prior to their detection by the TOF–MS. The resolution of an IMS, usually very low (10 to 12) can be increased to 200 to 400 by increasing the pressure of the buffer gas, connecting the ion source directly with the drift tube, increasing the length of the drift tube, and increasing the electric field gradient of the drift tube. Unlike LC/MS separation, an ion chromatogram can be obtained within 1 s.

1.2.4. High-Resolution Acquisition

High-resolution mass spectrometry (HRMS) strongly competes with the classical tandem MS, in particular in the field of quantitative methods (e.g., pesticides, drugs, etc). High-res analyses are commonly used for residues in food: the availability of new and more sensitive, selective instrumentation contributed to the development and the acceptance of such technology in the last few years. Many benefits are gained by using HRMS, including the collection of full-scan spectra, which provides greater insights into the composition of a sample. High-res instruments, including double focusing, FTICR, orbitrap mass spectrometers, are able to measure the mass of an ion and its associated isotope peaks with sufficient accuracy to allow the determination of its elemental composition. This is possible because each element has a slightly different characteristic mass defect. The accurate mass measurement shows the total mass defect and provides a means to determine its elemental composition. In the last years, the number of studies reposting HRMS-based approaches for food analysis applications had a significant increase. Even if laboratories still analyze pesticides, drugs and other microcomponents by using LC-MS/MS and other older approaches, HRMS instruments are used in both quantitative and confirmative works. High-res instruments allow not only for high resolution, superior sensitivity, larger dynamic range and improved speed. Generally, a high-resolution analysis provides more comprehensive information, e.g. identifying coeluted and therefore potentially suppressing matrix compounds. Selectivity of detection in complex matrices is only ensured when the HRMS instrument can physically resolve the analyte mass from that of other coeluted isobaric matrix compounds. HRMS permits selectivity in their elemental composition and therefore have a different exact mass. Many reports in the literature described the use of a narrow mass window to improve selectivity.[21-23] Other studies have addressed the issue of the mass resolution required for trace analysis in food matrices.[24] Resolutions of 50,000 or more FWHM were suggested for low levels of analytes in complex matrices such as animal feed. The same authors concluded that a resolving power of 25,000 FWHM is sufficient for less complex matrices. Their conclusion was based on experiments that involved analyzing matrix samples spiked at low levels with various veterinary drugs, mycotoxins, and pesticides. Moreover, the latest generation of Orbitrap instrumentation provides resolutions well above 100,000 FWHM (Figure 7). Such instrument provides a clearly higher selectivity than QqQ instrumentation.

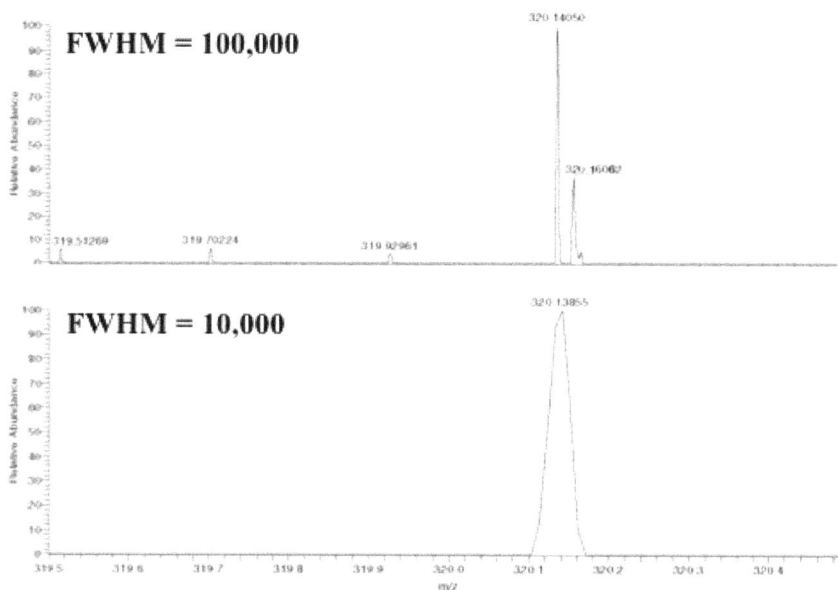

Figure 7. High resolution MS spectra registered by a Orbitrap instrument; the signal at m/z 320 has been resolved at 100,000 FWHM (top) and at 10,000 FWHM (bottom). The higher resolution allows for a clear separation of the analyte ion from an isobaric matrix compound. The lower resolution does not resolve these two ions.

The emergence of higher and higher resolving HRMS instruments is a significant advantage over QqQ technology. It is a fact that in the recent past every QqQ instrument generation showed an impressive increase in detector sensitivity as compared with the previous generation. This provided not only lower limits of detection, but also permitted the dilution of extracts in order to reduce signal suppression effects. The other side of the coin is the fact that detection selectivity remained virtually unchanged. This resulted in SRM traces which show many peaks corresponding to compounds being eluted before, after, or even together with the targeted analyte.[25] Further, most of the currently available literature for HRMS multiresidue screening is based on measurements made with instruments capable of providing resolutions between 10,000 and 20,000 FWHM.[26, 27] Basically, higher resolution leads to superior results when detecting low concentrations of analytes in complex matrices. However, depending on the technology used: TOF instruments, for instance, achieve higher resolution by employing very long flight tubes. This might be a physical limitation. Orbitraps provide a user-defined resolution. There is no relevant sensitivity drop when choosing a higher resolution. However, the resolution obtained is proportional to the measurement time. The resulting consequences are fewer data points per time unit. This is a limitation when using UPLC equipment or when attempting polarity-switching experiments. Furthermore, higher resolutions produce more data which have to be processed by the instrument and later accessed from the hard disk when extracting mass traces from the raw data.

1.3. Identification and assay of major components of olive oil

Triacylglycerols (TAG) are the most important lipid component of human diet and represent by far the main constituent of olive oil (96-99%), they differ in fatty acid composition as well as in their sequence within the glyceride. It is now well established that the biochemistry of lipids assumed with diet, hence those present in olive oil, is strictly related to the action of the digesting enzymes, such a pancreatic lipase, which are able to discriminate between the fatty acids etherifying glycerol at different sites (Figure 8).[28]

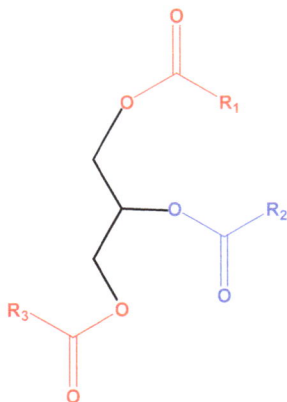

Figure 8. Schematic of TG structure. R, fatty acid, with R1, R2, and R3 indicating the fatty acid at the sn-1, -2, and -3 positions, respectively.

Therefore the classic analytical methods based on the identification and assay of methyl or ethyl fatty acid esters obtained by direct transesterification of TGAs, still the official method of analysis, lack by nutritional, hence quality, validation.[29] Chromatographic[30, 31] and ^{1}H NMR methods[32-34] have been introduced for a detailed analysis of TGA in olive oils. In a recent application of chromatographic methodologies combined with statistical analysis[35] an effective discrimination of extra virgin olive oils was achieved by direct injection of the samples without any preliminary chemical derivatization. Atmospheric pressure chemical ionization (APCI), linked to a triple quadrupole, mass spectrometer provided sensitivity to detect over 50 compounds in the sample. In figure 9 is reported the MS profile of a TGA mixture present in a given olive oil sample.

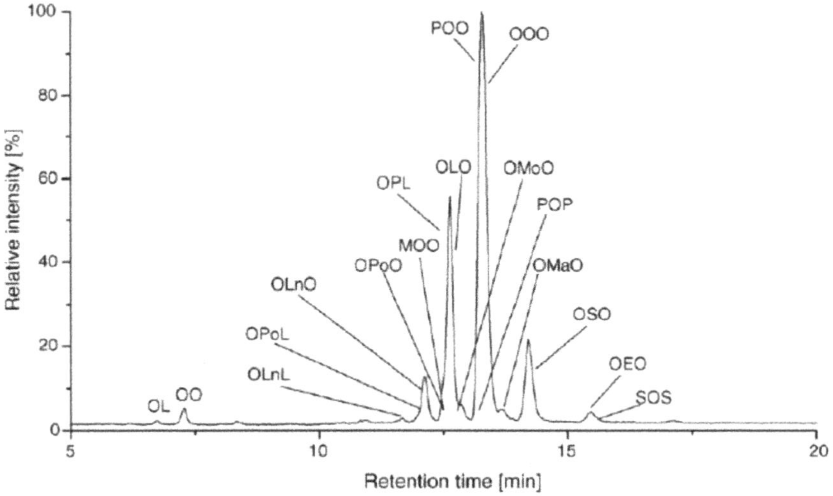

Figure 9. HPLC-MS profile of diluted extra virgin olive oil. The three letter acronyms have the conventional meaning, where O, M. Po, P, S, L, Ln, E. Ma, and Mo, correspond to asingle fatty acid respectively.

A peculiar application to the evaluation of TAG profiling[36] of Matrix assisted laser desorption (MALDI) time-of-flight (TOF) mass spectrometry was introduced (Figure 10). The method was applied to a peculiar type of olive oil produced in a desert, by means of a conventional 2,5-dihydroxybenzoic acid (DHB) matrix, using un-treated samples. The fatty acid composition of the analyzed foodstuff was in agreement with what could be anticipated by applying conventional GC techniques, with the main difference, however, that the presence of oleic acid, for instance, was evaluated through the identification of differently structured TAGs, such as Trioleyl (31.53%) dioleoyl-palmitoyl (23.06%) and dioleoyl-linoleoyl (14.31%) glycerol.

Microbiological studies have recently demonstrated the occurrence in the suspended fraction of freshly produced olive oil of a rich microflora mainly composed of yeasts[37] which, probably, induces the presence of lipase.[38] This fact can explain the many changes observed in the triglycerides profiles during aging as a consequence of the lipase action[39]

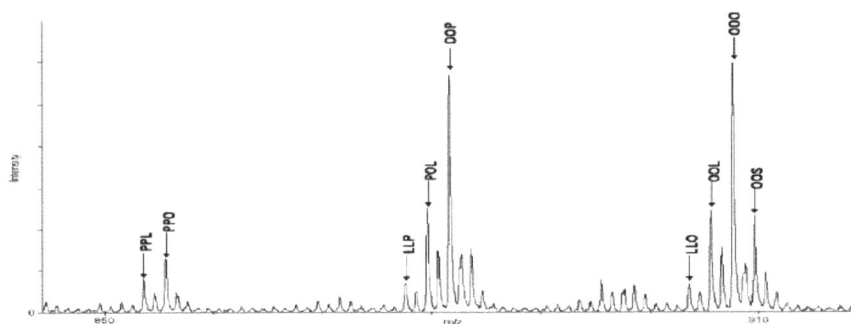

Figure 10. TAG spectrum obtained from MALDI-TOF/MS of an olive oils sample in positive ion mode.

and to the presence of several mono and di carboxylic acids in olive oil derived from many types of biotransformation.[40-42] The products resulted from this modifications are new type of triglyceride oligopolymers present in olive oil, as it was reported for Japanese wax.[43]

The relative activity of lipase to catalyze hydrolysis and transesterification of triglycerides was studied and confirmed either with model systems mimic olive oil, or directly with the foodstuff.[44] It was monitored, therefore, by mass spectrometry, the ability of lipase to catalyze the formation of new functionalized triglycerides from triglycerides and free mono and di-acids. The oligomerization of triglycerides has been monitored on real olive oil samples, and verified in solution under lipase catalysis, whose presence in olive oil was already documented. The hydrolytic pathways taken under enzymatic treatment is balanced by the formation of triglycerides oligomers that should not alter the nutritional value of the aliment. As a general view of the lipase action on triglyceride models in the presence of a carboxylic acid, Tripalmitoyl- (1) and trioleoyl (2)- glycerol were incubated with succinic under lipase catalysis at 70°C (Scheme 2). The choice of succinic acid was motivated by the known presence of this dienoic acid in olive oil.

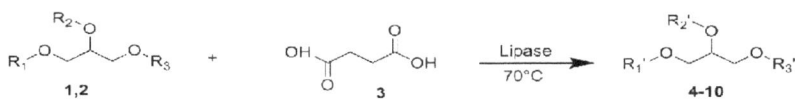

Scheme 2. Incubation of triglyceride mixture with succinic acid in the presence of lipase.

In Figure 11 is reported the partial full scan spectrum of the succinoyl derivative of of dioleylglycerol observed as sodiated species. The MS peak might correspond to other isomeric species, in the sense that the succinoyl moiety could be linked to different glycerol sites, according to the mechanism of lipase action. Nevertheless, it is clear that after thermal mild treatment, similar to the deodorizing protocol, the presence of lipase and of succinic acid in olive oil can cause, besides TAG isomerization, the formation of side products which could be useful as markers of the industrial procedure.

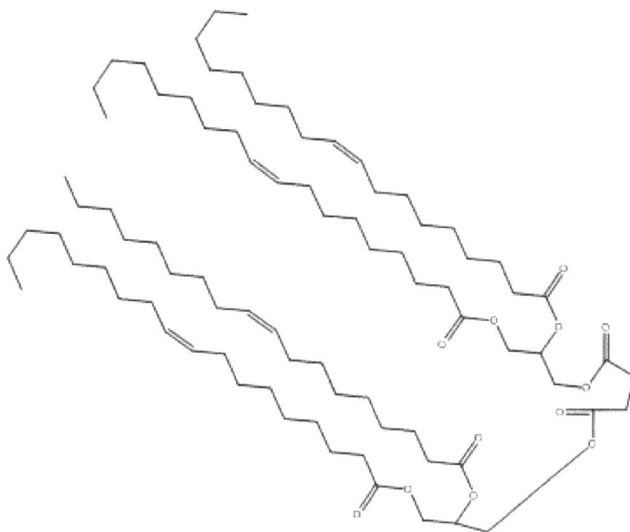

Formula. 1,2-dioleyl-3-succinoylglycerol (**6**). $C_{82}H_{146}O_{12}$

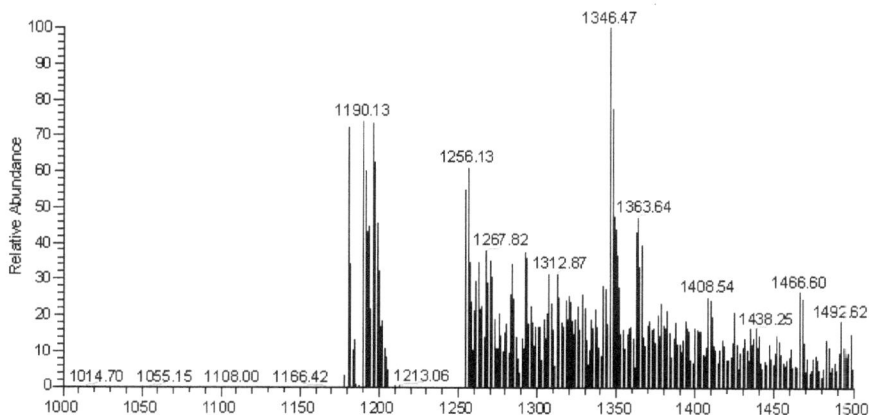

Figure 11. ESI-MS full scan spectrum of compound **6** identified as [M+Na$^+$] adduct.

The ideal objective of any extraction method is to extract the largest possible amount of matrix constituents without altering their identity. to measure "total fat" various methods have been approved by the regulatory agencies of most major countries.[45] Most of the older methods involve solvent extraction and gravimetric mass measurement of the lipid residue. Several apparatus have been developed since 1939 for automatic extraction (e.g., Soxhlet, Soxhterm, Soxtec, Butt, etc).[46] For total fat determination ethyl ether (diethyl ether) is often the solvent of choice as it is relatively non-polar and extracts mostly the non-polar lipids (triacylglycerols, sterols, tocopherols, etc) while poorly extracting the polar lipids (glycolipids and phospholipids).[47, 48] Methods have been developed to use supercritical fluid extraction for total fat determination.[49]

However, because of the diversity of the matrix content, methods are designed to efficiently extract specific molecule classes. The extraction of total lipids in foods is often considered to be a simple procedure; to extract total lipids (non-polar and polar) more polar solvents must be used as extractants (e.g., hexan or petroleum ether, chloroform, methanol, isopropanol, water). The most popular method for total lipid extraction is the Folch method.[50] Lipids are extracted from a sample by using chloroform-methanol (2:1 by volume). Many modifications to the Folch procedure have been published.[51]

1.4. Identification and assay of minor component of olive oil

The quality of olive oil is often associated with the presence of microcomponents whose healing effects have been proved in some special cases.[52, 53] Virgin olive oils contains also phenolic substances responsible for their stability against oxidation.[54] Phenolic fraction includes simple phenols, tyrosol, and hydroxytyrosol, derivatives of hydroxybenzoic and hydroxycinnamic acids, aglycons of some glucosides, namely oleuropein, demethyloleuropein, ligstroside, and verbascoside.[55, 56]

Absolute analytical methods for mass spectrometric detection as well as for quantification of such bio-active molecules, such as oleopentadialdehydes have been developed. As previously mentioned, the presence of dialdehydes in olive oil can be easily ascertained by a simple LC-MS approach (scheme and figure above). However *false positive* can be easily formed during the ionization process due to extreme reactivity of the sampled molecules with polar solvents such as water or methanol.[57] Such novel method provides an in situ chemical derivatization of the whole set of molecules into stable alkyloxime derivates, allowing the concomitant use of a stable isotope standard which improves both the precision and the accuracy of the measurements, thus reducing the drawbacks may arising from the calibration procedure, sample preparation, and matrix effects.

The method was applied to a set of different Italian virgin olive oils showing the presence of 5 and 6 in the range of 70÷166 and 23÷132 ppm, respectively. The same approach was exploited in the assay of hydroxytyrosol (1) and tyrosol (2), the strong antioxidant present in large amount in virgin olive oil, by LC–MS/MS under MRM condition and isotope dilution method, using d_2-labelled internal standards, 3 and 4, obtained by simple synthetic procedures (Chart 1). This active principles ranged from 10 to 47 and 5 to 25 ppm in experimental and commercial virgin olive oil, respectively.[58]

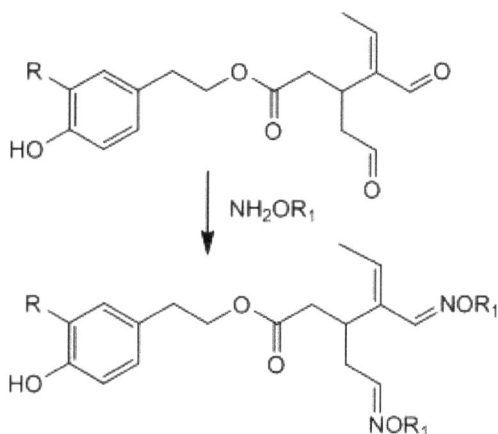

5 $R_1 = CH_3$, R=H O-methoxy-pentanedialdoxime-Tyr
6 $R_1 = CH_3$, R=OH O-methoxy-pentanedialdoxime-HTyr
7 $R_1 = CD_3$, R=OH d_6-O-methoxy-pentanedialdoxime-HTyr

Scheme 3. Derivatization of tyrosol (Tyr) and hydroxytyrosol dialdehydes with unlabelled and labeled methoxyamine.

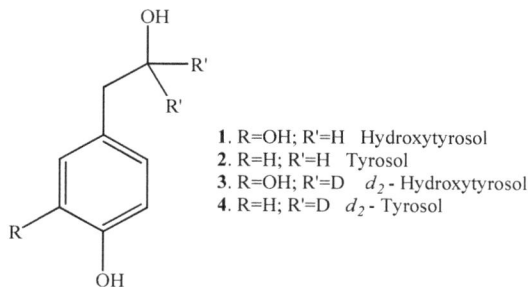

1. R=OH; R'=H   Hydroxytyrosol
2. R=H; R'=H   Tyrosol
3. R=OH; R'=D d_2- Hydroxytyrosol
4. R=H; R'=D d_2- Tyrosol

Scheme 4. Labelled and unlabelled hydroxytyrosol and tyrosol.

A renowned antioxidant present in olive oil is the secoiridoid oleuropein (OLP). A number of studies have recognized that a diet rich in olive oil, particularly unrefined oils, provides a healthy prevention of artery wall thickening as a consequence of low-density lipoprotein (LDL) oxidation process. This beneficial effect has been associated to the presence of oleic and linoleic monounsaturated fatty acids and to the action of potent antioxidants such as tocopherols and the "polyphenols".[59] OLP, a secondary metabolite of terpenoid origin, is the main iridoid of the "phenolic pool"[60] of *Olea europaea*, whose activity is likely associated to its o-dihydroxybenzene (catechol) moiety. The same moiety is shared by hydroxytyrosol which is formed by enzymatic degradation of the intact secoiridoid and exhibits similar redox activity. Other phenolic compounds, such as tyrosol, caffeic acid, etc., account for the radical scavenger effect of virgin olive oil;[61] however, attention has been paid to the actual content of oleuropein in foodstuffs due to its therapeutic action.[62] An ESR study performed on the pure molecule has demonstrated that OLP can likely be exhibited in particular physiological conditions a pro-oxidant activity.[63]

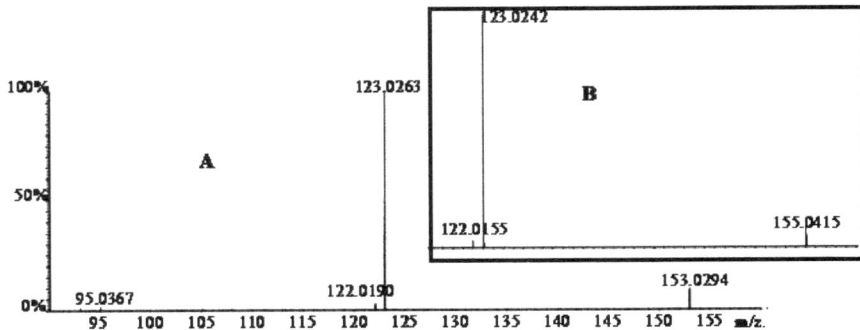

Figure 12. ESI (-) MSMS spectra of compound HTyr (A) and labelled HTyr (B,insert).

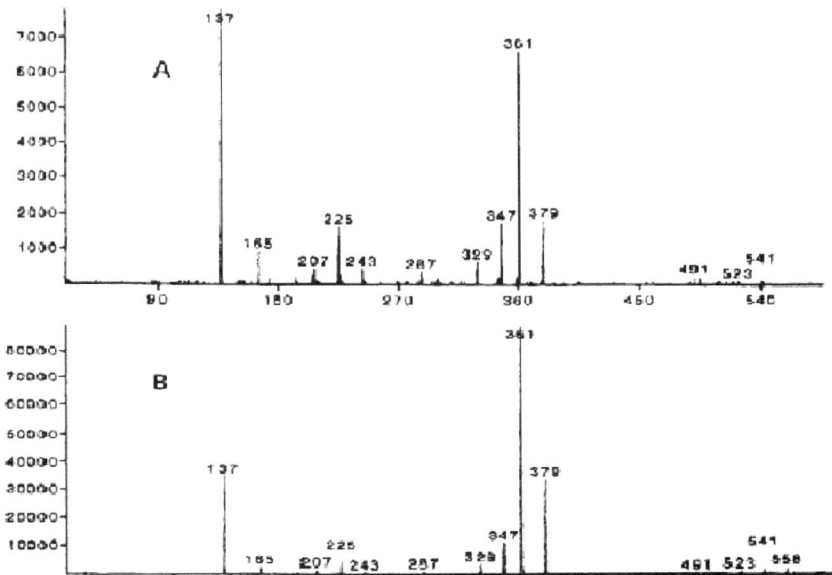

Figure 13. ESI tandem mass spectra of OLP commercial standard. Product ion spectra (A) from [M+H]+ and (B) from [M+NH4]+ precursors.

A comprehensive MS characterization of OLP, showed the advantage of sampling $[M+NH_4]^+$ ammonium adduct either for the sensitivity of the analytical method and for the peculiar gas-phase chemistry of the species as shown by low and high energy collision experiments. It was shown that the loss of ammonia neutral and the consequent formation of the $[M+H]^+$ species was controlling the unimolecular dissociations (Figure 13).[64]

The analytical data suggested, therefore, that the $[M+NH_4]^+$ species of oleuropein might correspond to a mixture of different structures and react, in the gas phase, through the elimination of neutral ammonia to give a transient $[M+H]^+$ which possesses enough internal energy to undergo further fragmentation. Oleuropein is extremely reactive in the experimental conditions leading to the preparation of olive oil from drupes, mainly due to the action of glycosidase and esterase enzymes. Therefore its assay in oils derived from different cultivars and different procedures could be considered a marker of process and of good practise. The validated process previous discussed was therefore applied to the monitoring of OLP in oils formed from different cultivars at different ripening stage. (Figure 14).[65]

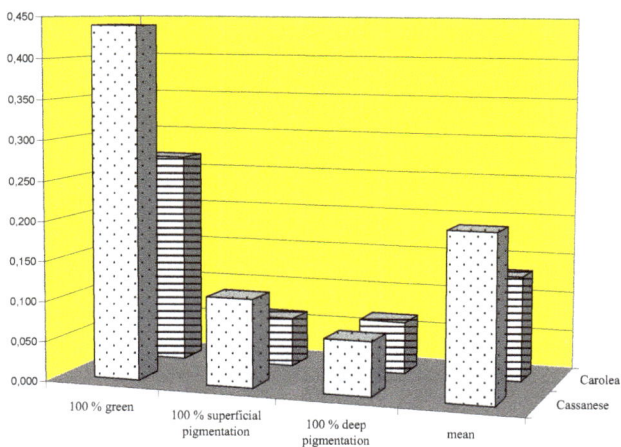

Figure 14. Ripening-phase-dependent oleuropein content in virgin olive oils of Cassanese and Carolea cultivars. Oleuropein assay in olive oil.

The importance of oleuropein has prompted the development of an absolute method for its assay in olive oil, based on APCI-MS/MS and isotope dilution.[66]

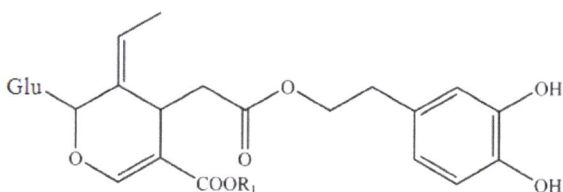

Scheme 5. R_1= CH$_3$- OLPd$_0$; R_1= CD$_3$- OLPd$_3$

The sensitivity and accuracy of the method allowed the discrimination of olive oil samples with very similar content of OLP. In particular filtered and not-filtered products from different southern Italian regions, located at latitude ranging between (38° 57′ N) and (42° 28′ N), were discriminated by the amount of OLP which was in the range of 0.093÷0.225 and 0.116÷0.344 ppm, respectively.

A similar approach was exploited in the characterization, at the molecular level, of olive oils produced from the whole fruits or from stoned olives obtained from a pilot plant.[67] The absolute amount of OLP was 3÷4 times higher in stoned oils, total phenol content showed a similar behaviour, whereas the amount of α-tocopherols was higher in oils obtained by conventional milling procedures. The total phenol content should be enhanced either when more lipophilic aglycones are formed, after the removal of the hydrophilic sugar moiety, and, likely, when oleocanthal[55] and its hydroxytyrosol homologue are obtained after deglycosylation and demethylation at position 11 of oleuropein.[68] This effect, indirectly, resembles the action of cell-wall-degrading enzymes added to the paste to improve the olive oil quality.[69]

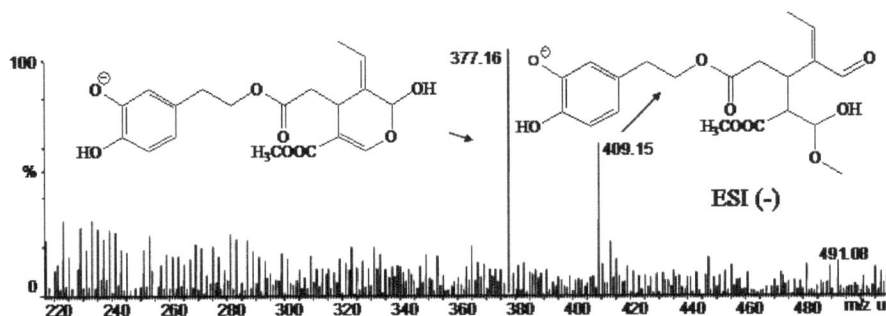

Figure 15. ESI-MS (-) spectra taken directly from the solution containing OLP and the water extracts of Carolea pits after 60 min. of incubation.

The main technological difference in the production of the sampled oils was represented by the presence or absence of stones during the manufacturing procedure. It was therefore checked, always by LC-MS/MS, the action of water extracts of crushed stones on pure samples of OLP was therefore planned. The extracted ion chromatogram showed, after 60 min of incubation, that elenoic acid and its open chain isomers were formed whereas hydroxytyrosol was not detected (Figure 15). This was considered an evidence of the preponderant presence of glucosidases in the stone enzyme pool. The quality of olive oils in relation to the identification and assay of the nutraceutical pool has been recently performed by the new tools of ambient ionization mass spectrometry. A low temperature plasma (LTP) ion source has allowed a rapid screening of free fatty acids, selected bioactive phenolic compounds, volatiles and adulterants in virgin olive oils (Figure 16).[70]

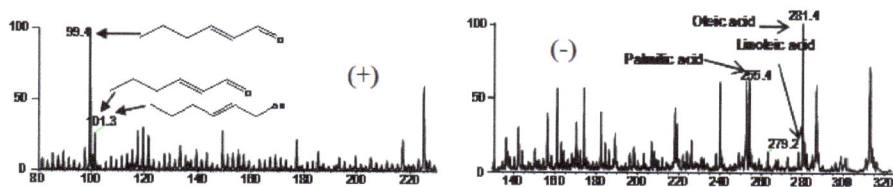

Figure 16. LTP-MS full scan spectrum of crude virgin olive oil in the positive (+) and negative ionization mode (-).

The peculiarity of the method is also represented by the possibility of analyzing each component present in the full scan spectrum with conventional MS/MS devices. Further, another ion source related to the ambient mass spectrometry techniques, named paper spray, has been used for olive oil rapid analysis. The latter, allowed for qualitative determination of several analytes as well as for their assay adding a suitable labeled internal standard. Currently, this technique is applyed also to the *is situ* derivatization of olive oil (e.g., by methoxyamine), allowing the formation of mass tags and increasing the ionization efficiency for such king of molecules (e.g., oleopentanedialdehydes) in less than one minute.

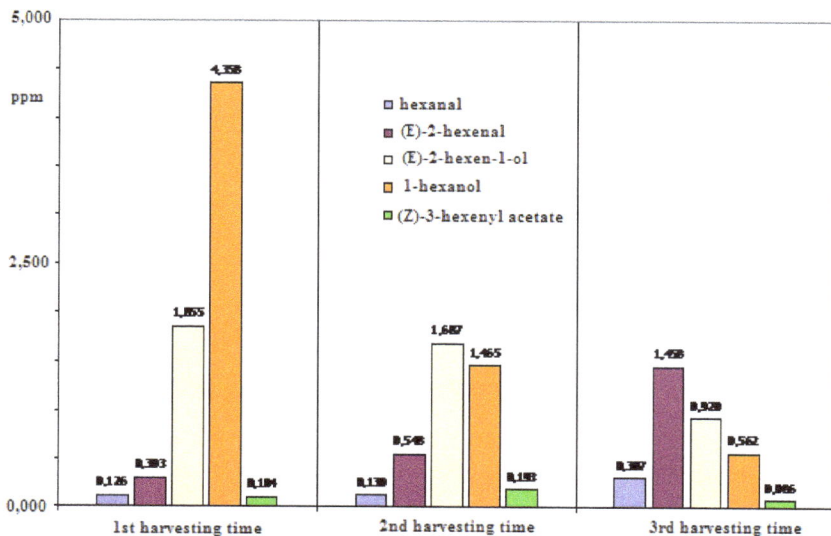

Figure 17. Variation of the distribution of the aroma markers with the harvesting time in the olive oils produced from two different southern Italy cultivar.

The volatile fraction of olive oil is considered as a source of quality data either for its organoleptic properties and for its correlation to genetic, agronomic and environmental factors. Aroma components of products of plant origin are dependent on genetic, agronomic, and environmental factors.[71] The identification and assay of the terminal species of the "lipoxygenase pathway" which are present in the volatile fraction of olive oils, has been performed by electron impact and/or chemical ionization mass spectrometry in a GC/MS Ion Trap apparatus. The quantitative data for each compound were subjected to principal component analysis to characterize the different cultivars of this work. PCA methodology was applied to confirm the hypothesis that the five selected markers of specific lipoxygenase oxidation could be used to differentiate the various cultivars (Figure 17).[72] The same method was applied to distinguish the origin of experimental oils produced from drupes harvested in different areas of Italian Calabria region and of Tunisia. An easy discrimination was achieved from each cluster of samples. Olive oils produced from irrigated and nonirrigated farms in Tunisia were also clearly distinguishable.[73] Quality and safety of oil is also associated to the presence of dangerous organic residues such as phthalates (PAEs). These compounds tend to be distributed mostly in fatty foods and this can cause the presence of remarkable amounts of PAEs in olive oil. Their determination in fatty matrices represents,therefore, a very important goal for the consumers' health and confidence. A rapid method for the analysis of phthalates in olive oil by GC-MS/MS after a GPC clean-up has been developed which exploits the capability of tandem mass spectrometry (GC-MS/MS) for the unequivocal confirmation and accurate quantification of PAEs at low limits of detection (LOD) levels in fatty matrices without the need for a liquid-liquid extraction prior to GPC and for SPE clean-up following GPC.[74] The interest in the identification of secondary metabolites by MS methods has been extended to different olive tissues. An extensive investigation, by means of high-resolution tandem mass spectrometry (MS/MS) has shown that some of these micro components are strongly cultivar dependent, and probably ripening dependent, while others are widespread and present in all the analyzed cultivars.[75] The new secoiridoid metabolites found in drupes reveal that the key molecules produced by secondary metabolism of terpenes can be conjugated with hydroxytyrosol, a secondary metabolite of phenol biosynthesis, through the formation of differently structured glucosides. The origin of this new species could be related to transport phenomena which can be different among the various tissues of a given plant. Moreover, more stringent evidence of the biogenetic similarity of the members of different oleaceae families is provided by the discovery of metabolites typical of ligustrum and fraxinus in olive tissues. Secondary metabolites of Olea europaea leaves have been selected as markers for the discrimination of cultivars and cultivation zones by multivariate analysis; moreover, the statistical approach has been exploited to correlated the identity and relative amounts of metabolites in leaves with the harvesting period. The mean values of the concentration of each compound, detected by tandem mass spectrometry were inputted into PCA analysis. The bidimensional plot (Figure 18) shows a shifting along PC1 going from March–April to January which indicates the increase of concentration of compounds on the left of plot and the decrease for methoxytyrosol glucoside and 2-methoxyhydroxytyrosol glucoside. Moreover, position of samples of July at highest score values on PC2 means a decrease of concentration of variables verbascoside and hydroxytyrosol glucoside in leaves harvested in this period.[76]

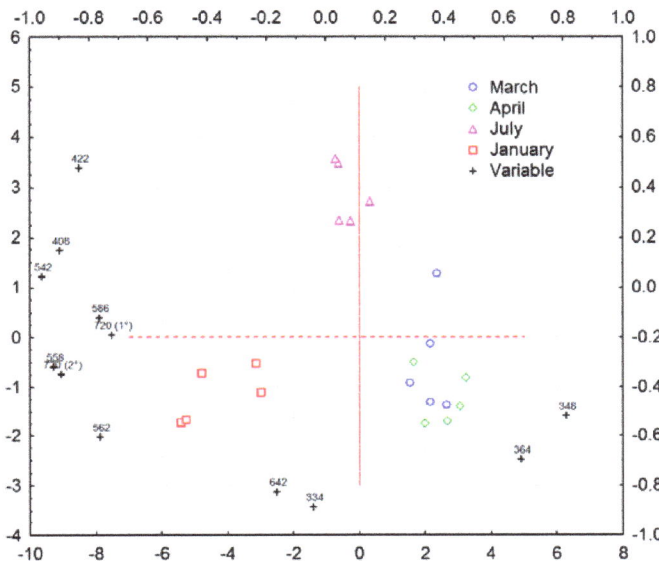

Figure 18. Biplot of principal component scores and loadings for leaves of Carolea cv harvested in March, April, July and January (variables are indicated as m/z values).

Author details

Giovanni Sindona and Domenico Taverna

University of Calabria, Dept. Chemistry, Arcavacata di Rende (CS), Italy

2. References

[1] Q.Q Wu, K.X Fan, H. Q Ruan, R. Zeng, C. H. Shieh. *Chinese Science Bulletin* 2009, 54,732-737.

[2] E.M. Usai , E. Gualdi , V. Solinas , E. Battistel *Bioresour. Technol.* 2010, *101* 7707–7712.

[3] G. Sindona, A. Caruso, A. Cozza, S. Fiorentini, E. Marini, A. Procopio, S. Zicari, *Curr. Med. Chem.* 2012, *19*, 4006-4013.

[4] A. De Nino, F. Mazzotti, E. Perri, A. Procopio, A. Raffaelli, G. Sindona, *J. Mass Spectrom.* 2000, *35*, 461–467.

[5] Di Donna L, Benabdelkamed H, Mazzotti F, Napoli A, Nardi M, Sindona G, Anal. Chem., 2011, 83, 6, 1990-1995.

[6] Cooks RG et al., Science, 2006, 311, 5767, 1566-70.

[7] Mazzotti F, Di Donna L, Benabdelkamel H, Gabriele B, Napoli A, Sindona, G, J. Mass Spectreom, 2010, 45, 4, 358-63.

[8] J. Charrow, S. I. Goodman, E. R.G. McCabe P. Rinaldo. *Genet. Med.* 2000, 2, 267-269.

[9]   J.B. Phílillips, J. Xu, *J.Chromatogr.*, 1995, 703, 327.

[10] Harynuk, James 2009, June 18.

[11] *SciTopics*. Retrieved July 25, 2012.

[12] F. Erni and R.W. Frei, *J.Chromatogr.* 149, 561 (1978).

[13] Schlosser A, Ripkorn R, Bossemeyer D, and Lehmann WD, *Anal. Chem.*, 2001, 73, 170.

[14] Eiceman GA and Stone JA, *Anal. Chem.*, 2004, 76, 391.

[15] Wu C, Siems WF, Asbury GR, and Hill HH, Jr., *Anal. Chem.*, 1998, 70, 4929.

[16] Clemmer DE and Jarrold MF, *J . Mass Spectrom.*, 1997, 32, 577.

[17] Laboda A, *J . Am. Soc. Mass Spectrom.*, 2006, 17, 691.

[18] Hoaglund-Hyzer CS and Clemmer DE, *Anal. Chem.*, 2001, 73, 177.

[19] Ruotolo BT, Gillig KJ, Stone EG, Russell DH, Fuhrer K, Gonin M, and Schultz JA, *Int. J. Mass Spectrom.*, 2002, 219, 253.

[20] Ruotolo BT, McLean JA, Gillig KJ, Stone EG, and Russell DH, *J. Am. Soc. Mass Spectrom.*, 2005, 16, 158; Srebalus Barnes CA and Clemmer DE, *Anal. Chem.*, 2001, 73, 424.

[21] Kellermann M, Muenster, Zomer P, Mol H, 2009, Am Soc Mass Spectrom 20:1464–1476.

[22] Kaufmann A, Butcher P, Maden K, Widmer M, 2010, Anal Chim Acta 673:60–72.

[23] Di Donna, L; Mazzotti, F; Salerno, R; Tagarelli, A; Taverna, D; Sindona, G, Characterization of new phenolic compounds from leaves of Olea europaea L. by high-resolution tandem mass spectrometry, RAPID COMMUNICATIONS IN MASS SPECTROMETRY, 21, 22, 3653-3657, 2007.

[24] Nielen MWF, Van Engelen MC, Zuiderent R, Ramaker R, 2007, Anal Chim Acta 586:122–129.

[25] Kaufmann A, The current role of high-resolution mass spectrometry in food analysis, Anal. Bioanal. Chem., 2012, 403:1233–1249

[26] Grimalt S, Sancho J, Pozo OJ, Hernández F, 2010, J Mass Spectrom 45:421–436.

[27] Kaufmann A, Butcher P, Maden K, Widmer M, 2007, Anal Chim Acta 586:13–21.

[28] S. M. Innis, *Adv. Nutr.* 2011, 2, 275–283.

[29] Amending Regulation (EEC) No 2568/91 on the characteristics of olive oil and olive-residue oil and on the relevant methods of analysis. COMMISSION REGULATION (EU) No 61/2011 of 24 January 2011.

[30] W. Moreda, M.C. Pérez-Camino and A. Cert.. *Grasas y Aceites* 2003, 54, 175-179; P. J. Kalo, A. Kemppinen,. *Eur. J. Lipid Sci. Technol.* 2012, 114, 399–411.

[31] O. Schreiberova, T. Krulikovska, K. Sigler, A. Cejkova and T Rezanka. *Lipids* 2010, 45,743756.

[32] L. Mannina, M. Patumi, N. Proietti, D. Bassi, A. L. Segre.. *J. Agric. Food Chem.* 2001, 49, 2687-2696.

[33] A. Ranalli, L. Pollastri, S. Contento, G. Di Loreto, E. Iannucci, L. Lucera, F. Russi, *J. Agric. Food Chem.* 2002, 50, 3775;G. Vigli, A. Philippidis, A. Spyros, P. Dais, *J. Agric. Food Chem.* 2003, 51 5715.

[34] P. Fronimaki, A. Spyros, S. Christophoridou, P. Dais, *J. Agric. Food Chem.* 2002, 50 2207.

[35] K. Nagy, D. Bongiorno, G. Avellone, P. Agozzino, L.Ceraulo, K. Vekey. *J.Chromatogr. A*, 2005, 1078, 90–97.

[36] P. B. Chapagain and Z. Wiesman. MALDI-TOF/MS. *J. Agric. Food Chem.* 2009, *57*, 1135–1142.

[37] G. Ciafardini, B.A. Zullo, A. Iride. *Food Microbiology*. 2006, *15*, 2360.

[38] B.A. Zullo, G. Cioccia, G. Ciafardini. *Food Microbiology*. 2010, *27*,1035.

[39] P. Agozzino, G. Avellone, D. Bongiorno, L. Ceraulo, S. Indelicato, K. Vèkey.. *J Mass Spectrom*. 2010, *45*, 989.

[40] C.M. Kalua, M.S. Allen, D.R. Bedgood Jr, A.G. Bishop, P.D. Prenzler, K. Robards *Food Chemistry*. 2007, *100*, 273.

[41] H. L. Ngo, K. Jones, T.A. Foglia.. *JAOCS*, 2006, *83*, 629.

[42] C. Saiz-Jimenez, B. Hermosin. *J Anal Appl Pyrol*, 1999, *49*, 349 - 357.

[43] D. Tsuchiy, Y. Murakamib, Y. Ogomaa, Y. Kondob, R. Uchio, S. Yamanaka. *J. Mol. Catal. B-Enzym.* 2005, *35*, 52.

[44] M.Attya, A. Russo, E. Perri, and G. Sindona *J. Mass Spectrom*, 2012, in press.

[45] USDA National Nutrient Database for Standard Reference, Release 15, 2003, www.nal.usda.gov/fnic/foodcomp/Data/SR15/sr15/.html.

[46] Matthaus B, Bruhl L, Comparison of the different methods for the determination of oil content in oilseeds, JAOCS, 78, 95-102, 2001.

[47] Official Methods and Recognized Practices of the American Oil Chemists' Society, 5th ed., Firestone D, Eds., AOCS Press., Champaign, 1998.

[48] Official Methods of Analysis, 17th ed., AOCS International, Gaithersburg, MD, 2000.

[49] Official Methods and Recognized Practices of the American Oil Chemists' Society, 5th ed., Firestone D, Eds., AOCS Press., Champaign, 1998.

[50] Folch J, Lees M, Stanley GHS, A simple method for the isolation and purification of total lipids from animal tissue, J. Biol. Chem., 226, 497-509, 1957.

[51] Ways P, Hanahan DJ, Characterization and quantification of red cell lipids in normal man, J. Lipid Res., 5, 318-328, 1964.

[52] A. De Nino, L. Di Donna, F. Mazzotti, E. Perri, A. Raffaelli, G. Sindona and E. Urso in *Biologically-active Phytochemicals in Food Analysis, Metabolism, Bioavailability and Function*, Royal Society of Chemistry, Cambridge 2001, 131.

[53] A. Procopio , C. Celia , M. Nardi ,M. Oliverio , D. Paolino , G. Sindona, *J Nat Prod.* 2011;*74*, 2377-81.

[54] Servili and Montedoro, 2002, Europ. J. of Lipid Science and Tech., 104, 602-613.

[55] Montedoro, G.; Servili, M.; Baldioli, M.; Selvaggini, R.; Miniati, E.; Macchioni, A. *J. Agr. Food Chem.* 1993, *41*, 2228-2234.

[56] Attya et al, Molecules, 15, 12, 8734-8746, 2010.

[57] Di Donna et al. Anal. Chem., 2011, 83, 1990-1995.

[58] F. Mazzotti, H. Benabdelkamel, L. Di Donna, L. Maiuolo, A. Napoli and G. Sindona. Assay of tyrosol and hydroxytyrosol in olive oil by tandem mass spectrometry and isotope dilution method *Food Chem.* 2012, 135, 1006–1010.

[59] Halliwell, B.; Aeschbach, R.; Loliger, J.; Aruoma, O. I. *Food Chem. Toxicol.* 1995, *33*, 601.

[60] Servili, M.; Salvaggini, R.; Esposto, S.; Taticchi, A.; Montedoro,G.; Marozzi, G. *J. Chromatogr. A* 2004, *1054*, 113-127.

[61] Roche, M.; Dufour, C.; Mora, N.; Dangles, O. *Org. Biomol. Chem.* 2005, *3*, 423-430.

[62] Lee-Huang, S.; Zhang, L.; Huang, P. L.; Chang, Y. T.; Huang, P. L. *Biochem. Biophys. Res. Commun.* 2003, *307*, 1029-1037.

[63] A. Mazziotti, F. Mazzotti, M. Pantusa, L. Sportelli and G. Sindona J. Agric. Food Chem. 2006, *54*, 7444-7449(15).

[64] A. De Nino, N. Lombardo, E. Perri A. Procopio, A. Raffaelli and G.Sindona. *J Mass Spectrom.* 1997, *32*, 533-541.

[65] E. Perri, A. Raffaelli, and G. Sindona *J. Agric. Food Chem.* 1999, *47*, 4156-4160.

[66] A. De Nino, L. Di Donna, F. Mazzotti; I. Muzzalupo, E. Perri,; A. Tagarelli, *Anal. Chem.* 2005, *77*, 5961-5964.

[67] A. De Nino, L. Di Donna, F. Mazzotti, A. Sajjad, G. Sindona, E. Perri, A. Russo, L. De Napoli, L. Filice. *Food Chem.* 2008, *106*, 677–684.

[68] Beauchamp, G. K.; Keast, R.S. J.; Morel, D.; Lin, J.; Pika, J.; Han, Q.; Lee, C. H.; Smith, A. B.; Breslin, P. A. S. *Nature*, 2005, *437*, 45-46.

[69] Vierhuis, E.; Servili, M.; Baldioli, M.; Schols, H.A.; Voragen, A.G.H.; *J Agric Food Chem.* 2001, *49*, 1218-1223.

[70] J. F. Garcıa-Reyes, F. Mazzotti. J. D. Harper, N. A. Charipar, S. Oradu, Z. Ouyang, G. Sindona and R. G. Cooks. *Rapid Commun. Mass Spectrom.* 2009; *23*, 3057–3062.

[71] F. Angerosa, N. d'Alessandro, C. Basti, R. Vito, *J. Agric. Food Chem.* 1998, *46*, 2940-2944; F Angerosa; C. Basti, R. Vito, *J. Agric. Food Chem.* 1999, *47*, 836-839.

[72] C. Benincasa, A. De Nino, N. Lombardo, E. Perri, G. Sindona, and A. Tagarelli. *J. Agric. Food Chem.* 2003, *51*, 733-741.

[73] B. Cavaliere, A. De Nino, F. Hayet, A. Lazez, B. Macchione, C. Moncef, E.Perri, G. Sindona, A. Tagarelli. *J. Agr. Food Chem.* 2007, *55*, 1454.

[74] B. Cavaliere, B. Macchione, G. Sindona and A Tagarelli. J. *Chromatogr. A*, 2008, *1205*, 137–143)

[75] L. Di Donna, F. Mazzotti, A. Napoli, R. Salerno, As. Sajjad and G. Sindona *Rapid Commun. Mass Spectrom.* 2007; *21*, 273–27.

[76] L. Di Donna, F. Mazzotti, A. Naccarato, R. Salerno, A. Tagarelli, D. Taverna and G. Sindona. *Food Chem.* 2010, *121*, 492–496.

Chapter 12

Sensory Analysis of Virgin Olive Oils

Innocenzo Muzzalupo, Massimiliano Pellegrino and Enzo Perri

Additional information is available at the end of the chapter

1. Introduction

The evaluation of the sensory properties and determination of the importance of these properties in consumer product acceptance represent a major accomplishment in sensory analysis.

Most of the sensory evaluation carried out by a trained *panel* involves measurement in two main areas, difference testing and descriptive analysis. Sensory difference tests are procedures used to determine whether judges can distinguish between two similar stimuli. In terms of food, the two stimuli are two very similar food samples. These evaluations are used to determine whether slight changes occur due to either product reformulation or the change in technological processing. Difference tests are particularly well adapted to the assessment of vegetable oils during their processing, being used to control refining efficiency.

Furthermore, they may be used to measure slight flavour variation determined by changes in storage or packaging. There are many different methods for sensory difference testing. They are described in detail in several texts (Amerine *et al.*, 1965; Lawless and Heymann, 1998; Meilgaard *et al.*, 1999).

Descriptive analysis, the other main area of sensory evaluation, precisely describes and measures the sensory attributes of food. It follows that this can be done only with trained judges. There are various methodologies that can be used. One procedure is the Quantitative Descriptive Analysis (QDA) developed by the Tragon Corporation of Palo Alto, California (Stone *et al.*, 1974). Judges examine the food and list all the relevant sensory properties. They then use standards to obtain agreement on the descriptive terms being used. By using a scaling procedure, the judge estimates the intensity of each sensory attribute considered. The raw data obtained are then evaluated through common statistic techniques.

There are several applications for descriptive analysis: product development and assessment, comparison of the sensory properties of products which cannot be compared simultaneously (e.g., fresh olive virgin oil from consecutive years), shelf-life studies, correlation of instrumental and sensory properties, studies of the effects of technological processing on the sensory characteristics of a product, quality assurance, certification of a preset quality standard, etc.

Another branch of sensory analysis is represented by consumer science. Generally, people eat and drink food because they like it. But, food intake is not completely driven by hedonic motives. Other factors such as market segmentation, advertising, price, packaging, opinions and beliefs play a role too. The sensory perception of sensory quality is an important factor in motivating consumer choice. It derives from the intrinsic properties of the product, perceived by the consumer at the moment of buying (colour, shape, aspect, etc) and subsequently from direct individual experiences (odour, taste). The measurement of liking is necessary before a product is launched onto the market, with substantial capital being needed. This can therefore save investing in a product that may not be liked due to a deficiency of sensory qualities.

The most frequently used method to measure acceptability and preference is the 9-point hedonic scale, developed in 1955 (Jones *et al.*, 1955). The subjects that take part in the sensory testing are not trained, and should be relatively naive to this kind of task.

2. Sensory analysis of olive oil

The European Commission (EC) upon issuing the following regulations:

- Commission Regulation n. 2568/91 of 11 July 1991 (EC Regulation, 1991);
- Commission Regulation n. 2472/1997 of 11 December 1997 (EC Regulation, 1997);
- Commission Regulation n. 796/2002 of 06 May 2002 (EC Regulation, 2002);
- Commission Regulation n. 1989/2003 of 06 November 2003 (EC Regulation, 2003);
- Commission Regulation n. 1640/2008 of 04 July 2008 (EC Regulation, 2008);
- Commission Regulation n. 61/11 of 24 January 2011 (EC Regulation, 2011);

set the objective of establishing and developing the criteria needed to evaluate the chemical and sensory characteristics of oil and virgin olive oil as well as the opportune methodology. The introduction of the *Panel Test* has led to an evolution in the concept of oil quality. The necessary cognitions are set in order to carry out the sensory analysis of virgin olive oil. It tries to standardise the behaviour and procedures of the tasters, who should take into consideration not only the more general indications but also those specific for tasting olive oil. These procedures have led to an evaluation sheet being drawn up by the International Olive Oil Council (IOOC), leading to results being obtained from different *Panels*, in different areas of the same country as well as different countries to be compared.

Since 1991, this methodology has been part of the regulations of the European Commission for classifying oils and is described in detail in Appendix II of the EU Regulation n. 2568/91.

The method, Quantitative Descriptive Analysis (QDA), defines the main attributes of oil, both positive as well as negative. Appearance as colour was not selected as a quality parameter of virgin olive oil, as a specific dark-coloured glass is used. The official evaluation sheet used within the European Union to establish the sensory profile of virgin olive oil is shown below (Fig. 1). The values, expressed as centimetres, are statistically processed to calculate the median of each positive and negative characteristic.

COI/T.20/Doc. No 15/Rev. 3
page 11

Figure 1

PROFILE SHEET FOR VIRGIN OLIVE OIL

INTENSITY OF PERCEPTION OF DEFECTS:

Fusty/
muddy sediment

Musty-humid-
earthy

Winey-vinegary -
acid-sour

Metallic

Rancid

Others (specify)

INTENSITY OF PERCEPTION OF POSITIVE ATTRIBUTES:

Fruity

greenly   ripely

Bitter

Pungent

Name of taster:

Sample code:

Date:

Comments:

Figure 1. Sheet for sensory analysis

Sensory analysis is an essential technique to characterize food and investigate consumer preferences. International cooperative studies, supported by the International Olive Oil Council (IOOC) have provided a sensory codified methodology for virgin olive oils, known as the "COI *Panel test*". Such an approach is based on the judgments of a panel of technicians, conducted by a *panel* leader, who has sufficient knowledge and skills to prepare sessions of sensory analysis, motivate judgement, process data, interpret results and draft the report.

The *panel* generally consists of a group of 8 to 12 persons, selected and trained to identify and measure the intensity of the different positive and negative sensations perceived.

Sensory assessment is carried out according to codified rules, in a specific tasting room, using controlled conditions to minimize external influences, using a proper tasting glass and adopting both a specific vocabulary and a profile sheet that includes positive and negative sensory attributes. Collection of the results and statistical elaboration must be standardized. The colour of virgin olive oil, which is not significantly related to its quality, may produce expectations and interferences in the flavour perception phase. In order to eliminate any prejudices that may affect the smelling and tasting phases, *panel* lists use a dark-coloured (blue or amber-coloured) tasting glass.

The *panel* leader is the person responsible for selecting, training and monitoring tasters to ascertain their level of aptitude according to (IOOC/T.20/Doc.15/rev.3). The number of candidates is generally greater than that needed in order to select people that have a grater sensitivity and discriminatory capability. Screening criteria of candidates are founded on sensory capacity, but also on some personal characteristics of candidates. Given this, the *panel* leader will personally interview a large number of candidates to become familiar with their personality and understand habits, hobbies, and interest in the food field. He uses this information to screen candidates and rejects those who show little interest, who are not readily available or who are incapable of expressing themselves clearly.

The determination of the detection threshold of the group of candidates for characteristic attributes is necessary because the "threshold concentration" is a point of reference common to a "normal group" and may be used to form homogeneous *panels* on the basis of olfactory-gustatory sensitivity.

A selection of tasters is made by the intensity rating method, as described by Gutiérrez Rosales (Gutiérrez Rosales *et al.*, 1984). A series of 12 samples is prepared by diluting a virgin olive oil characterized by a very high intensity of a given attribute in an odourless and tasteless medium (refined oil or paraffin). The *panel* leader sends out the candidate, removes one of the 12 tasting glasses from the series, and places the remaining together; the candidate is called back in the room and is asked to correctly replace the testing glass withdrawn from the series by comparing the intensity of this last with that of the others. The test is carried out for fusty, rancid, winey and bitter attributes to verify the discriminating capacity of the candidate on the entire scale of intensities.

The stage training of technicians is necessary to familiarize tasters with the specific sensory methodology, to heighten individual skill in recognizing, identifying and quantifying the sensory attributes and to improve sensitivity and retention with regards to the various attributes considered, so that the end result is precise and consistent. In addition, they learn to use a profile sheet.

The maintenance of the *panel* is made through continuous training during the entire life of the same *panel*, the check of the sensory acuity of tasters, and exercises that allow the measurement of *panel* performance.

Every year, every *panel* must assess a number of reference samples in order to verify the reliability of the results obtained and to harmonize the perception criteria; they must also update the Member State on their activity and on composition changes of their group.

3. Test conditions

The test conditions for tasting are described in COI/T.20/Doc. No. 15/Rev. 3:

"The oil sample for analysis shall be presented in standardised tasting glasses conforming to the standard COI7T.20/Doc. No. 5 (Fig. 2).

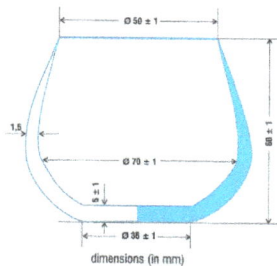

Figure 2. Tasting glass

The glass shall contain 14-16 ml of oil, or between 12.8 and 14.6 g if the samples are to be weighed, and shall be covered with a watch-glass. Each glass shall be marked with a code made up of digits or a combination of letters and digits chosen at random.

Figure 3. Heating the samples when in the glass

The oil samples intended for tasting shall be kept in the glass at 28 °C ± 2 °C throughout the test (Fig. 3).

This temperature has been chosen because it makes it easier to observe organoleptic differences than at ambient temperature and because at lower temperature the aromatic compounds peculiar to these oils volatilize poorly while higher temperatures lead to the formation of volatile compounds peculiar to heated oils. The test room must be at a temperature between 20 and 25 °C.

The morning is the best time for tasting oils. It has been proved that there are optimum perception periods with regards to taste and smell during the day. Meals are preceded by a period in which olfactory-gustatory sensitivity increases, whereas afterwards this perception decreases. However, this criterion should not be taken to the extreme where hunger may distract the tasters, thus decreasing their discriminatory capacity; therefore, it is recommended to hold the tasting session between 10:00 in the morning and 12:00 noon (Fig. 4).

Figure 4. Cabins of the test room

The following recommendations apply to the conduct of the tasters during their work.

When called by the panel leader to participate in an organoleptic test, tasters should be able to attend at the time set beforehand and shall observe the following:

-   they shall not smoke or drink coffee at least 30 minutes before the time set for the test;
-   they must not have used any fragrance, cosmetic or soap whose smell could linger until the time of the test;

- they shall fast at least one hour before the tasting is carried out;
- should they feel physically unwell, and in particular if their sense of smell or taste is affected, or if they are under any psychological effect that prevents them from concentrating on their work, the tasters shall refrain from tasting and shall inform the panel leader accordingly;
- when they have complied with the above, the tasters shall take up their place in the booth allotted to them in an orderly, quiet manner;
- they shall carefully read the instructions given on the profile sheet and shall not begin to examine the sample until fully prepared for the task they have to perform (relaxed and unhurried). If any doubts should arise, they should consult the panel leader in private;
- they must remain silent while performing their tasks;
- they must keep their mobile phone switched off at all times to avoid interfering with the concentration and work of their colleagues.

4. Technique for tasting of virgin olive oil

The technique applied for tasting is the one described in annex XXI of (EEC) Reg. N. 2568/91:

"the taster shall pick up the glass (Fig. 2), keeping it covered with the watch glass, and shall bend it gently fully in this position so as to wet the inside as much as possible.

Once this stage is completed, he shall remove the watch-glass and smell the sample taking even, slow deep breaths until he has formed a criterion on the oil under assessment. Smelling shall not exceed 30 seconds. If no conclusion has been reached during this time, he shall take a short rest before trying again (Fig. 5).

Figure 5. The olfactory test during

When the olfactory test has been performed, the taster shall then judge the flavour (overall olfactory-gustatory-tactile sensation). To do so, he shall take a small sip of approximately 3 ml of oil. It is very important to distribute the oil throughout the whole of the mouth cavity, from the front part of the mouth and tongue along the sides to the back part and to the palate support, since it is a known fact that the perception of the four primary tastes, sweet, salty, acid and bitter varies in intensity depending on the area of the tongue and palate.

It should be stressed that it is essential for a sufficient amount of the oil to be spread very slowly over the back of the tongue towards the throat while the taster concentrates on the order in which the bitter and pungent stimuli appear; if this is not done, both of these stimuli may escape notice in some oils or else the bitter stimulus may be obscured by the pungent stimulus.

Taking short, successive breaths, drawing in air through the mouth, enables the taster not only to spread the sample extensively over the whole of the mouth but also to perceive the volatile aromatic components via the back of the nose.

Tactile sensation shall also be taken into consideration. Consequently, fluidity, stickiness and sharpness or sting shall be noted down when detected, and if required for the test, their intensity shall be quantified.

When organoleptically assessing a virgin olive oil, only one sample shall be evaluated in each session to avoid the contrast effect that could be produced by immediately tasting other samples.

As successive tasting produces fatigue or loss of sensitivity, it is important to use a product that can eliminate the remains of the oil from the preceding tasting from the mouth.

The use of a small slice of green apple (about 15 g) is recommended which, after being chewed, can be spat out into a spittoon. Then the mouth should be rinsed out with a little water at ambient temperature. At least 15 minutes should lapse between the end of one session and the start of the next.

The panel leader will collect the profile sheets completed by each taster and review the intensities assigned to the different attributes. Should they find any anomaly, they shall invite the taster to revise his or her profile sheet and, if necessary, to repeat the test.

The panel leader will enter the assessment data of each panel member in a computer program like that appended to this method with a view to statistically calculating the results of the analysis, based on their median.

5. Definition and classification of olive oil.

Virgin olive oil is the oil obtained from the fruit of the olive tree either by mechanical or other physical means under conditions, particularly thermal conditions, that do not lead to alterations in the oil, and which has not undergone any treatment other than washing, decantation, centrifugation and filtration.

In Table 1, the various classes of olive oil as divided by European legislation are reported. The subdivisions in different classes are based on the degree of acidity as well as other analytical parameters and sensory evaluation indices.

1. *Extra virgin olive oil*: is the virgin olive oil which has a free acidity, expressed as oleic acid, of not more than 0.8 gram per 100 grams, and the sensory characteristics with median defects is 3.5 and the median of fruitiness greater than 0.
2. *Virgin olive oil*: is the virgin olive oil which has a free acidity, expressed as oleic acid, of not more than 2 grams per 100 grams and the sensory characteristics with median defects greater than 0, but less than or equal to 2.5 and the median of fruitiness greater than 0.
3. *Lampante olive oil*: is the virgin olive oil which has a free acidity, expressed as oleic acid, of more than 2 grams per 100 grams and the sensory characteristics with median defects greater than 3.5 or if the median defects is less than or equal to 3.5 and the median of fruitiness is 0. Such olive oil is intended for refining purposes.
4. *Refined olive oil*: is the olive oil obtained from lampante olive oils by refining methods which do not lead to alterations in the initial glyceridic structure, which has a free acidity, expressed as oleic acid, of less than or equal to 0.3 grams per 100 grams.
5. *Blended olive oil*: is the oil consisting of a blend of refined olive oil and virgin olive oil which has a free acidity, expressed as oleic acid, of less than or equal to 1 grams per 100 grams. It can be used for human consumption.
6. *Crude olive-pomace oil*: is the oil obtained by treating olive pomace with solvents, to the exclusion of oils obtained by re-esterification processes and of any mixture with oils of other kinds. This oil is intended for refining with a view to its use in food for human consumption.
7. *Refined olive-pomace oil*: is the oil obtained from crude olive-pomace oil by refining methods which do not lead to alterations in the initial glyceridic structure, which has a free acidity, expressed as oleic acid, of less than or equal to 0.3 grams per 100 grams.
8. *Olive-pomace oil*: is the oil comprising the blend of refined olive-pomace oil and virgin olive oil which has a free acidity, expressed as oleic acid, of less than or equal to 1 gram per 100 grams. It can be used for consumption, but in no case should this blend be called "olive oil".

One important consideration is that the first two oils are the best for human consumption due to them being within the set parameters, when obtained directly from the olive press. The third type of oil cannot be consumed until it has been rectified, which gives the fourth type of oil. A small amount of extra-virgin olive oil or virgin olive oil is added to this rectified oil and is known as "Olive oil".

ANNEX I

ANNEX I

OLIVE OIL CHARACTERISTICS

Category	Fatty acid methyl esters (FAME) and fatty acid ethyl esters (FAEE)	Acidity (%) (*)	Peroxide index mEq O₂/kg (*)	Waxes mg/kg (**)	2-glyceryl monopalmitate (%)	Stigmastadiene mg/kg (*)	Difference: ECN 42 (HPLC) and ECN 42 (theoretical calculation)	K₂₃₂ (*)	K₂₇₀ (*)	Delta-K (*)	Organoleptic evaluation Median defect (Md) (*)	Organoleptic evaluation Fruity median (Mf) (*)
1. Extra virgin olive oil	Σ FAME + FAEE ≤ 75 mg/kg or 75 mg/kg < Σ FAME + FAEE ≤ 150 mg/kg and (FAEE/FAME) ≤ 1.5	≤ 0.8	≤ 20	≤ 250	≤ 0.9 if total palmitic acid % ≤ 14 %	≤ 0.10	≤ 0.2	≤ 2.50	≤ 0.22	≤ 0.01	Md = 0	Mf > 0
					≤ 1.0 if total palmitic acid % > 14 %							
2. Virgin olive oil	—	≤ 2.0	≤ 20	≤ 250	≤ 0.9 if total palmitic acid % ≤ 14 %	≤ 0.10	≤ 0.2	≤ 2.60	≤ 0.25	≤ 0.01	Md ≤ 3.5	Mf > 0
					≤ 1.0 if total palmitic acid % > 14 %							
3. Lampante olive oil	—	> 2.0	—	≤ 300 (*)	≤ 0.9 if total palmitic acid % ≤ 14 %	≤ 0.50	≤ 0.3	—	—	—	Md > 3.5 (*)	—
					≤ 1.1 if total palmitic acid % > 14 %							
4. Refined olive oil	—	≤ 0.3	≤ 5	≤ 350	≤ 0.9 if total palmitic acid % ≤ 14 %	—	≤ 0.3	—	≤ 1.10	≤ 0.16	—	—
					≤ 1.1 if total palmitic acid % > 14 %							
5. Olive oil composed of refined and virgin olive oils	—	≤ 1.0	≤ 15	≤ 350	≤ 0.9 if total palmitic acid % ≤ 14 %	—	≤ 0.3	—	≤ 0.90	≤ 0.15	—	—
					≤ 1.0 if total palmitic acid % > 14 %							
6. Crude olive-residue oil	—	—	—	> 350 (*)	≤ 1.4	—	≤ 0.6	—	—	—	—	—
7. Refined olive-residue oil	—	≤ 0.3	≤ 5	> 350	≤ 1.4	—	≤ 0.5	—	≤ 2.00	≤ 0.20	—	—
8. Olive-residue oil	—	≤ 1.0	≤ 15	> 350	≤ 1.2	—	≤ 0.5	—	≤ 1.70	≤ 0.18	—	—

(*) Total isomers which could (or could not) be separated by capillary column.
(*) Or where the median defect is less than or equal to 3.5 and the fruity median is equal to 0.
(*) Oils with a wax content of between 300 mg/kg and 350 mg/kg are considered to be lampante olive oil if the total aliphatic alcohol content is less than or equal to 350 mg/kg or if the erythrodiol and uvaol content is less than or equal to 3.5 %.
(*) Oils with a wax content of between 300 mg/kg and 350 mg/kg are considered to be crude olive-residue oil if the total aliphatic alcohol content is above 350 mg/kg and if the erythrodiol and uvaol content is greater than 3.5 %.

Category	Acid content (*)						Total mono-oleic trans-linolenic isomers (%)	Total trans-linoleic + trans-linolenic isomers (%)	Sterol composition						Total sterols mg/kg	Erythrodiol and uvaol (%) (**)
	Myristic (%)	Linolenic (%)	Arachidic (%)	Eicosenoic (%)	Behenic (%)	Lignoceric (%)			Cholesterol (%)	Brassicasterol (%)	Campesterol (%)	Stigmasterol (%)	Beta-sitosterol (%)	Delta-7-stigmastenol (%)		
1. Extra virgin olive oil	≤ 0.05	≤ 1.0	≤ 0.6	≤ 0.4	≤ 0.2	≤ 0.2	≤ 0.05	≤ 0.05	≤ 0.5	≤ 0.1	≤ 4.0	< Camp.	≥ 93.0	≤ 0.5	≥ 1 000	≤ 4.5
2. Virgin olive oil	≤ 0.05	≤ 1.0	≤ 0.6	≤ 0.4	≤ 0.2	≤ 0.2	≤ 0.05	≤ 0.05	≤ 0.5	≤ 0.1	≤ 4.0	< Camp.	≥ 93.0	≤ 0.5	≥ 1 000	≤ 4.5
3. Lampante olive oil	≤ 0.05	≤ 1.0	≤ 0.6	≤ 0.4	≤ 0.2	≤ 0.2	≤ 0.10	≤ 0.10	≤ 0.5	≤ 0.1	≤ 4.0	—	≥ 93.0	≤ 0.5	≥ 1 000	≤ 4.5 (*)
4. Refined olive oil	≤ 0.05	≤ 1.0	≤ 0.6	≤ 0.4	≤ 0.2	≤ 0.2	≤ 0.20	≤ 0.30	≤ 0.5	≤ 0.1	≤ 4.0	< Camp.	≥ 93.0	≤ 0.5	≥ 1 000	≤ 4.5
5. Olive oil composed of refined and virgin olive oils	≤ 0.05	≤ 1.0	≤ 0.6	≤ 0.4	≤ 0.2	≤ 0.2	≤ 0.20	≤ 0.30	≤ 0.5	≤ 0.1	≤ 4.0	< Camp.	≥ 93.0	≤ 0.5	≥ 1 000	≤ 4.5
6. Crude olive-residue oil	≤ 0.05	≤ 1.0	≤ 0.6	≤ 0.4	≤ 0.3	≤ 0.2	≤ 0.20	≤ 0.10	≤ 0.5	≤ 0.2	≤ 4.0	—	≥ 93.0	≤ 0.5	≥ 2 500	> 4.5 (*)
7. Refined olive-residue oil	≤ 0.05	≤ 1.0	≤ 0.6	≤ 0.4	≤ 0.3	≤ 0.2	≤ 0.40	≤ 0.35	≤ 0.5	≤ 0.2	≤ 4.0	< Camp.	≥ 93.0	≤ 0.5	≥ 1 800	> 4.5
8. Olive-residue oil	≤ 0.05	≤ 1.0	≤ 0.6	≤ 0.4	≤ 0.3	≤ 0.2	≤ 0.40	≤ 0.35	≤ 0.5	≤ 0.2	≤ 4.0	< Camp.	≥ 93.0	≤ 0.5	≥ 1 600	> 4.5

(*) Other fatty acids content (%): palmitic 7.5-20.0; palmitoleic 0.3-3.5; heptadecanoic ≤ 0.3; heptadecenoic ≤ 0.3; stearic 0.5-5.0; oleic 55.0-83.0; linoleic 3.5-21.0
(*) Total Delta-5,23-stigmastadiene+chlerosterol+beta-sitosterol+sitostanol+delta-5-avenasterol+delta-5,24-stigmastadienol.
(*) Oils with a wax content of between 300 mg/kg and 350 mg/kg are considered to be lampante olive oil if the total aliphatic alcohol content is less than or equal to 350 mg/kg or if the erythrodiol and uvaol content is less than or equal to 3.5 %.
(*) Oils with a wax content of between 300 mg/kg and 350 mg/kg are considered to be crude olive-residue oil if the total aliphatic alcohol content is above 350 mg/kg and if the erythrodiol and uvaol content is greater than 3.5 %.

Notes:
(a) The results of the analyses must be expressed to the same number of decimal places as used for each characteristic.
The last digit must be increased by one unit if the following digit is greater than 4.
(b) If just a single characteristic does not match the values stated, the category of an oil can be changed or the oil declared impure for the purpose of this Regulation.
(c) If a characteristic is marked with an asterisk (*), referring to the quality of the oil this means the following:
  — for lampante olive oil, it is possible for both the relevant limits to be different from the stated values at the same time
  — for virgin olive oil, if at least one of these limits is different from the stated values, the category of the oil will be changed, although they will still be classified in one of the categories of virgin olive oil.
(d) If a characteristic is marked with two asterisks (**), referring to the quality of the oil, this means that for all types of olive-residue oil, it is possible for both the relevant limits to be different from the stated values at the same time.

Table 1. Analytical and sensory properties of olive oils

6. Olive oil tasting attributes

Olive oil as judged by experts shows a multitude of either positive or negative characteristics.

6.1. Positive attributes

Almond: light smell recalling that of fresh or dried almond.

Apple: a sensation recalling this fruit.

Artichoke: a smell recalling raw artichoke.

Astringent: a puckering sensation in the mouth created by tannin.

Bitter: this is a preferred characteristic taste of olive oils, if it is not too highly intense.

Fruity: range of smells (dependent on variety) characteristic of oil from healthy fresh fruit, green or ripe, perceived directly and/or retro nasally. Fruitiness is qualified as green if the range of smells is reminiscent of green grass. Fruitiness is qualified as ripe if the range of smells is reminiscent of ripe fruit and is characteristic of oil from green and ripe fruit.

Green grass: a sensation recalling that of freshly cut grass.

Hay: a smell recalling that of dried grass.

Spicy: a tactile sensation similar to that of a light chilli pepper, especially in the back of the throat, which can force a cough.

6.2. Negative attributes

Brine: salty taste of oil made from brined olives.

Coarse: a tactile sensation in the mouth due to texture of oil.

Cucumber: off flavour from prolonged storage, particularly in tin.

Dreggiest: odour of warm lubricating oil and is caused by the poor or lack of the decanting process.

Earthy: this term is used when oil has acquired a musty humid odour because it has been pressed from unwashed, muddy olives.

Esparto: hemp-like smell acquired when olive paste has been spread on Esparto mats. Smells may differ according to whether the mates are green or dried.

Hemp: caused by the use of filtering *panels*, which are not perfectly clean, and recalls hemp.

Flat: oils which have lost their characteristic aroma and have neither taste nor smell.

Frozen: due to olives which have been exposed to freezing temperatures. When cooked, this oil gives off very unpleasant odours.

Fusty: due to olives fermenting in piles while in storage waiting to be pressed.

Grubby: smell imparted by grubs of the olive fly. The smell is both rotten and putrid at the same time.

Heated: prolonged heating during extraction processing.

Muddy: typical odour of oil that has been stored to long on its own sediment.

Musty: mouldy smell from olives being stored too long before pressing.

Metallic: oils processed or stored with extended contact to metal surfaces.

Rancid: old oils that have started oxidizing due to exposure to light or air.

Vegetable water: oils that have absorbed the unpleasant odours and flavours of the vegetable water after pressing that they have remained in contact for too long.

Wine-vinegar: typical odour of wine or vinegar due to fermentation of olives.

7. Factors that affect the sensory quality of virgin olive oil

The evaluation of exactly how various factors affect the sensory quality of the final product is essential in order to distinguish among different types of oil. The following factors all play a role in producing high quality olive oil (Angerosa, 2002; Angerosa *et al.*, 2004, Benincasa *et al.*, 2011).

Cultivar. The numerous varieties constitute an important element for the production of extra virgin olive oils, characterised by different organoleptic characteristics.

Cultivation Techniques (irrigation, fertilization, treatment of plants, diseases etc). Among the environmental factors that influence the quality of the olives and therefore the oil, both the temperature and the amount of water available have an important role, with the first affecting the acidic composition of the olives, while the latter the amount of phenolic substances.

Maturation of the olive. An early harvest generally gives a more bitter and spicy oil due to the high phenol content.

Harvest and Storage of the olive. The quality of the oil is highly conditioned by the state of integrity of the olive. Traditional manual harvesting techniques avoid damaging the fruit in comparison to mechanical methods. Storage of the olives in not very big crates, avoids an excessive mass of olives that could either become crushed or overheated, facilitating attacks from micro-organisms as well as oxidation and fermentation.

De-leafing and washing of the olive. Before being processed, the olives must be cleaned of any superfluous material, including leaves, and branches. These are all elements that can negatively influence the quality of the oil.

Pressing. The olives are broken during the pressing phase with the skin and the pulp being lacerated as well as the stone crushed. The press can be a traditional "pan-mill" one, either in a discontinuous system or combined with an extraction system in order to carry out continual centrifugation. These presses can either be hammers or disks. Metal presses have a more violent pressing of the olive (above all hammered ones) as well as a greater laceration of the skin, giving a higher extraction of the phenolic composites and therefore a more bitter and spicy oil that lasts longer.

Kneading. Prolonged kneading and high temperatures could increase the activity of pectoltic and proteolitic enzymes, negatively modifying the chemical-physical characteristics and therefore the quality of the oil.

Extraction. The systems to separate the liquid from the solid can be divided into two groups:

a.   mechanical pressure on the paste through a series of operations that make the process discontinuous,

b.   by centrifugation that is continuous. The extraction process by pressure has the risk of contaminating the oil due to wear and dirtying of the filters. The continual centrifugation system guarantees greater hygiene and therefore gives oil with elevated qualitative characteristics.

Centrifugation. The oily liquid contains a certain amount of water (called "of vegetation") that is eliminated by centrifuging the product. This operation allows the suspended solid substances to also be eliminated. Water is often added in order to rid the oil of the watery impurities. However, this reduces the phenolic substances content.

Clarification and filtration. The oil obtained from centrifugation still contains mucilage, water and small pieces of the fruit. It is also turbid and opalescent. A clarification process is then carried out in order to eliminate these substances that can favour hydrolysis and/or oxidation. Traditional clarification methods include sedimentation, and have now been substituted by filtration. "Light" filtration systems are preferable rather than more drastic ones that can provoke a reduction in the anti-oxidants and subsequently a reduced shelf-life with the possibility of turning rancid.

Conservation of the oil. In order to maintain both the chemical-physical and organoleptic properties of the oil, the conservation conditions must be controlled. The main factors affecting the conservation of oil are:

a. the temperature (12-15°C),

b. light (oil should be stored in the dark otherwise the photo-oxidization process on the polyunsaturated fatty acids can determine the rancid defect),

c. oxygen in the air (a series of oxidation reactions occur when the oil comes into contact with the air, modifying the chemical composition and subsequently the colour, smell and flavour). It is therefore good practice to store extra virgin olive oil in a sealed environment.

8. Conclusion

Virgin olive oil is the only product that since 1991 has undergone sensory analysis regulated by European norms. It can be noted that the methodology for the sensory evaluation of virgin olive oil is well documented and standardized. Sensory analysis of olive oil is the evaluation of an oil's organoleptic attributes, which are appreciated through the senses of smell and taste. Sensory analysis is an essential part of evaluating olive oil quality and complements chemical analysis, both of which are requirements for determining the quality of olive oil according to International Olive Council (IOC) and new USDA standards.

The studies on olive oil have highlighted that the sensory qualities are affected by several agronomic and climatic parameters as well as extraction process. Due to here being a variation in the chemical composition of the oil, there is a possibility that these chemical compounds could be considered markers to be used to differentiate the various factors that affect overall quality.

The results of the quantitative descriptive analysis carried out by 9 members of staff of the CRA-OLI on the virgin olive oils produced by single varieties cultivated in the collection field of CRA-OLI are given in the elaiographic cards attached to Chapter *"Description of varieties"*

Author details

Innocenzo Muzzalupo*, Massimiliano Pellegrino and Enzo Perri
Agricultural Research Council - Olive Growing and Oil Industry Research Centre, Rende (CS), Italy

Acknowledgement

The authors thank all tasters and the panel leader of the panel at the CRA-OLI. The author thank the Projects: CERTOLIO, GERMOLI and OLIOPIU' for financial support.

---

* Corresponding Author

9. References

Amerine, M.A.; Pangborn, R.M. & Roessler, E.B. (1965). Principles of sensory evaluation of food. Academic Press, New York, ISBN 0120561506.

Angerosa F. (2002). Influence of volatile compounds on virgin olive oil quality evaluated by analytical approaches and sensor panels. *European Journal of Lipid Science and Technology*, Vol. 104, pp. 639–660, ISSN 1438-9312.

Angerosa, F.; Servili, M; Selvaggini, R.; Taticchi, A.; Esposto, S. & Montedoro, G.F. (2004). Volatile compounds in virgin olive oil: occurance and their relationship with the quality. *Journal of Chromatography A*, Vol. 1054, pp. 17-31, ISSN 0021-9673.

Benincasa, C.; Russo, A.; Romano, E.; Elsorady, M.E.; Perri, E. & Muzzalupo I. (2011). Chemical and sensory analysis of some egyptian virgin olive oils. *Journal Nutrition & Food Sciences*, Vol. 1, pp. 1-8, ISSN 0034-6659.

European Community, Commission Regulation (1991) 2568/91. On the characteristics of olive oil and olive residue oil and on the relevant methods of analysis. *Official Journal of the European Communities*, July 11, L248, 1-83.

European Community, Commission Regulation (1997) 2472/97. Amending Regulation No 2568/91/EEC. *Official Journal of the European Communities*, December 11, L341, 25-39.

European Community, Commission Regulation (2002) 796/2002. Amending Regulation No 2568/91/EEC. *Official Journal of the European Communities*, May 6, L128, 8-28.

European Community, Commission Regulation (2003) 1989/2003. Amending Regulation No 2568/91/EEC. *Official Journal of the European Communities*, November 6, L295, 57-77.

European Community, Commission Regulation (2008) 640/2008. Amending Regulation No 2568/91/EEC. *Official Journal of the European Communities*, July 4, L178, 11-16.

European Community, Commission Regulation (2011) 61/2011. Amending Regulation No 2568/91/EEC. *Official Journal of the European Communities*, January 24, L23, 1-14.

Gutiérrez-Rosales, F.; Risco, M.A. & Gutiérrez González-Quijano, R. (1984). Selección de catadores mediante el método de clasificación por intensidad. *Grasas y Aceites*, Vol.35, pp. 310-314, ISSN 0017-3495.

International Olive Oil Council (2007). Glass for oil tasting. IOOC/T.20/Doc. 5/rev.1

International Olive Oil Council (2010). Sensory analysis of olive oil. Method for the organoleptic assessment of virgin olive oil. IOOC/T.20/Doc. No 15/Rev. 3.

Jones, L.V.; Peryam, D.R. & Thurstone, L.L. (1955). Development of a scale for measuring soldiers' food preferences. *Food Research*, Vol. 20, pp. 512-520.

Lawless, H.T. & Heymann, H. (1998). Sensory evaluation of food: principle and practices. Ed. Chapman & Hall. New York, ISBN 0-412-99441-0.

Meilgaard, M.; Civille, G.V. & Carr, B.T. (1999). Sensory evaluation techniques. 3rd ed. CRC Press, Boca Raton, Florida, ISBN 0-8493-0276-5.

Stone, H. & Sidel, J.L. (2004). Sensory evaluation practise. 3nd ed., Academic Press, San Diego, CA, ISBN 0-12-672690-6.

Chapter 13

Olive Oil Traceability

Enzo Perri, Cinzia Benincasa and Innocenzo Muzzalupo

Additional information is available at the end of the chapter

1. Introduction

The market for imported, premium priced foods has increased dramatically over recent years, as consumers become ever more aware of products originating from around the world. There are many food products that are of superior quality (taste, texture, fragrance etc.) because of the locale in which they are cultivated. Environmental conditions, such as local climate and soil characteristics, combine to yield crops that exhibit specific traits. Clearly, higher quality products demands higher market prices, therefore unscrupulous traders may attempt to increase profits by deliberately mislabelling foods, or by increasing the volume of a good quality batch through adulteration with sub-standard produce.

In recent years, there has been increasing legislation to protect the rights of both the consumer and honest producers. On 2009 the European Union (EU) Member States agreed to require origin labelling for virgin and extra virgin olive oils (EC Regulation 182/2009) to defend consumers need about true characteristics and origin. To enforce these laws, a measure of the authenticity of samples must be made, most often in the form of proving the presence/absence of adulterants, or verifying geographical or cultivar origin by comparison with known and reliable samples.

The term traceability is often used in modern language, however its usage is applied for different meanings. In fact, in certain cases it is used to describe the origins of the raw materials, whilst in others to describe from which geographical region a foodstuff originates. It is also used more generically to identify a particular producer or from where it has been purchased. Many consider it as a guarantee of food safety and therefore obligatory for all food products; others however as a guarantee of quality and therefore a voluntary value.

Also the applied legislation is highly complex, with lawmakers giving different functions to traceability depending on the application and environment of the law. With the event of the food crisis in the 90's it was evident that new guidelines were urgently required to simplify and streamline the processes which could also be applied to food contamination. For this reason, the new laws of the European Union were considered as an umbrella and termed

"General Laws for food safety" and was not applied only to food safety but also: to introduce the concept of traceability where food companies (producers, manufacturers and importers) need to guarantee to be able to demonstrate the traceability of every food, animal feed and ingredient, showing the chain from producer to consumer; create the " European Food Safety Authority" (EFSA) to unify the various commities and to make public the scientific process and risks; to strenghthen the ealry warning system adopted by European governments and the European Commission to enable rapid interventions in cases of food safety in the human and animal food chains.

In this chapter, we will only consider the "traceability" as a process that allows us to track "from downstream to upstream" the informations distributed along the olive-oil production chain, from producer to consumer (from farm to fork), and to include a evaluation of analytical methods useful in ascertaining what is claimed on the label, in according to the Regulation (CE) 178/2002.

2. Traceability in regulation (EC) 178/2002

Since 1 January 2005, a rule has required that EU countries implement labelling and identification procedures for products sold by farmers, producers and first importers to the EU to enable and facilitate their traceability when they are put on the market. The main purpose is to be able to initiate a withdrawal and/or recall procedure for products in the event of a food crisis. The quality of traceability will enable targeted and precise withdrawals. It will also limit the extent of recalls and ensure the removal of holds on products that are not involved.

The law (EC Regulation No. 178/2002) defines traceability as: "The ability to reconstruct and follow a food, feed, a food-producing animal or substance intended to be, or to join a food or feed, through all stages of production, processing and distribution" (Article 3, paragraph 15).

Many use the terms tracking and traceability synonymously. In reality, these two terms identify two inverse processes:

- tracking identifies the location of a product from upstream to downstream in the chain and at every stage of the journey;
- traceability or tracing is the inverse process, which allow us to gather the information previously issued. It is evident that the two processes are strongly related and based on the same system.

Traceability does not refer to the production of a generic good. It makes each unit of production physically identifiable, managing production processes that are determined by "lots", and manage traceability means identifying each group of products and following the path.

It is necessary to record information relating to inputs (products and companies), processing (product lots, which lots and what end products) and outflows (products/companies /clients). The key is to define the composition of a set of products that have undergone the same process of transformation.

Moreover, the amount of information that identifies a batch may vary and of course the complexity of the whole system increases with what information the company chooses to include in the identification of a lot.

Internal traceability, then, helps to express the internal procedures of each company to trace the origin of materials used.

The label is the instrument through which information is transferred to consumers. It's easy to understand why this essential requirement is effective.

Finally, the information given in the production lot should be able to trace all the links along the chain, back to the first producer and/or supplier of the product or substance to whom it belongs. Traceability chain is an inter-company process, resulting from internal processes of every operator in the industry. These systems should be linked by efficient information systems. It is not governed by a single person in the chain, the relations between the operators allow the tracing of the chain.

3. Analytical methods for the traceability of extra virgin olive oil

The increase in the demand for high-quality *olive oil*s has led to the appearance in the market of *olive oil*s elaborated with specific characteristics. They include oils of certain regions possessing well-known characteristics, that is, *olive oil*s with a denomination of origin, or with specific olive variety composition, that is, coupage or monovarietal *olive oil*s. Monovarietal *olive oil*s have certain specific characteristics related to the olive variety from which they are elaborated (Montealegre *et al.*, 2010). However, coupage *olive oi*'s are obtained from several olive varieties to achieve a special flavor or aroma.

The appearance of denominations and protected indications of origin has promoted the existence of oils labeled according to these criteria. Regulation 2081/92, to promote and protect food products, created the systems known as:

- Protected Designation of Origin (PDO);
- Protected Geographical Indication (PGI);
- "Traditional Speciality Guaranteed" (TSG).

An *olive oil* with a PDO denomination requires meeting precise definition of several parameters such as cultivar, geographical origin, agronomic practice, production technology, and organoleptic qualities (Giménez *et al.*, 2010), and all of these parameters have to be investigated to study its traceability and to certify its quality. The introduction of certifications of origin and quality for virgin *olive oil* as PDO makes necessary the implementation of traceability procedures.

Any research dealing with *olive oil* traceability is focused on investigating the botanical or geographical origin. In both cases, the selection of the markers (compounds with discriminating power) to be studied is complicated because the composition of extra virgin *olive oil*s is the result of complex interactions among olive variety, environmental conditions, fruit ripening, and oil extraction technology (Araghipour *et al.*, 2006). The verification of the cultivars employed to produce an *olive oil* sample may contribute to address the oil origin. This fact may have commercial interest in the case of monovarietal *olive oils* or *olive oils* with PDO because these high-quality *olive oils* may be adulterated by other oils of lower quality, using anonymous or less costly cultivars (Sanz-Cortes *et al.*, 2003).

Unfortunately, morphological traits have been difficult to evaluate, are affected by subjective interpretations, and are severely influenced by the environment and plant developmental stage (Japon-Lujan *et al.*, 2006). Nowadays, several efforts have been focused on the investigation of one or several compounds present in *olive oils* usable to differentiate olive varieties (Sanz-Cortes *et al.*, 2003).

Compositional markers (substances that take part of the composition of the olive oils) include major and minor components. Major components such as sterols, phenolic compounds, volatile compounds, pigments, hydrocarbons, tocopherols, fatty acids and triglycerides may provide basic information on olive cultivars. Minor components can provide more useful information and have been more widely used to differentiate the botanical origin of *olive oils* (Howarth and Vlahov, 1996; Lanteri *et al.*, 2002).

In recent years, there has been increasing legislation to ensure consumer confidence and to protect the rights of both the consumer and honest producers.

It is of great importance the development of analitycal methods to verify the correspondence between what is stated on the label and what is contemplated in the documents (EC Regulation 182/2009 and Reg. 834/2007) in relation to the production of olives and extra virgin olive from organic farming. Moreover, to enforce these laws, a measure of the authenticity of samples must be made, most often in the form of proving the presence/absence of adulterants, or verifying geographical or cultivar origin by comparison with known and reliable samples. The latter method often includes the use of multivariate statistical techniques such as principal components analysis, linear discriminant analysis, canonical variance analysis and partial least squares regression to investigate sample data (Giménez *et al.*, 2010).

3.1. Chromatographic techniques for the traceability of olive oil

In the last decade, gas chromatography (GC), high performance liquid chromatography (HPLC), supercritical fluid chromatography (SFC), and capillary electrophoresis (CE) have been widely used in the authentication analysis of olive oil. The fat, sterols, protein, carbohydrate or other natural compound profiles (Aparicio-Ruiz, 2000; Benincasa *et al.*, 2003; Vichi et al., 2005; Lopez Ortiz et al., 2006; Temime et al., 2006; Canabate-Diaz et al., 2007; Cavaliere et al., 2007; Haddada et al., 2007; Vichi et al., 2007; Lazzez et al., 2008) have been used to provide species and geographical differentiation. In several cases, enantiomeric composition has been utilised (Rossmann et al., 2000). These methods of discrimination rely on samples of different species and/or different geographical origin having different chemical compositions. This is not always the case, samples from the same location have been found to contain different components and conversely samples from different regions may display identical chemical composition.

Detectors usually used, in combination with chromatographic techniques (GC and HPLC), may be more or less selective and sensitive, but lack information about the identity of compounds. Therefore, the coupling of chromatographic techniques and mass spectrometry (MS) overcomes this drawback (Cavaliere *et al.*, 2007; Lazzez *et al.*, 2008).

MS is a sensitive and selective detector, sometimes allowing preparation steps to be avoided. GC–MS is a robust technique, used routinely in many laboratories for food analysis; for example for the determination of aroma compounds and pesticide analysis. More recently, LC coupled to quadrupoles, magnetic sectors or time-of-flight (TOF) detectors, has also had a great expansion into the field of food analysis.

The recently introduced spray methods (ESI) have fostered qualitative and quantitative analysis of medium to high polar analytes by mass spectrometry. The designing of ion source houses for ESI has also fostered the rediscovery of atmospheric pressure ionization (API) methods such APCI where the chemical ionization (CI) is achieved at atmospheric pressure. Both techniques produce soft ionization, but additional fragmentation can be achieved by performing in-source collision induced dissociation (CID) in tandem or trap instruments. MS/MS in space (tandem, sectors, quadrupoles, TOFs, etc.) and in time (traps) provide additional and unique information on the structure of analytes. ESI is useful for polar and ionic solutes ranging in molecular weight from 100 to 150×10^3 dalton. APCI is applicable to non-polar and medium polarity molecules with a molecular weight from 100 to 2000 dalton. Although the choice of the right interface, as well as the detection polarity are based mostly on the compounds polarity and thermal stability, and the HPLC operating conditions, many classes of compounds can give good response with both ionization techniques. In certain circumstances both positive and negative ionization modes are needed, while in most of the cases the choice of only one operation mode is enough. The number of applications of HPLC–API-MS to food analysis has rapidly increased in recent years. ESI is much more widespread than APCI, but for both techniques the trend is towards an increase in the number of applications (Sindona et al., 1999; 2000; Di Donna et al., 2001).

3.2. Stable isotope techniques for the traceability of olive oil

Stable isotope techniques enable differentiation of chemically identical substances through alterations in their isotopic fingerprint, and have been used in authenticity studies for many food products. The isotopic composition of light elements (lighter than calcium) in plant material can vary depending on location but, the dominant factor is the influence of latitude on the fractionation of the elements in groundwater.

Fractionation occurs during physical processes such as evaporation. Lighter isotopes evaporate very slightly faster than their heavier counterparts, therefore in warmer regions where the amount of evaporation is higher, the isotopes are fractionated to a greater degree. The discrimination between isotopes in such physical processes is only significant for light elements, with a high relative mass difference between the isotopes. Thus hydrogen ratios, measured by site-specific natural isotope fractionation nuclear magnetic resonance (SNIF-NMR), and carbon, nitrogen, oxygen and sulphur isotope ratios measured by isotope ratio mass spectrometry (IRMS) have been applied to the authentication of foods.

The elemental composition of vegetation reflects (to a certain extent) that of the soil in which it has grown (Kelly *et al.*, 2002) which in turn will depend on the topography, geology and soil characteristics. Therefore, no two countries will have identical soil maps, and the concentration of elements in food product, can then be used to assign their geographical origin. As with other methods of authentication, a database of samples of known origin must be available, against which unknown samples can be compared. The most useful elements for the assignment of origin are those that are not homeostatically controlled. Elements such as K, Ca and Zn are actively absorbed by organisms and will therefore be present in samples at similar levels, regardless of the environmental conditions experienced.

Elements that have no role in normal physiological processes, such as the rare earth elements (REEs) and the heavy metals, are passively absorbed and the concentration of these elements in an organism will strongly reflect the environmental levels to which the organism has been exposed.

3.2.1. Inductively coupled plasma mass spectrometry technique

Several studies have used multi-element concentration profiles in the determination of food authenticity, either alone or in combination with chromatographic or stable isotope ratio data. Elemental concentrations were determined by atomic absorption spectroscopy (AAS), inductively coupled plasma - atomic emission spectroscopy (ICP-AES) or inductively coupled plasma – mass spectrometry (ICP-MS) (Dugo *et al.*, 2004; Benincasa *et al.*, 2007; Cindric *et al.*, 2007).

Several trace elements have variable natural isotopic abundance due to the decay of radioactive isotopes. These include Li, B, Cu, Sr, Nd, Hf, Pb and U. The composition and age of the local rocks dictates the abundance of the radioactive precursors and their daughter species. Elements are taken up into plants in the same isotopic proportions as they occur in the soil and in precipitation.

Therefore isotope ratios in plant-derived food products depend on the geology of the region in which the source crop was grown, and are different in produce of different geographical origin. There have, however, been relatively few reports of the use of heavier stable isotope ratio measurements for the authentication of foodstuffs. For many years thermal ionization mass spectrometry (TIMS) was the only technique capable of performing isotope ratio measurements with sufficient precision to allow geographical assignment of food products based on trace element isotopic composition. Samples for TIMS analysis must be loaded in the form of the pure element, meaning that extensive sample preparation is required if this technique is to be applied.

ICP-MS is now a well established technique for isotopic trace element determinations. ICP-MS allows rapid analysis of a large range of sample types, requires minimal sample preparation and due to the ionising power of the ICP, can be applied to a wider range of elements than TIMS. The precision of isotope ratio analyses by ICP-MS has only recently matched that achievable by TIMS, through the application of double focusing mass analysers coupled to multi-collector detection arrays. The increasing availability of multi-collector ICP-MS is likely to lead to wider application of heavier stable isotope ratio measurements to the authentication of food products.

3.2.2. ^{13}C compound-specific isotope analysis

One of the most powerful techniques to be used in food authenticity studies is stable isotope ratio mass spectrometry (SIRMS). SIRMS has found application in the authentication of a wide range of foodstuffs, especially for the detection of added cane sugar in fruit juices (Bricout and Koziet, 1987), wines (Dunbar and Schmidt, 1984), spirits (Parker *et al.*, 1998), honey (White *et al.*, 1998) and to detect the adulteration of flavour compounds with synthetic analogues (Culp and Noakes, 1990). The majority of the reported research has focused on bulk or global stable carbon isotope ratios ($^{13}C/^{12}C$).

Approximately 98.89% of all carbon in nature is ^{12}C and 1.1% is ^{13}C. The amount of ^{13}C present in a sample is expressed as a ratio to the amount of ^{12}C present, relative to an international standard using the δ notation. This notation has the units per mil (‰). The $\delta^{13}C$ value of plants varies because of isotopic fractionation during physical, chemical and biological processes.

Photosynthetic fixation of CO_2 by plants takes place by three different routes, depending on the nature of the plant. Most terrestrial plants photosynthesise using the Calvin or C3 pathway in which CO_2 is fixed via the carbon cycle to from a three-carbon compound. Some plants, mainly tropical grasses, such as maize and sugar cane use the Hatch-Slack or C4 pathway in which the initial "Hatch-Slack" step is via a dicarboxylic acid, a four carbon compound. A third photosynthetic class of plants uses the CAM (Crassulacean acid metabolism) pathway.

Typical CAM plants are succulents and are of minor importance in the oleochemical industry. In the C3 pathway carbon dioxide is fixed via the carboxylation of ribulose-1,5-diphosphate to form phosphoglyceric acid (a 3-carbon molecule). This enzyme catalysed reaction discriminates against ^{13}C and so proceeds with a relatively large isotope effect (O'Leary, 1981). This means that less $^{13}CO_2$ is incorporated into C3 plants than into C4 plants. C3 plants have bulk $\delta^{13}C$ values in the range -24 to -30‰ whereas C4 plants have bulk $\delta^{13}C$ values from -9 to –14‰.

Olive oil shows a $\delta^{13}C$ values of 28.7‰.

The most widely used technique to assess the authenticity of edible oils is to measure the fatty acid composition. However it is not always possible to detect adulteration by this technique because of the natural variation in fatty acid profiles and because blends of oils with fatty acid compositions similar to the authentic oil may be prepared relatively easily. A more sophisticated approach is to determine the fatty acid composition at the 2 position of the triacylglycerol, since this is known to differ from the overall fatty acid composition. Woodbury *et al.*, (1998) used this technique to obtain the compound-specific $\delta^{13}C$ values of the fatty acids specifically at the 2-position of the triacylglycerol.

Royer *et al.*, (1999) examined 188 olive oils produced mainly in Greece during 1993 to 1996. The concentration and $\delta^{13}C$ value of individual fatty acids present in the olive oils were determined by gas chromatography and GC-C-IRMS respectively. The results were examined in terms of geographical, temporal, and botanical factors. French and Italian olive oils were securely classified at the 99.9% confidence interval using the $\delta^{13}C$ values of the principal fatty acids palmitic (C16:0), oleic (C18:1) and linoleic (C18:2). Regional classifications for the Greek olive oils were also achieved on the basis of differences in the ^{13}C abundance of oleic acid compared to linoleic acid and palmitic acid.

Glycerol is a primary metabolite in plants. It is nominally present in its ester form as glycerolipids in fats and oils (Kiritsakis and Christie, 1999). Glycerol is bio-synthesised relatively early in the lipidic metabolic pathway compared to fatty acids (Weber *et al.*, 1997; Harwood and Sánchez, 1999).

Consequently, it may be expected that the isotopic distribution in glycerol is a better indicator of the botanical and environmental influences on any given plant. A number of compound specific IRMS studies on glycerol have been performed, some of which include data derived from vegetable oils.

3.2.3. ^{18}O and ^{2}H pyrolysis CF-IRMS

The abundance of the stable isotopes oxygen-18 (^{18}O) and deuterium (^{2}H) are particularly interesting isotopic probes for both botanical and geographical identification of a variety of different food products. The primary source of all organic hydrogen and oxygen is the hydrosphere. The meteoric water that has passed through the meteorological cycle of evaporation, condensation and precipitation finally constitutes the groundwater and exhibits a systematic geographical isotope variation. Decreasing temperatures cause a progressive heavy-isotope depletion of the precipitation when the water vapour from oceans in equatorial regions moves to higher latitudes and altitudes. Evaporation of water from the oceans is a fractionating process that decreases the concentration of the heavy isotopomers of water ($^{1}H^{2}H^{16}O$, $^{1}H^{1}H^{18}O$) in the clouds compared to the sea. As the clouds move inland and gain altitude further evaporation, condensation and precipitation events occur decreasing the concentration of deuterium and oxygen-18.

Consequently, the ground water reflects this isotopic gradient from the coast to inland areas. For land plants, a further pre-assimilation affects the isotopic composition of the water substrate. The hydrogen and oxygen present in plant material originates from the water taken up by the roots. The water is transported through the plants xylem system. The isotopic composition of the xylem water is the same as that of water taken in by the roots, and the water is taken into the leaves without a change in isotopic composition. Evapo-transpiration of water through the leaf enriches the remaining water in the heavier isotopomers. Therefore, it is expected that growing regions with relatively low humidity, where the rate of evaporation from the leaf is higher, result in plant materials with relatively enriched $\delta^{2}H$ and $\delta^{18}O$ values.

Over the past 5 years there has been a marked increase in the use and application of ^{2}H and ^{18}O stable isotopes in many areas of food research. This has been facilitated by recent developments in on-line gas preparation devices that proceed by high temperature pyrolysis of organic products and the availability of commercial IRMS analysers capable of measuring ^{2}H/^{1}H ratios in the presence of a helium carrier gas. These innovations have, to a large extent, overcome the difficulties associated with offline gas preparation for DI-MS and greatly increased the applicability of this measurement. It is now possible to routinely measure ^{2}H and ^{18}O abundances in organic samples by Pyrolysis-Continuous Flow-Isotope Ratio Mass Spectrometry (Py-CF-IRMS).

Therefore, measurements of stable isotope ratios of the light elements (H, C ,N ,O, ₃ and bioelements) and of the heavy element stronzium, in natural cycles, have provided geographical fingerprints (Roßmann *et al.*, 2000).

Preliminary investigations into the application of ^{18}O-pyrolysis continuous-flow IRMS to obtain information about the geographical origin of olive oil samples has been conducted by Angerosa *et al.* (1999). They measured the δ^{13}C and δ^{18}O values of whole olive oil, sterols and aliphatic alcohol fractions from fruits of *Olea europaea* L. produced in Greece, Italy, Morocco, Spain, Tunisia, and Turkey. The results permitted provincial classification of the oils. However, there were some misclassifications observed for oil samples coming from neighbouring countries with similar climates.

A secure geographical classification of an olive oil, in order to ensure that the consumer is not defrauded and that the honest trader is not disadvantaged by having their PDO oils misrepresented by inferior products, can be achieved by performing heavy isotope ratios (e.g. ^{88}Sr/^{86}Sr) and multi-element analysis.

3.2.4. Nuclear magnetic resonance spectroscopy

During the last ten years, nuclear magnetic resonance spectroscopy (NMR) (Del Coco *et al.*, 2012; Mannina & Segre, 2002), has played an ever-increasing role in the study olive oil characterization and autentication. In particular, it has been shown that high-resolution NMR together with statistical analysis constitutes a powerful tool for the geographical characterization of olive oils on Mediterranean, national, regional and PDO scales. On this regard, innovative techniques like NMR spectroscopy seem to be able to distinguish olive oils on the basis of their geographical origin, whereas the conventional analyses suitable for the determination of quality and genuineness seem not to be so appropriate for this type of discrimination (Frankel, 2010; Guillen & Ruiz, 2001).

Important information on the fatty acid distribution on the glycerol moiety can be obtained by ^{13}C NMR (Rezzi *et al.*, 2005; Petrakis *et al.*, 2008; Alonso-Salces *et al.*, 2010b; Mannina *et al.*, 2010; Alonso-Salces *et al.*, 2011b). Two groups of resonances are observed in the carbonyl region of the ^{13}C NMR spectrum of an olive oil: one group is due to fatty chains in position sn-1,3 of the glycerol moiety, the other one is due to fatty acids in position sn-2.

It must be noted that although gas chromatographic methods give the full composition of fatty chains, no information is given about the fatty chains distribution on glycerol. Thus in this case, gas chromatography and ^{13}C NMR methodologies must be considered complementary and not alternative. ^{13}C NMR spectroscopy together with discriminant analysis has been proposed by Mavromoustakos et al., to detect the presence of soybean oil, cottonseed oil, corn oil and sunflower seed oil in virgin olive oils. Only double bonds signals have been used for the analysis. Fatty acids composition and fatty chain positional distribution on glycerol moiety determined by ^{13}C NMR spectroscopy allow, in combination with the multivariate statistical procedure, the classification of olive oils according to their variety (Marini et al., 2004, 2006).

^{31}P NMR spectroscopy has been employed for the detection and quantification of the minor compounds mono- and diacylglycerols, polyphenols, sterols, phospholipids and for the determination of free acidity and moisture (Petrakis et al., 2008). This analytical technique requires the derivatization of the labile hydrogens of the hydroxyl and carboxyl groups of the olive oil constituents using the phosphorous reagent 2-chloro-4,4,5,5- tetramethyl dioxaphospholane and the use of ^{31}P chemical shifts of the phosphitylated compounds to identify the labile centers. The main phospholipids found in olive oil were phosphatidic acid, lyso-phosphatidic acid and phosphatidylinosotol, although very small amounts of other phospholipids were detected as well. In this way, the content of free fatty acids has been determined showing a good correlation with official titration method. Besides, using the same derivatization method, sn-1,2, sn-1,3-diacylglycerides and monoacylglycerides have been quantified and have showed a characteristic trend during olive oil storage.

The composition of diacylglycerols (sn-1,2- 1,3-diacylglycerols and total) determined by ^{31}P NMR derivatization method has been used to characterize olive oils from different Greek areas (Petrakis et al., 2008). Some preliminary correlations between the diacylglycerols content and the geographical regions have been observed.

The geographical characterization of monovarietal virgin olive oils from three regions of Southern Greece, namely Peloponnesus, Crete and Zakynthos, has been performed applying both ^1H and ^{31}P NMR (Petrakis et al., 2008). The correct geographical prediction at the level of three regions, based on discriminant analysis, has been rather high (87%). Phenolic compounds and free acidity determined by ^{31}P NMR spectroscopy have enabled to classify Koroneiki, Athinolia and Kolovi monovarietal Greek extra virgin olive oils using ANOVA and discriminant analysis.

^1H NMR spectrum of an olive oil does not provide this information directly; however, it is possible to have an indirect measurement of acidity using the measurable amount of diglycerides and monoglycerides.

Another study carried out by using ^1H NMR involved the characterization of the phenolic fraction of olive oil from three different areas of the Apulia region (Sacco et al., 2000). Statistical evaluation has showed discrimination between Coast, Hinterland and North samples.

Extra virgin olive oils sampled in three harvesting years and coming from different Italian regions (Tuscany, Lazio, Lake of Garda) have been analyzed by ^{1}H methodology. The statistical elaboration applied on the intensity of b-sitosterol, aldehydes and some other volatile compounds has allowed the classification of the olive oils according to the geographical regions.

^{1}H-NMR fingerprinting of olive oil is a valuable analytical tool for the traceability of virgin olive oils from different points of view, i.e. food authentication and food quality.   As described before, ^{1}H and ^{13}C NMR techniques provide different information: ^{1}H NMR spectrum allows the measurement of minor components of olive oils such as β-sitosterol, hexanal, 2-(E)-hexenal, formaldehyde, squalene, cycloartenol and linolenic acid; the ^{13}C NMR spectrum detects major components such as glycerol tri-esters of olive oils, defining also acyl composition and positional distribution on the glycerol moiety. The main sensitivity concern about ^{1}H NMR analysis is the dynamic range, which makes the detection of signals below the threshold imposed by the ADC (analogue to digital converter) impossible. Therefore, in a standard extra virgin oil spectrum, trace component detection is limited to signals more intense than 10−5 times the intensity of the fatty acid CH_2 signals. Weak signals above the dynamic range threshold are in any case affected by severe baseline distortion. Selective pulses coupled to gradient spin-echo refocusing such as DPFGSE (69), in NMR modern instruments, allow these limitations to be circumvented, paving the way for the easy detection of minor components, especially those presenting resonances away from other interfering resonances.

The power of these new sequences could be demonstrated by the detection of aldehydes, carotenoids or other specific minor components whose resonances fall in a relatively free region. It was performed a specific DPFGSE analysis on the aldehydic components of some extra virgin olive oils, mostly responsible of the sensorial properties of extra virgin olive oil. Comparison with standard ^{1}HNMR spectra show tremendous improvement of quantitative and qualitative information with drastic instrumental time reduction.

Fingerprinting techniques such as NMR, NIR (Galtier et al., 2007; Mignani et al., 2011), MIR (Reid et al., 2006), fluorescence (Kunz et al., 2011), FT-IR, FT-MIR and FT-Raman (Baeten et al., 2005; Lopez-Diez et al., 2003; Yang et al., 2005) spectroscopies, have been used for the determination of food authenticity (Reid et al., 2006).

These types of techniques are particularly attractive since they are non selective, require little or no sample pre-treatment; use small amounts of organic solvents or reagents; and the analysis takes only a few minutes per sample.

3.3. Molecular markers for the traceability of olive oil

The food crisis situation seen in last years and the controversy about genetically modified organisms (GMO), with a sharp increase in basic food prices, highlights the extreme susceptibility of the current agricultural and food model and the need for more strict food quality control, which should include determination of the origin of the product and the raw materials used in it. That's why a well documented traceability system has become a requirement for quality control in the food chain.

Almost 84% from the total olive oil production derives from the European Union, especially from Spain, Italy and Greece. The olive oil is a main constituent of the Mediterranean diet. However there has recently been an increase in olive oil consumption internationally, due to greater availability and the current consideration of high nutritive and health benefits, including a qualified health claim from Food and Drug Administration (FDA, USA).

Molecular markers allow the detection of DNA polymorphisms and enable to effectively distinguish different varieties in an effective way, without any environmental influence.

Olive oil extraction is the process of extracting the oil present in the olive drupes for food use. Olive oil extraction is the process of separating the oil from the other fruit contents. It is possible to attain this separation by physical means alone, *i.e.* oil and water do not mix, so they are relatively easy to separate. The modern method of olive oil extraction uses an industrial decanter to separate all the phases by centrifugation. In this method the olives are crushed to a fine paste. This can be done by a hammer crusher, disc crusher, depicting machine or knife crusher. This paste is then malaxed for about 30 minutes in order to allow the small olive droplets to agglomerate.

When we blend olive oils of the same category, but from different provenances, most chemical analyses are of limited significance. Due to their high variability according to environmental conditions, neither morphological characteristics of different groups, nor the analyses of chemical composition of fatty acid and secondary metabolites can provide reliable results for oil traceability (Ben Ayed *et al.*, 2009; Papadia *et al.*, 2011). For this reason, genetic identity seems to be the most appropriate method for identifying the variety from which the olive oil under study derives.

Molecular markers allow the detection of DNA polymorphisms and enable to effectively distinguish different varieties in an effective way, without any environmental influence. In fact, DNA in oil is not affected by the environment and is identical to the mother tree DNA since the oil containing tissues are formed by diploid somatic cells of the tree (Muzzalupo *et al.*, 2007). However, depending on the molecular markers used correctly, extra alleles can be detected in the oil that do not correspond to the mother tree allele but to the pollinator alleles contained in the embryo, itself located inside the seed (Muzzalupo and Perri, 2002; Ben Ayed *et al.*, 2010).

The use of DNA based technology in the field of food authenticity is gaining increasing attention. This technique makes use of molecular markers that mostly use polymerase chain reaction (PCR) and are thus easy to genotype. Even in a complex matrix such as olive oil, molecular marker techniques such as RAPDs (random amplified polymorphic DNA, Pasqualone *et al.*, 2001; Muzzalupo and Perri, 2002).), AFLPs (amplified fragment length polymorphism, Pafundo *et al.*, 2005; Montemurro *et al.*, 2007) and SSRs (simple sequence repeat, Muzzalupo *et al.*, 2007; Breton *et al.*, 2004; Bracci *et al.*, 2011; Ben Ayed *et al.*, 2009) are very useful in the study of the traceability of olive oil. Single nucleotide polymorphism

(SNPs) markers have been recently developed in olive and utilized to study the genetic diversity of olive trees (Reale *et al.*, 2006). SNP detection can be delivered in a number of ways, but the simultaneous detection of multiple SNPs from a single DNA sample is of particular interest (Consolandi *et al.*, 2008). PDO oils are typically not monovarietal, so a method for quantifying the components of the mixture is essential if conformity with certification depends on a prescribed proportion of varietal types. So far, application of Real-Time as a tool for olive oil authentication has been explored by Giménez *et al.*, (2010). The authors evidenced that Real-Time PCR is useful to quantify DNA extracted from oil, and thus to assess the yields of different methods of extraction. But the size of amplicon, is critical for the success of analysis. A possibility of utilising qRT-PCR to quantify varieties in PDO oils rests on the use of taqMan probes designed on SNPs specific of varieties entering in the oil composition (Marmiroli *et al.*, 2009). Several sequences from noncoding spacer region between *psbA-trnH* and partial coding region of *matK* of plastid genome provided a good discrimination of pure *olive oil* and its admixture by other vegetable oils such as canola and sunflower. The plastid based molecular DNA technology has a great potential to be used for rapid detection of adulteration easily up to 5% in olive oil (Kumar *et al.*, 2011).

The knowledge of genome nucleotide sequences also could be useful to identify new sequence polymorphisms, which will be very useful in the development of many new variety-specific molecular markers and in the implementation of more efficient protocols for tracking and protect olive oil origin.

A recent publication reported by Papadia *et al.*, 2011 a systematic effort to obtain genetic characterization by SSR amplification, soil analyses, and 1H-NMR spectra, is carried out in order to make a direct connection between the olive tree variety (genetic information) and the NMR spectra (chemical information) of the extra virgin olive oil produced. The results reported show that a multidisciplinary approach, through the application of multivariate statistical analysis, could be used to set up a method for variety and/or geographic origin certification, based on the construction of a suitable database. Further research will be directed to the growth of an organic genetic/NMR/soil database, in order to improve the prediction ability of the LDA, and furthermore to develop a way to correlate ^{1}H-NMR spectra of commercial extra virgin olive oils with their geographical and genetic origin.

3.4. Electrochemical sensor analysis for the traceability of olive oil

Electronic noses (e-noses) and electronic tongues (e-tongues) for liquid analysis, based on the organizational principles of biological sensory systems, developed rapidly during the last decade. E-noses and e-tongues crudely mimic the human smell and taste sensors (gas and liquid sensors) and their communication with the human brain. The human olfactory system is by far the more complex and contains thousands of receptors that bind odor molecules and can detect some odors at parts per trillion levels (Breer, 1997) and include between 10 and 100 million receptors (Deisingh *et al.*, 2004). Apparently some of the receptors in the olfactory mucus can bind more than one odor molecule and in some cases one odor molecule can bind more than one receptor. This results in a mind-boggling amount of combinations that send unique signal patterns to the human brain. The brain then interprets these signals and makes a judgment and/or classification to identify the substance consumed, based in part, on previous experiences or neural network pattern recognition. The electronic nose often consists of non-selective sensors that interact with volatile molecules that result in a physical or chemical change that sends a signal to a computer which makes a classification based on a calibration and training process leading to pattern recognition. The non-selectivity of the sensors results in many possibilities for unique signal combinations, patterns or fingerprints. The human tongue contains sensors, in the form of 10,000 taste buds of 50–100 taste cells each (Deisingh *et al.*, 2004) for sweet, sour, bitter, salty and umami and is much less complicated than the human olfactory system. The e-tongue then uses a range of sensors that respond to salts, acids, sugars, bitter compounds, etc. and sends signals to a computer for interpretation. The interpretation of the complex data sets from e-nose and e-tongue signals is accomplished by use of multivariate statistics. For non-linear responses, artificial neural networks (NAA) can be used for modeling the data. Biosensors are also being developed, but are not yet commercialized. In contrast to chemical sensing materials, that are broad spectrum to generate characteristic response patterns, there are biological systems. The problem with chemical sensors is that these systems are extensive, require large sample sizes for analysis, have low sensitivity and poor specificity compared with the human nose. The bioelectronic nose utilizes olfactory receptors as sensing mechanisms and are cell or protein-based to mimic a mammalian olfactory system (Lee and Park, 2010). Another type of sensing system is based on colorimetric sensor array built in disposable chips (Suslick *et al.*, 2010). These arrays are based on the chemical interactions between the analyte and a chemical dye. They are being developed for volatile and non volatile molecules (Zhang *et al.*, 2006; Zhang and Suslick, 2007; Musto *et al.*, 2009) for applications by the food industry.

The advantage of the human sensory system is that the brain can receive signals from both olfactory and tongue receptors and integrate both sets of data to form classifications and/or judgments. The e-nose and e-tongue are not integrated since each has its own software package, but the data from both instruments could be imported into another program and integrated. The disadvantage of the human sensory system is that no two brains are alike (of course from another point of view, this is a good thing), and the same brain may react differently from one day to the next, depending on an individual's health, mood or environment, making the data subjective. On the contrary, e-nose and e-tongue instruments can be calibrated to be reliably consistent and can give objective data for important functions like quality and safety control. These instruments can also test samples that are unfit for human consumption. A disadvantage for the e-nose and e-tongue systems (as with humans) is that they are also affected by the environment including temperature for both e-nose and e-tongue and humidity for e-nose, which can cause sensor drift, although calibration systems and built-in algorithms help compensate for this. There are more or at least not less different types of sensing materials for e-tongue (liquid sensors) compared to e-nose systems, and liquid sensors often possess higher selectivity and significantly lower detection limits compared to the gas sensors (e-nose).

There are several reviews on the subject of e-nose and e-tongue technology, including reviews on e-noses (Di Giacinto et al., 2010; Wilson and Baietto, 2009), biomimetic/biotechnology e-nose and/or e-tongue sensing systems (Rudnitskaya and Legin, 2008; Ghasemi-Varnamkhasti *et al.*, 2010)], applications for e-noses and e-tongues (Scampicchio et al., 2008), neural networks for e-noses (Lu *et al.*, 2000), pattern recognition techniques (Berrueta *et al.*, 2007); meat quality assessment by e-nose (Ghasemi-Varnamkhasti *et al.*, 2009) and computational methods for analysis of e-nose data (Jurs *et al.*, 2000). This review will concentrate on the recent literature on applications of e-noses and e-tongues in the food industry.

4. Conclusions

Some varieties of olive oil are recognized as being of higher quality because they derive from well-defined geographical areas, command better prices and generally are legally protected. Indeed, the aim of Protected Designations of Origin (PDO), Protected Geographical Indication (PGI) and Traditional Specialty Guaranteed (TSG) is to add value to certain specific high quality products from a particular origin.

The development of accurate analytical fingerprinting methods for the authentication of olive oils and for the certification of the geographical origin is an actual issue and an important challenge. In fact, to protect the rights of both the consumer and honest producers and, to enforce the laws, it is important to develop analytical methods to measure the authenticity of the samples, to verify the geographical or cultivar origin and to provide the presence/absence of adulterants.

Author details

Enzo Perri[*], Cinzia Benincasa and Innocenzo Muzzalupo
Agricultural Research Council - Olive Growing and Oil Industry Research Centre, Rende (CS), Italy

Acknowledgement

Financial support for this study was provided by Italian Ministry of Agriculture, Food and Forestry through the project GERMOLI "Salvaguardia e valorizzazione del GERMoplasma OLIvicolo delle collezioni del CRA-OLI"

5. References

Alonso-Salces, R. M., Moreno-Rojas, J. M., Holland, M. V., & Guillou, C. (2011b). Authentication of Virgin Olive Oil using NMR and isotopic fingerprinting. In series: Food Science and Technology, Nova Science Publishers, ISBN 978-1-61122-309-5, New York.

---

[*] Corresponding Author

Alonso-Salces, R. M., Moreno-Rojas, J. M., Holland, M. V., Reniero, F., Guillou, C., & Heberger, K. (2010b). Virgin Olive Oil Authentication by Multivariate Analyses of 1H NMR Fingerprints and δ13C and δ2H Data. Journal of Agricultural and Food Chemistry, Vol.58, No.9, pp. 5586-5596, ISSN 0021-8561.

Angerosa, F., Breas, O., Contento, S., Guillou, C., Reniero, F. & Sada, E, (1999). *J. Agric. Food Chem.*, Vol.47, pp. 1013-1017.

Aparicio, R., & Aparicio-Ruiz, R. (2000). Authentication of vegetable oils by chromatographic techniques. Journal of Chromatography A, Vol.881, No.1-2, pp. 93-104, ISSN 0021-9673.

Araghipour, N., Colineau, J., Koot, A., Akkermans, W., Rojas, J. M. M., Beauchamp, J., Wisthaler, A., Märk, T. D., Downey, G., Guillou, C., Mannina, L., van Ruth, S. (2008). Geographical origin classification of olive oils by PTR-MS. *Food Chem.*, Vol.108, pp. 374–383.

Baeten, V., Pierna, J. A. F., Dardenne, P., Meurens, M., Garcia-Gonzalez, D. L., & Aparicio-Ruiz, R. (2005). Detection of the presence of hazelnut oil in olive oil by FT-Raman 204 Olive Oil – Constituents, Quality, Health Properties and Bioconversions and FT-MIR spectroscopy. Journal of Agricultural and Food Chemistry, Vol.53, No.16, pp. 6201-6206, ISSN 0021-8561.

Ben Ayed, R., Grati-Kamoun, N., Moreau, F. & Rebai, A. (2009). Comparative study of microsatellite profiles of DNA from oil and leaves of two Tunisian olive cultivars. *European Food Research and Technology*, Vol.229, pp. 757–762, ISSN 1438-2377.

Benincasa, C., Lewis, J., Perri, E., Sindona, G. & Tagarelli, A. (2007), Determination of Trace Element in Italian Virgin Olive Oils and their Characterization According to Geographical Origin by Statistical Analysis. *Analitica Chimica Acta*, Vol.585, pp. 366-370.

Benincasa, C., De Nino, A., Lombardo, N., Perri, E., Sindona, G. & Tagarelli, A. (2003), Assay of aroma active components of virgin olive oils from southern italian regions by SPME-GC/ion trap mass spectrometry. *J. Agric. Food Chem.*, Vol.51, pp. 733-741.

Bracci, T.; Busconi, M.; Fogher C. & Sebastiani L. (2011). Molecular studies in olive (*Olea europaea* L.): overview on DNA markers applications and recent advances in genome analysis. *Plant Cell Reports*, Vol.30, pp. 449–462, ISSN 0721-7714.

Breton, C.; Claux, D.; Metton, I.; Skorski, G. & Berville, A. (2004). Comparative study of methods for DNA preparation from olive oil samples to identify cultivar SSR alleles in commercial oil samples: possible forensic applications. *Journal of Agriculture and Food Chemistry*, Vol.52, pp. 531-537, ISSN 0021-8561.

Bricout, J., Koziet, J. (1987). Control of the authenticity of orange juice by isotopic analysis. *J. Agric. Food Chem.*, Vol.35, pp. 758-760.

Canabate-Diaz, B., Segura Carretero, A., Fernandez-Gutierrez, A., Belmonte Vega, A., Garrido Frenich, A., Martinez Vidal, J. L., & Duran Martos, J. (2007). Separation and determination of sterols in olive oil by HPLC-MS. Food Chemistry, Vol.102, No.3, pp. 593-598, ISSN 0308-8146.

Cavaliere B., De Nino A., Hayet F., Lazez A., Macchione B., Moncef C., Perri E., Sindona G., Tagarelli A.(2007), A metabolomic approach to the evaluation of the origin of extra virgin olive oil: a conventional statistical treatment of mass spectrometric analytical data. *J. Agric. Food Chem.*, Vol.55, pp. 1454-1462.

Cindric, I. J., Zeiner, M., & Steffan, I. (2007), Trace elemental characterization of edible oils by ICP–AES and GFAAS. Microchemical Journal, 85, 136–139. doi:10.1016/j.microc.2006.04.011.

Commission Regulation (EC) No. 178/2002 of the European Parliament and of the Council of 28 January 2002 laying down the general principles and requirements of food law, establishing the European Food Safety Authority and laying down procedures in matters of food safety.

Commission Regulation (EC) No. 182/2009 of 6 March 2009 amending Regulation (EC) No 1019/2002 on marketing standards for olive oil.

Consolandi, C., Palmieri, L., Severgnini, M., Maestri, E.. Marmiroli, N., Agrimonti, C., Baldoni, L., Donini, P., Bellis, G. & Castiglioni, B. (2008). A procedure for olive oil traceability and authenticity: DNA extraction, multiplex PCR and LDR-universal array analysis. *European Food Research and Technology*, Vol.227, pp. 1429–1438, ISSN 1438-2377.

Council Regulation (EEC) No. 2081/92 on the protection of geographical indications and designations of origin for agricultural products and foodstuffs. Off. J. European Union 1992, L 208, 1–15.

Culp, R. A., Noakes, J. E. (1990). Identification of isotopically manipulated cinnamic aldehyde and benzaldehyde. *J. Agric. Food Chem.*, Vol.38, pp. 1249-1255.

Del Coco, L., Schema, F. P. & Fanizzi, F. P. (2012). 1H Nuclear Magnetic Resonance Study of Olive Oils Commercially Available as Italian Products in the United States of America. *Nutrients*, Vol. 4(5), pp. 343-355; doi:10.3390/nu4050343

Di Donna, L., De Nino, A., Mazzotti, F., Perri E., Raffaelli, A., & Sindona, G. (2001). Quantitative determination of oleuropein in organic olive oil. In Biologically-active Phytochemicals in Food (a cura di), Cambridge: Royal society of chemistry, pp. 131-132.

Sindona, G., Perri, E., Mazzotti, F., & Raffaelli, A. (2000). High-throughput screening of tocopherols in natural extracts". *J. Mass Spectr*, Vol. 35, pp. 1360-1361.

Sindona, G., De Nino, A., Mazzotti, F., Perri, E., Procopio, A., & Raffaelli, A. (2000). Virtual Freezing of the Hemiacetal-Aldehyde Equilibrium of the Aglycones of Oleuropein and Ligstroside present in olive oils from Carolea and Coratina Cultivars by Ion-spray Ionisation Tandem Mass Spectrometry. J. Mass Spectr., Vol.35, pp. 461-467.

Kelly, S. D., Parker, I., Sharman, M. and Dennis, M. J. (1997) Assessing the authenticity of single seed vegetable oils using fatty acid stable carbon isotope ratios (13C/12C), *Food Chemistry*, Vol.2, pp. 181-186.

Berrueta, L. A., Alonso-Salces, R. M. & Heberger, K. (2007). Supervised pattern recognition in food analysis. J. Chromatogr. A , Vol.1158, pp. 196-214.

Breer, H. (1997). Sense of smell: Signal recognition and transductions in olfactory receptor neurons. In Handbook of Biosensors and Electronic Noses: Medicine, Food and Environment; Kress-Rogers, E., Ed.; CRC Press: Boca Raton, FL, USA, 1997; pp. 521-532.

Deisingh, A. K., Stone, D. C. & Thompson, M. (2004). Applications of electronic noses and tongues in food analysis. Int. J. Food Sci. Technol, Vol.39, pp. 587-604.

Dugo, G., La Pera, L., Giuffrida, D., Salvo, F., Lo Turco, V. (2004), Influence of the olive variety and the zone of provenience on selenium content determined by cathodic stripping potentiometry (CSP) in virgin olive oils. *Food Chemistry*, Vol.88, pp. 135–140. doi:10.1016/j.foodchem.2003.12.036.

Di Giacinto, L., Di Loreto, G., Di Natale, C., Gianni, G., Guasti, S., Migliorini, M., Pellegrino, M., Perri, E., Santonico, M., 2010, Caratterizzazione analitica degli attributi sensoriali degli oli vergini di oliva, Ed. Camera di Commercio di Firenze.tività Dunbar, J. & Schmidt, H-L. (1984). Measurement of the 2H/1H ratios of the carbon bound hydrogen-atoms in sugars. *Fres, Z. Anal. Chem.*, Vol.317, pp. 853-857.

Frankel, E. N. (2010). Chemistry of extra virgin olive oil: Adulteration, oxidative stability, and antioxidants. Journal of Agricultural and Food Chemistry, Vol.58, No.10, pp. 5991-6006, ISSN 0021-8561.

Galtier, O., Dupuy, N., Le Dréau, Y., Ollivier, D., Pinatel, C., Kister, J., & Artaud, J. (2007). Geographic origins and compositions of virgin olive oils determinated by chemometric analysis of NIR spectra. Analytica Chimica Acta, Vol.595, No.1-2, pp. 136-144, ISSN 0003-2670.

Ghasemi-Varnamkhasti, M., Mohtasebi, S. S. & Siadat, M. (2010). Biomimetic-based odor and taste sensing systems to food quality and safety characterization: An overview on basic principles and recent achievements. J. Food Eng., Vol.100, pp. 377-387.

Ghasemi-Varnamkhasti, M., Mohtasebi, S. S., Siadat, M. & Balasubramanian, S. (2009). Meat quality assessment by electronic nose (Machine Olfaction Technology). Sensors, Vol.9, pp. 6058-6083.

Giménez, M.J.; Pistón, F.; Martín, A. & Atienza, G.S. (2010). Application of real-time PCR on the development of molecular markers and to evaluate critical aspects for olive oil authentication. *Food Chemistry*, Vol.118, pp. 482–487, ISSN 0308-8146.

Guillen, M. D., & Ruiz, A. (2001). High resolution H-1 nuclear magnetic resonance in the study of edible oils and fats. Trends in Food Science & Technology, Vol.12, No.9, pp. 328-338, ISSN 0924-2244.

Haddada, F. M., Manai, H., Daoud, D., Fernandez, X., Lizzani-Cuvelier, L., & Zarrouk, M. (2007). Profiles of volatile compounds from some monovarietal Tunisian virgin olive oils. Comparison with French PDO. Food Chemistry, Vol.103, No.2, pp. 467- 476, ISSN 0308-8146

Harwood, J. & Sánchez, J. (1999). Lipid biosynthesis in olives in Handbook of olive oil, p. 129-158. J. Harwood & R. Aparicio (Eds.), Aspen publisher Inc., Maryland, USA.

Howarth, O. W., & Vlahov, G. (1996). C-13 Nuclear -Magnetic-Resonance study of cyclopropenoid triacylglycerols. *Chemistry and physics of lipids*, Vol.81, no.1, pp. 81-85.

Japon-Lujan, R., Luque-Rodriguez, J. M. & Luque de Castro, M. D. (2006). Dynamic ultrasound-assisted extraction of oleuropein and related biophenols from olive leaves. *J. Chrom. A*, Vol.1108, pp. 76-82.

Jurs, P. C., Bakken, G. A. & McClelland, H. E. (2000). Computational methods for the analysis of chemical sensor array data from volatile analytes. Chem. Rev., Vol.100, pp. 2649-2678.

Kiritsakis, A. & Christie, W. W. (1999). Analysis of edible oils in Handbook of olive oil, p. 129-158. J. Harwood & R. Aparicio (Eds.), Aspen publisher Inc., Maryland, USA

Kumar, S.; Kahlon, T. & Chaudhary, S. (2011). A rapid screening for adulterants in olive oil using DNA barcodes, *Food Chemistry*. Vol.127, pp. 1335–1341, ISSN 0308-8146.

Kunz, M. R., Ottaway, J., Kalivas, J. H., Georgiou, C. A., & Mousdis, G. A. (2011). Updating a synchronous fluorescence spectroscopic virgin olive oil adulteration calibration to a new geographical region. Journal of Agricultural and Food Chemistry, Vol.59, No 4, pp. 1051-1057, ISSN 1520-5118.

Lanteri S., Armanino C., Perri E., Palopoli A. (2002). Study of oils from calabrian olive cultivars by chemometric methods. *Food Chemistry*, Vol.76, pp. 501-507.

Lanteri, S., Armanino, C., Perri, E., & Palopoli, A. (2002). Study of oils from Calabriar olive cultivars by chemometric methods. Food Chemistry, Vol.76, No.4, pp. 501-507, ISSN 0308-8146.

Lazzez, A., Perri, E., Caravita, M. A., Khlif, M., and Cossentini, M. (2008). Influence of Olive Maturity Stage and Geographical Origin on Some Minor Components in Virgin Olive Oil of the Chemlali Variety. *J. Agric. Food Chem.*, Vol.56, pp. 982–988.

Lee, S. & Park, T. (2010). Recent advances in the development of bioelectronic nose. Biotechnol. Bioprocess Eng., Vol.15, pp. 22-29.

Lopez Ortiz, C. M., Prats Moya, M. S., & Berenguer Navarro, V. (2006). A rapid chromatographic method for simultaneous determination of ⦵-sitosterol and tocopherol homologues in vegetable oils. Journal of Food Composition and Analysis, Vol.19, No.2-3, pp. 141-149, ISSN 0889-1575.

Lu, Y., Bian, L. & Yang, P. (2000). Quantitative artificial neural network for electronic noses. Anal. Chim. Acta, Vol.417, pp. 101-110.

Mannina, L., & Segre, A. (2002). High resolution nuclear magnetic resonance: From chemical structure to food authenticity. Grasas y Aceites, Vol.53, No.1, pp. 22-33, ISSN 0017-3495.

Mannina, L., Marini, F., Gobbino, M., Sobolev, A. P., & Capitani, D. (2010). NMR and chemometrics in tracing European olive oils: the case study of Ligurian samples. Talanta, Vol.80, No.5, pp. 2141-2148, ISSN 1873-3573.

Marini, F., Balestrieri, F., Bucci, R., Magrì, A. D., Magrì, A. L., & Marini, D. (2004). Supervised pattern recognition to authenticate Italian extra virgin olive oil varieties. Chemometrics and Intelligent Laboratory Systems, Vol.73, No.1, pp. 85-93, ISSN 0169-7439.

Marini, F., Magrì, A. L., Bucci, R., Balestrieri, F., & Marini, D. (2006). Class-modeling techniques in the authentication of Italian oils from Sicily with a Protected Denomination of Origin (PDO). Chemometrics and Intelligent Laboratory Systems, Vol.80, No.1, pp. 140-149, ISSN 0169-7439.

Marmiroli, N.; Maestri, E.; Pafundo. S. & Vietina, M. (2009). Molecular traceability of olive oil: from plant genomics to Food Genomics. In: Berti L, Maury J (eds) *Advances in olive resources*. Research Signpost, pp. 1–16, ISBN 978-81-7895-388-5, Kerala, India.

Mavromoustakos, T., Zervou, M., Bonas, G., Kolocouris, A., & Petrakis, P. (2000). J. Am. Oil Chem. Soc., Vol. 77, pp. 405–411.

Mignani, A. G., Ciaccheri, L., Ottevaere, H., Thienpont, H., Conte, L., Marega, M., Cichelli, A., Attilio, C., & Cimato, A. (2011). Visible and near-infrared absorption spectroscopy by an integrating sphere and optical fibers for quantifying and discriminating the adulteration of extra virgin olive oil from Tuscany. Analytical and Bioanalytical Chemistry, Vol.399, No.3, pp. 1315-1324, ISSN 1618-2650.

Montealegre, C, Marina Alegre, M. L., García-Ruiz, C. (2010). Traceability markers to the botanical origin in olive oils. *J Agric Food Chem.* Vol.58, no.1, pp. 28-38.

Montemurro, C.; Pasqualone, A.; Simeone, R.; Sabetta, W. & Blanco, A. (2007). AFLP molecular markers to identify virgin olive oils from single Italian cultivars. *European Food Research and Technology*, Vol.226, pp. 1439–1444, ISSN 1438-2377.

Musto, C. J., Lim, S. H. & Suslick, K. S. (2009). Colorimetric detection and identification of natural and artificial sweeteners. Anal. Chem., Vol.81, pp. 6526-6533.

Muzzalupo, I. & Perri, E. (2002). Recovery and characterization of DNA from virgin olive oil. *European Food Research and Technology*, Vol.214, pp. 528–531, ISSN 1438-2377.

Muzzalupo, I.; Pellegrino, M. & Perri, E. (2007). Detection of DNA in virgin olive oils extracted from destoned fruits. *European Food Research and Technology*, Vol.224, pp.469–475, ISSN 1438-2377.

O'Leary MH (1981). Carbon siotope fractionation in plants, *Phytochemistry*, Vol.20, pp. 553-567.

Pafundo, S.; Agrimonti, C. & Marmiroli, N. (2005). Traceability of plant contribution in olive oil by amplified fragment length polymorphisms. *Journal of Agricultural and Food Chemistry*, Vol.53, pp. 6995–7002, ISSN 0021-8561.

Papadia, P.; Del Coco, L.; Muzzalupo, I.; Rizzi, M.; Perri, E.; Cesari, G.; Simeone, V.; Mondelli, F.; Schena, F.P. & Fanizzi, F.P. (2011). Multivariate analysis of 1H-NMR spectra of genetically characterized extra virgin olive oils and growth soil correlations. *Journal of the American Oil Chemists Society*, Vol.88, N. 10, pp. 1463-1475, ISSN 0003-021X.

Parker, I G., Kelly, S.D., Sharman, M., Dennis M.J., & Howie D. (1998) Investigation into the use of Stable Carbon Isotope Ratios (13C/12C) of Scotch Whisky Congeners to establish Brand Authenticity. *Food Chem.*, Vol.63, pp. 423-428.

Pasqualone, A.; Caponio, F. & Blanco, A. (2001). Inter-simple sequence repeat DNA markers for identification of drupes from different *Olea europaea* L. cultivars. *European Food Research and Technology*, Vol.213, pp. 240-243, ISSN 1438-2377.

Petrakis, P. V., Agiomyrgianaki, A., Christophoridou, S., Spyros, A., & Dais, P. (2008). Geographical characterization of Greek virgin olive oils (cv. Koroneiki) using ^{1}H and ^{31}P NMR fingerprinting with canonical discriminant analysis and classification binary trees. Journal of Agricultural and Food Chemistry, Vol.56, No.9, pp. 3200-3207, ISSN 0021-8561.

Reale, S.; Doveri, S.; Díaz, A.; Angiolillo, A.; Lucentini, L.; Pilla, F.; Martín, A.; Donini, P. & Lee, D. (2006). SNP-based markers for discriminating olive (*Olea europaea* L.) cultivars. *Genome*, Vol.15, pp. 1193-1209, ISSN 0831-2796.

Reid, L. M., O'Donnell, C. P., & Downey, G. (2006). Recent technological advances for the determination of food authenticity. Trends in Food Science & Technology, Vol.17, No.7, pp. 344-353, ISSN 0924-2244.

Rezzi, S., Axelson, D. E., Heberger, K., Reniero, F., Mariani, C., & Guillou, C. (2005). Classification of olive oils using high throughput flow 1H-NMR fingerprinting with principal component analysis, linear discriminant analysis and probabilistic neural networks. Analytica Chimica Acta, Vol.552, No.1-2, pp. 13-24, ISSN 0003-2670.

Roßmann, A. (2001). Private communication from Dr Andreas Roßmann Technical University of Munich, FRG.

Rossmann, A., Haberhauer, G., Holzl, S., Horn, P., Pichlmayer, F. & Voerkelius, S. (2000). The potential of multielement stable isotope analysis for regional origin assignment of butter. Eur. Food Res. Technol., 211, 32-40.

Royer, A., Gerard, C., Naulet, N., Lees, M. & Martin G. J. (1999) Stable isotope characterization of olive oils. I - Compositional and carbon-13 profiles of fatty acids. JAOCS., Vol.76, pp. 357-363.

Rudnitskaya, A. & Legin, A. (2008). Sensor systems, electronic tongues and electronic noses, for the monitoring of biotechnological processes. J. Ind. Microbiol. Biotechnol., Vol.35, pp. 443-451.

Sacco, A., Brescia, M. A., Liuzzi, V., Reniero, F., Guillou, C., Ghelli, S., & van der Meer, P. (2000). Characterization of Italian olive oils based on analytical and nuclear magnetic resonance determinations. JAOCS, Vol.77, No.6, pp. 619-625, ISSN 0003-021X.

Sanz-Cortes, F., Parfitt, D. E., Romero, C., Struss, D., Llacer, G. & Badenes, M. L. (2003). Intraspecific olive diversity assessed with AFLP. Plant Breeding, Vol.122, no.2, pp. 173-177.

Scampicchio, M., Ballabio, D., Arecchi, A., Cosio, S. M. & Mannino, S. (2008). Amperometric electronic tongue for food analysis. Microchim. Acta, Vol.163, pp. 11-21.

Shaw, A. D., di Camillo, A., Vlahov, G., Jones, A., Bianchi, G., Rowland, J., & Kell, D. B. Anal. Chim. Acta, 1997, Vol.348, pp. 357–374.

Suslick, B. A., Feng, L. & Suslick, K. S. (2010). Discrimination of complex mixtures by a colorimetric sensor array: Coffee aromas. Anal. Chem., Vol.82, pp. 2067-2073.

Temime, S. B., Campeol, E., Cioni, P. L., Daoud, D., & Zarrouk, M. (2006). Volatile compounds from Chétoui olive oil and variations induced by growing area. Food Chemistry, Vol.99, No.2, pp. 315-325, ISSN 0308-8146.

Vichi, S., Pizzale, L., Conte, L. S., Buxaderas, S., & López-Tamames, E. (2005). Simultaneous determination of volatile and semi-volatile aromatic hydrocarbons in virgin olive oil by headspace solid-phase microextraction coupled to gas chromatography/mass spectrometry. Journal of Chromatography A, Vol.1090, No.1-2, pp. 146-154, ISSN 0021-9673.

Vichi, S.; Pizzale, L.; Conte, L. S.; Buxaderas, S.; Lopez-Tamames, E. J. Agric. Food Chem. 2003, 51, 6572-6577.

Weber, D., Kexel, H. & Schmidt, H-L. (1997). 13C-pattern of natural glycerol: Origin and practical importance. J. Agric. Food Chem., Vol.45, pp. 2042-2046.

White, J.; Winters, K.; Martin, P. (1998) Stable carbon isotope ratio analysis of honey: Validation of internal standard procedure for worldwide application. JAOAC. Int., Vol.81, pp. 610-619.

Wilson, A. & Baietto, M. (2009). Applications and advances in electronic-nose technologies. Sensors, Vol.9, pp. 5099-5148.

Woodbury, S. E., Evershed, R. P. & Rossell, J. B. (1998) $\delta^{13}C$ analyses of vegetable oil fatty acid components, determined by gas chromatography-combustionisotope ratio mass spectrometry, after saponification or regiospecific hydrolysis. *J. Chromatography*, Vol.805, p. 249.

Yang, H., Irudayaraj, J., & Paradkar, M. M. (2005). Discriminant analysis of edible oils and fats by FTIR, FT-NIR and FT-Raman spectroscopy. Food Chemistry, Vol.93, No.1, pp. 25-32, ISSN 0308-8146.

Zhang, C. & Suslick, K. S. (2007). Colorimetric sensor array for soft drink analysis. J. Agr. Food Chem.,Vol. 55, pp. 237-242.

Zhang, C., Bailey, D. P. & Suslick, K. S. (2006). Colorimetric sensor arrays for the analysis of beers: A feasibility study. J. Agr. Food Chem., Vol. 54, pp. 4925-4931.

Chapter 14

Oleuropein an Olive Oil Compound in Acute and Chronic Inflammation Models: Facts and Perspectives

Domenico Britti, Daniela Impellizzeri,
Antonio Procopio and Salvatore Cuzzocrea

Additional information is available at the end of the chapter

1. Introduction

A large mass of research has been accumulating to provide evidence for the health benefits of olive oil feeding and to scientifically support the widespread adoption of traditional Mediterranean diet as a model of healthy eating (Menendez et al., 2007). This evidence has been attributed to the fact that olive oil, the predominant source of fat in the Mediterranean diet (Petroni et al., 1995), contains several minor non-nutrients chemicals such as α- and γ-tocopherols and β-carotene, phytosterols, pigments, terpenic acids, flavonoids such as luteolin and quercetin, squalene, and phenolic compounds, usually and incorrectly termed polyphenols (Menendez et al., 2007; Visioli et al., 2002; Trichopoulou et al., 2003; Tripoli et al., 2005; Servili et al., 2004). The main phenolic compounds in virgin olive oil are secoiridoid derivatives of 2-(3,4-dihydroxyphenyl)ethanol (3,4-DHPEA) and 2-(4-hydroxyphenyl)-ethanol (p-HPEA) that occur as either simple phenols or esterified with elenolic acid to form, respectively, oleuropein and its derivative demethyloleuropein, and ligstroside, their aglycones 3,4-DHPEA-EA and p-HPEA-EA (Figure 1) (Bendini et al., 2007; Suárez et al., 2009). The aglyconic form of oleuropein and ligstroside, 3,4-DHPEA-EA and p-HPEA-EA respectively, were reported for the first time by Montedoro et al, who also assigned their chemical structures, later confirmed by other (Montedoro et al., 1992; Angerosa et al., 1996; Owen et al., 2000). 3,4-DHPEA-EA and p-HPEA-EA, associated to the intense sensory of bitterness and pungency respectively attribute in VOO (Gutierrez-Rosales et al., 2003), is endowed with numerous beneficial effects on human health (Servili et al., 2004; Bendini et al., 2007; Esti et al., 1998; Della Ragione et al., 2000; Paiva-Martins et al., 2001; Fabiani et al., 2002; Carasco-Pancorbo et al., 2005; Artajo et al., 2006; Fabiani et al., 2006; Fabiani et al., 2008; Paiva-Martins et al., 2009).

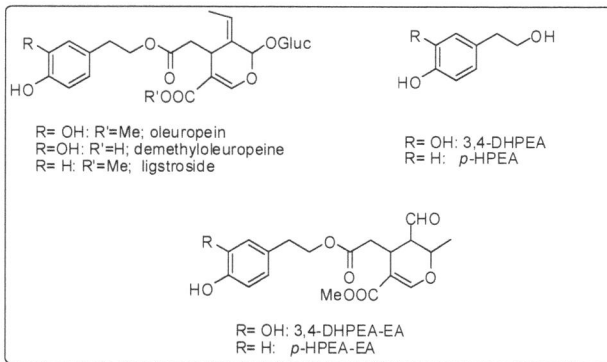

Figure 1. Chemical structures.

In particular, the anti-inflammatory properties of olive oil phenolic compounds seem to overlap with those attributed to non-steroidal anti-inflammatory drugs (Procopio A, *et al.*, 2009). The majority of phenolic compounds found in olive oil or table olives are derived from the hydrolysis of oleuropein, the major phenolic constituent of the leaves and unprocessed olive drupes of *Olea europaea* and responsible for the bitter taste of immature and unprocessed olives. Concentrations of up to 9.0 mg/l of oleouropein and 5.6 mg/l of its hydrolysis product hydroxytyrosol, have been detected in some preparations of olive oil (Montedoro *et al.*, 1992). Oleuropein, a glucoside with hydroxyaromatic functionality, has recognized several pharmacological properties including antioxidant, anti-inflammatory, anti-atherogenic, anti-cancer, antidiabetic, antimicrobial, and antiviral, and for these reasons, it is commercially available as food supplement in Mediterranean countries (Miles *et al.*, 2005; Covas, 2008; Omar, 2010). A more efficient anti-inflammatory role of the aglyconic 3,4-DHPEA-EA compared with the glycosidic form of oleuropein possibly derives from the greater lipophilicity of the former, a property that should allow better cell membrane incorporation and/or interaction with other lipids (Saija *et al.*, 1998). We focused on the antinflammatory properties of 3,4-DHPEA-EA, a hydrolysis product obtained from Oleuropein by the action of β-glucosidase on the parent glucoside, has been evaluated in a mice model of acute inflammation (carrageenan-induced pleurisy) and in a mice model of chronic inflammation (collagen-induced arthritis) (Impellizzeri *et al.*, 2011a-b).

2. 3,4-DHPEA-EA and acute inflammation

Premise - The inflammatory reaction is characterized by an initial increase in blood flow to the site of injury, enhanced vascular permeability, production of mediators such us prostaglandins, leukotrienes, histamine, bradykinin, platelet-activating factor (PAF) and the ordered and directional influx and selective accumulation of different effector cells from the peripheral blood at the site of injury. Influx of antigen non-specific but highly destructive cells (neutrophils) is one of the earliest stages of the inflammatory response. Carrageenan-induced local inflammation is commonly used to evaluate anti-inflammatory effects of non-steroidal drugs (NSAIDs). Therefore, carrageenan-induced local inflammation (pleurisy) is a useful model to assess the contribution of mediators involved in cellular alterations during the inflammatory process. In particular, the initial phase of acute inflammation (0-1h) which is not inhibited by NSAIDs such as indomethacin or aspirin, has been attributed to the release of histamine, 5-hydroxytryptamine and bradykinin, followed by a late phase (1-6 h) mainly sustained by prostaglandin release and attributed to the induction of inducible cyclo-oxygenase (COX-2) in the tissue (Nantel *et al.*, 1999). It appears that the onset of the carrageenan-induced acute inflammation has been linked to neutrophil infiltration and the production of neutrophil-derived free radicals, such as hydrogen peroxide, superoxide and hydroxyl radical, as well as the release of other neutrophil-derived mediators. Free radicals are produced in small amounts by normal cellular processes as part of the mitochondrial electron transport chain and the microsomal cytochrome P-450 system. They are formed during traumatic or hypoxic injuries as a consequence of insufficient oxygenation. Reactive oxygen species (ROS) and reactive nitrogen species (RNS) can react with and subsequently damage proteins, nucleic acids, lipids, and extracellular matrix proteins. During the inflammatory response, ROS and RNS modulate phagocytosis, secretion, gene expression, and apoptosis. Indeed, under pathological circumstances such as acute lung injury and sepsis, excess production of neutrophil-derived ROS and RNS may influence neighbouring endothelial or epithelial cells, contributing to the amplification of inflammatory tissue injury (Fialkow *et al.*, 2007). Furthermore, oxidative stress elicits the activation of the redox-sensitive transcription factors such as nuclear factor-κB (NF-κB) and AP-1, that play a central and crucial role in inducing the expression of inflammatory cytokines and intercellular adhesion molecule (ICAM-1) (Chen *et al.*, 2004) and the activation of the redox-sensitive protein kinases such as the mitogen-activated protein kinase (MAPK) superfamily (Li *et al.*, 2002). Thus, the study model was designed to evaluate the effects of 3,4-DHPEA-EA in a mice model of acute inflammation (0.1 ml of saline containing 2% λ-carrageenan was injected into the pleural cavity). In particular, we investigated the effects of 3,4-DHPEA-EA on the lung injury associated with carrageenan induced pleurisy. In order to gain a better insight into the mechanism of action of 3,4-DHPEA-EA, we have also investigated the effects on: 1) lung damage (histology), 2) polymorphonuclear (PMN) infiltration (myeloperoxidase [MPO] activity), 3) ICAM-1 and platelet-adhesion-molecule (P-selectin) expression, 4) nitrotyrosine and poly-ADP-ribose (PAR) formation, 5) pro-inflammatory cytokines production, tumor necrosis factor-α (TNF-α) and interleukin-1β (IL-1β), 6) lipid peroxidation (malondialdehyde [MDA] levels), and 7) nitric oxide (NO) synthesis (nitrite-nitrate concentration).

2.1. Materials and methods

Animals - Male CD mice, weight 20-25 g; Harlan Nossan, Milan, Italy, were used in these studies. The animals were housed in a controlled environment and provided with standard rodent chow and water. Animal care was in compliance with Italian regulations on the protection of animals used for experimental and other scientific purposes (D.lgs 116/92) as well as with EEC regulations (O.J. of E.C. L358/1 12/18/1986).

Carrageenan-induced pleurisy - Carrageenan-induced pleurisy was induced as previously described (Cuzzocrea *et al.*, 1999). Mice were anaesthetized with isoflurane and subjected to a skin incision at the level of the left sixth intercostals space. The underlying muscle was dissected and saline (0.1 ml) or saline containing 2% λ-carrageenan (0.1 ml) was injected into the pleural cavity. The skin incision was closed with a suture and the animals were allowed to recover. At 4 h after the injection of carrageenan, the animals were killed by inhalation of CO_2. The chest was carefully opened and the pleural cavity rinsed with 1 ml of saline solution containing heparin (5 U ml^{-1}) and indomethacin (10 μg ml^{-1}). The exudate and washing solution were removed by aspiration and the total volume measured. Any exudate, which was contaminated with blood, was discarded.

Experimental Design - Mice were randomly allocated into the following groups: (i) CAR + saline group. Mice were subjected to carrageenan-induced pleurisy (N = 10); (ii) CAR + 3,4-DHPEA-EA group (100 μM/kg). Same as the CAR + saline group but 3,4-DHPEA-EA (100 μM/kg, i.p.) were administered 30min after to carrageenan (N = 10); (iii) CAR + 3,4-DHPEA-EA group (40 μM/kg). Same as the CAR + saline group but 3,4-DHPEA-EA (40 μM/kg, i.p.) were administered 30min after to carrageenan (N = 10); (iv) Sham + saline group. Sham-operated group in which identical surgical procedures to the CAR group was performed, except that the saline was administered instead of carrageenan (N = 10); (v) Sham + 3,4-DHPEA-EA group. Same as the Sham+saline group but 3,4-DHPEA-EA (100 μM /kg, i.p.) were administered 30min after to carrageenan (N = 10). The doses of 3,4-DHPEA-EA (40 and 100 μ M /kg, i.p.) used here were based on previous in vivo studies (Procopio A, *et al.*, 2009).

Histological examination - Lung tissues samples were taken 4 h after injection of carrageenan. Lung tissues samples were fixed for 1 week in 10 % (w/v) PBS-buffered formaldehyde solution at room temperature, dehydrated using graded ethanol and embedded in Paraplast (Sherwood Medical, Mahwah, NJ, USA). Sections were then deparaffinized with xylene, stained with hematoxylin and eosin. All sections were studied using Axiovision Zeiss (Milan, Italy) microscope.

Measurement of cytokines - TNF-α and IL-1β levels were evaluated in the exudate 4 h after the induction of pleurisy by carrageenan injection as previously described (Cuzzocrea *et al.*, 1999). The assay was carried out using a colorimetric commercial ELISA kit (Calbiochem-Novabiochem Corporation, Milan, Italy).

Measurement of nitrite-nitrate concentration - Total nitrite in exudates, an indicator of NO synthesis, was measured as previously described (Cuzzocrea *et al.*, 2001). Briefly, the nitrate in the sample was first reduced to nitrite by incubation with nitrate reductase (670 mU/ml) and β-nicotinamide adenine dinucleotide 3'-phosphate (NADPH) (160 μM) at room temperature for 3 h. The total nitrite concentration in the samples was then measured using the Griess reaction, by adding 100 μl of Griess reagent (0.1% w/v) naphthylethylendiamide dihydrochloride in H_2O and 1% (w/v) sulphanilamide in 5% (v/v) concentrated H_3PO_4; vol. 1:1) to the 100 μl sample. The optical density at 550 nm (OD_{550}) was measured using ELISA microplate reader (SLT-Lab Instr., Salzburg, Austria). Nitrite concentrations were calculated by comparison with OD_{550} of standard solutions of sodium nitrite prepared in H_2O.

Immunohistochemical localization of ICAM-1, P-selectin, nitrotyrosine and PAR - At the end of the experiment, the tissues were fixed in 10% (w/v) PBS-buffered formaldehyde and 8 μm sections were prepared from paraffin embedded tissues. After deparaffinization, endogenous peroxidase was quenched with 0.3% (v/v) hydrogen peroxide in 60% (v/v) methanol for 30 min. The sections were permeablized with 0.1% (w/v) Triton X-100 in PBS for 20 min. Non-specific adsorption was minimized by incubating the section in 2% (v/v) normal goat serum in PBS for 20 min. Endogenous biotin or avidin binding sites were blocked by sequential incubation for 15 min with biotin and avidin, respectively. Sections were incubated overnight with anti-nitrotyrosine rabbit polyclonal antibody (Upstate, 1:500 in PBS, v/v), anti-PAR antibody (BioMol, 1:200 in PBS, v/v), anti-ICAM-1 antibody (Santa Cruz Biotechnology, 1:500 in PBS, v/v) or with anti-P-selectin polyclonal antibody (Santa Cruz Biotechnology, 1:500 in PBS, v/v). Sections were washed with PBS, and incubated with secondary antibody. Specific labeling was detected with a biotin-conjugated goat anti-rabbit IgG and avidin-biotin peroxidase complex (Vector Laboratories, DBA). In order to confirm that the immunoreaction for the nitrotyrosine was specific, some sections were also incubated with the primary antibody (anti-nitrotyrosine) in the presence of excess nitrotyrosine (10 mM) to verify the binding specificity. To verify the binding specificity for PAR, ICAM-1, P-selectin, some sections were also incubated with only the primary antibody (no secondary) or with only the secondary antibody (no primary). In these situations no positive staining was found in the sections indicating that the immunoreaction was positive in all the experiments carried out.

MPO activity - MPO activity, an indicator of PMN accumulation, was determined as previously described (Mullane *et al.*, 1985). At the specified time following injection of carrageenan, lung tissues were obtained and weighed, each piece homogenized in a solution containing 0.5% (w/v) hexadecyltrimethyl-ammonium bromide dissolved in 10 mM potassium phosphate buffer (pH 7) and centrifuged for 30 min at 20,000 x g at 4°C. An aliquot of the supernatant was then allowed to react with a solution of tetramethylbenzidine (1.6 mM) and 0.1 mM H_2O_2. The rate of change in absorbance was measured spectrophotometrically at 650 nm. MPO activity was defined as the quantity of enzyme degrading 1 μmol of peroxide min^{-1} at 37°C and was expressed in milliunits per gram weight of wet tissue.

MDA measurement - MDA levels in the lung tissue were determined as an indicator of lipid peroxidation as previously described (Ohkawa *et al.*, 1979). Lung tissue collected at the specified time, was homogenized in 1.15% (w/v) KCl solution. A 100 μl aliquot of the homogenate was added to a reaction mixture containing 200 μl of 8.1% (w/v) SDS, 1.5 ml of 20% (v/v) acetic acid (pH 3.5), 1.5 ml of 0.8% (w/v) thiobarbituric acid and 700 μl distilled water. Samples were then boiled for 1 h at 95°C and centrifuged at 3,000 x g for 10 min. The absorbance of the supernatant was measured using spectrophotometry at 650 nm.

Materials - Unless otherwise stated, all compounds were obtained from Sigma-Aldrich Company Ltd. (Poole, Dorset, U.K.). 3,4-DHPEA-EA was obtained from the controlled hydrolysis of oleuropein extracted from olive leaves by means the patented method reported by Procopio *et al.* (2009). All other chemicals were of the highest commercial grade available. All stock solutions were prepared in non-pyrogenic saline (0.9% NaCl; Baxter, Italy, UK).

Statistical evaluation - All values in the figures and text are expressed as mean ⊚ standard error (s.e.m.) of the mean of n observations. For the in vivo studies n represents the number of animals studied. In the experiments involving histology or immunohistochemistry, the figures shown are representative of at least three experiments (histological or immunohistochemistry coloration) performed on different experimental days on the tissue sections collected from all the animals in each group. The results were analyzed by one-way ANOVA followed by a Bonferroni post-hoc test for multiple comparisons. A p-value less than 0.05 were considered significant and individual group means were then compared with Student's unpaired t test. A p-value of less than 0.05 was considered significant.

2.2. Results

Effects of 3,4-DHPEA-EA on carrageenan-induced pleurisy – When compared to lung sections taken from saline-treated animals (sham group Fig. 2A, see histological score 2D) histological examination of lung sections taken from mice treated with carrageenan revealed significant tissue damage and edema (Fig. 2B, see histological score 2D) as well as infiltration of neutrophils (PMNs) within the tissues (Fig. 2B). 3,4-DHPEA-EA (100 µM/kg) reduced the degree of lung injury (Fig. 2C, see histological score 2D). The pleural infiltration with PMN appeared to correlate with an influx of leukocytes into the lung tissue, thus we investigated the effect of 3,4-DHPEA-EA on neutrophil infiltration by measurement of MPO activity. MPO activity was significantly elevated at 4 h after carrageenan administration in vehicle-treated mice (Fig. 2E). Treatment with 3,4-DHPEA-EA significantly attenuated neutrophil infiltration into the lung tissue (Fig. 2E).

Effects of 3,4-DHPEA-EA on the expression of adhesion molecules (ICAM-1, P-selectin) - Staining of lung tissue sections obtained from saline-treated mice with anti-ICAM-1 antibody showed a specific staining along bronchial epithelium demonstrating that ICAM-1 is constitutively expressed (Fig. 3A). No positive staining for P-selectin was found in lung tissue sections from saline-treated mice (Fig. 3D). At 4 h after carrageenan injection, the ICAM-1 staining intensity increased in the vascular endothelium (Fig. 3B). Lung tissue sections obtained from carrageenan-treated mice showed positive staining for P-selectin localized in the vessels (Fig. 3E). No positive staining for ICAM-1 or P-selectin was observed in the lungs of carrageenan-treated mice treated with 3,4-DHPEA-EA (Fig. 3C and 3F respectively).

Effects of 3,4-DHPEA-EA on the release of pro-inflammatory cytokine and nitrite–nitrate concentration - When compared to sham animals, injection of carrageenan resulted in an increase in the levels of TNF-α and IL-1β in the pleural exudates (Fig. 4A,B). The release of TNF-α and IL-1β was significantly attenuated by treatment with 3,4-DHPEA-EA (Fig. 4A,B).

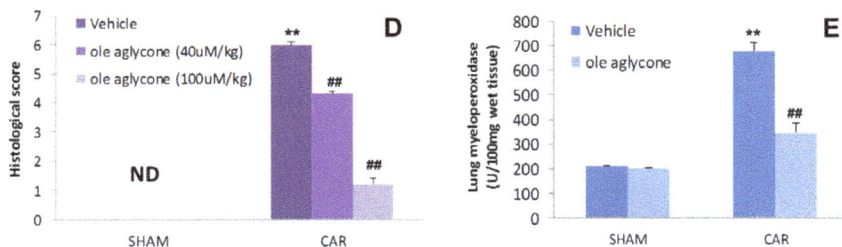

Figure 2. Effect of 3,4-DHPEA-EA (Ole aglycone) on histological alterations of lung tissue 4 h after carrageenan-induced injury and on PMN infiltration in the lung. Lung sections taken from carrageenan-treated mice treated with vehicle demonstrated edema, tissue injury (B) as well as infiltration of the tissue with neutrophils (B). Carrageenan-treated animals treated with 3,4-DHPEA-EA (C) demonstrated reduced lung injury and neutrophil infiltration. Section from sham animals demonstrating the normal architecture of the lung tissue (A). The histological score (D) was made by an independent observer. MPO activity, index of PMN infiltration, was significantly elevated at 4 h after carrageenan (CAR) administration in vehicle-treated mice (E), if compared with sham mice (E). 3,4-DHPEA-EA significantly reduced MPO activity in the lung (E). The figure is representative of at least 3 experiments performed on different experimental days. Data are expressed as mean ± S.E.M. from n = 10 mice for each group. **, P < 0.01 versus sham group. ##, P < 0.01 versus carrageenan.

No significant increase of TNF-α and IL-1β exudates levels was found in the sham animal (Fig. 4A,B). NO levels were also significantly increased in the exudate obtained from mice administered carrageenan (Fig. 4C). Treatment of mice with 3,4-DHPEA-EA significantly reduced NO exudates levels (Fig. 4C). No significant increase of NO exudates levels was found in the sham animal (Fig. 4C).

Effects of 3,4-DHPEA-EA on carrageenan-induced nitrotyrosine formation, lipid peroxidation and poly-ADP-ribosyl polymerase (PARP) activation - Immunohistochemical analysis of lung sections obtained from mice treated with carrageenan revealed positive staining for nitrotyrosine (Fig. 5B). In contrast, no positive staining for nitrotyrosine was found in the lungs of carrageenan-treated mice, which had been treated with 3,4-DHPEA-EA (100 μM/kg) (Fig. 5C). In addition, at 4 hours after carrageenan-induced pleurisy, MDA levels were also measured in the lungs as an indicator of lipid peroxidation. As shown in Figure 5D, MDA levels were significantly increased in the lungs of carrageenan-treated mice. Lipid peroxidation was significantly attenuated by the intraperitoneal injection of 3,4-DHPEA-EA (Fig. 5D). At the same time point (4 h after carrageenan administration), lung tissue sections were taken in order to determine the immunehistological staining for poly ADP-ribosylated proteins (an indicator of PARP activation). A positive staining for the PAR (Fig. 5F) was found primarily localized in the inflammatory cells present in the lung tissue from carrageenan-treated mice. 3,4-DHPEA-EA treatment reduced the degree of PARP activation (Fig. 5G). Please note that there was no staining for either nitrotyrosine (Fig. 5A) or PAR (Fig. 5E) in lung tissues obtained from the sham group of mice.

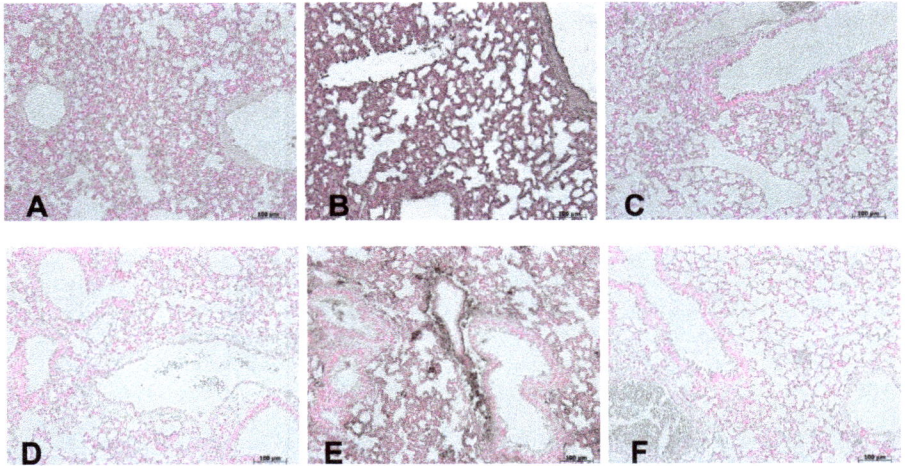

Figure 3. Effect of 3,4-DHPEA-EA (Ole aglycone) on the immunohistochemical localization of ICAM-1 and P-selectin in the lung after carrageenan injection. No positive staining for ICAM-1 was observed in lung sections taken from sham mice (A). Lung sections taken from carrageenan-treated mice showed intense positive staining for ICAM-1 along the vessels (B). The degree of positive staining for ICAM-1 was markedly reduced in lung sections obtained from mice treated with 3,4-DHPEA-EA (C). No positive staining for P-selectin was observed in lung sections taken from sham mice (D). Lung sections taken from carrageenan-treated mice treated with vehicle showed intense positive staining for P-selectin along the vessels (E). The degree of positive staining for P-selectin was markedly reduced in tissue sections obtained from mice treated with 3,4-DHPEA-EA (F). The figure is representative of at least three experiments performed on different experimental days.

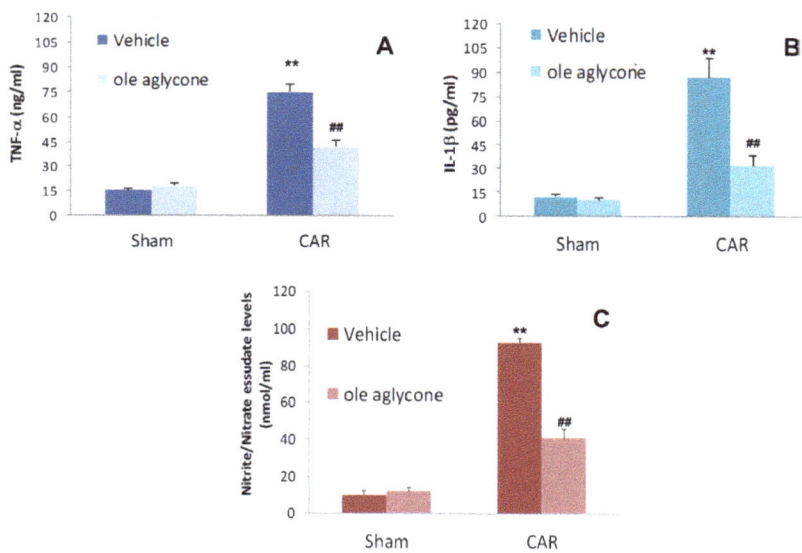

Figure 4. Effect of 3,4-DHPEA-EA (Ole aglycone) on carrageenan-induced pro-inflammatory cytokine release and NO formation in the lung. TNF-α and Il-1β levels were significantly elevated at 4 h after carrageenan administration in vehicle-treated mice (A and B respectively), if compared with sham mice (A and B respectively). 3,4-DHPEA-EA significantly reduced TNF-α and Il-1β levels (A and B respectively). Moreover nitrite and nitrate levels, stable NO metabolites, were significantly increased in the pleural exudates at 4 h after carrageenan administration (C) if compared with sham mice (C). 3,4-DHPEA-EA significantly reduced the carrageenan-induced elevation of nitrite and nitrate exudates levels (C). Data are expressed as mean ± S.E.M. from n 10 mice for each group. **, P < 0.01 versus sham group. ##, P < 0.01 versus carrageenan.

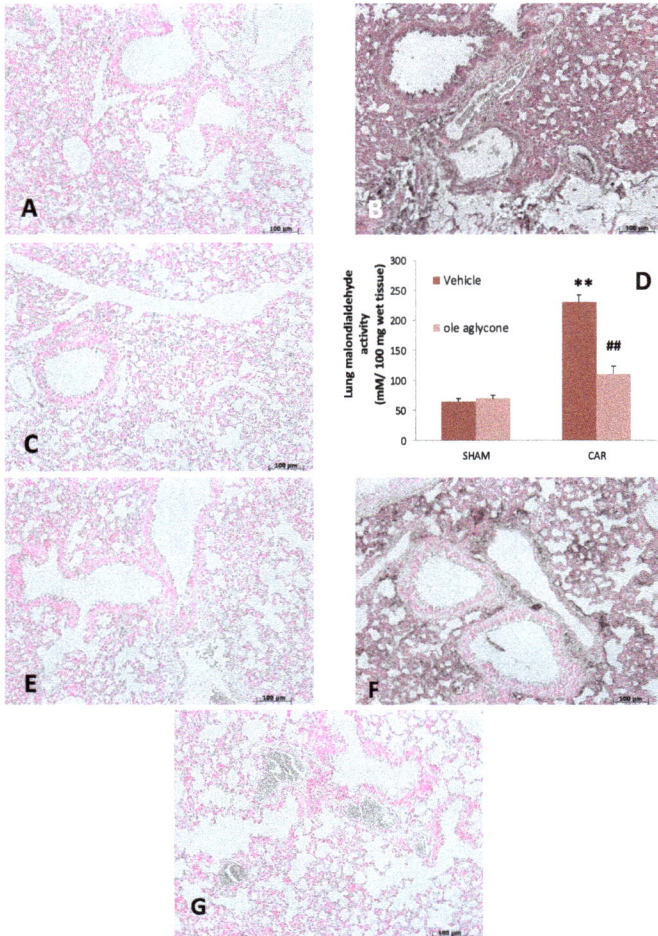

Figure 5. Effect of 3,4-DHPEA-EA (Ole aglycone) on carrageenan-induced nitrotyrosine formation, lipid peroxidation and PARP activation in the lung. No staining for nitrotyrosine is present in lung section from sham mice (A). Lung sections taken from carrageenan-treated mice treated with vehicle showed positive staining for nitrotyrosine, localized mainly in inflammatory cells (B). There was a marked reduction in the immunostaining for nitrotyrosine in the lungs of carrageenan-treated mice treated with 3,4-DHPEA-EA (C). Malondialdehyde (MDA) levels, an index of lipid peroxidation, were significantly increased in lung tissues 4 h after carrageenan administration (D), if compared with lung from sham mice (D). 3,4-DHPEA-EA significantly reduced the carrageenan-induced elevation of MDA tissues levels (D). Lung sections taken from carrageenan-treated mice showed positive staining for PAR (F). There was a marked reduction in the immunostaining for PAR in the lungs of carrageenan-treated mice treated with 3,4-DHPEA-EA (G). Lung section from sham mice showed no staining for PAR (E). The figure is representative of at least 3 experiments performed on different experimental days. Data are expressed as mean ± S.E.M. from n 10 mice for each group. **, $P < 0.01$ versus sham group. ##, $P < 0.01$ versus carrageenan.

2.3. Discussion

All of the above findings are in support of the view that 3,4-DHPEA-EA attenuates the degree of acute inflammation in the mouse. What, then, is the mechanism by which ole reduces acute inflammation? One consequence of increased oxidative stress is the activation and inactivation of redox-sensitive proteins (Bowie & O'Neill, 2000). Recent studies have observed that the acute consumption of olive oil decreased the activation of NF- κB system on mononuclear cells from healthy men (Perez-Martinez *et al.*, 2007) and that 3,4-DHPEA-EA, trans-resveratrol, and hydroxytyrosol incubated with human umbilical vein endothelial cells inhibit LPS-triggered NF- κB and AP-1 activation (Carluccio *et al.*, 2003). Moreover, various experimental evidence have clearly suggested that NF- κB plays a central role in the regulation of many genes responsible for the generation of mediators or proteins in acute lung inflammation associated with carrageenan administration (Cuzzocrea *et al.*, 2006) such us TNF-α, IL-1β, nitric oxide synthase inducible (iNOS) and COX-2. By inhibiting the activation of NF- κB, the production of joint destructive inflammatory mediators may be reduced as well. In this regard, Miles *et al.*, demonstrated that ole glycoside significantly decreased the concentration of IL-1β in LPS-stimulated human whole blood cultures. Therefore, this study also demonstrates that 3,4-DHPEA-EA attenuates the TNF-α and IL-1β production in the lung of carrageenan-treated mice. In addition, recent studies also showed that the potential cardioprotective activity of oleuropein in acute cardiotoxicity induced by doxorubicin treatment was determined in vivo in rats (Andreadou *et al.*, 2007) by inhibiting lipid peroxidation products, decreasing oxidative stress and reducing iNOS in cardiomyocytes. We show here that NO levels, evaluated as NO_2/NO_3 and MDA levels which is the products of lipid peroxidation, were increased at 4h after carrageenan injection while 3,4-DHPEA-EA decreased the levels of NO and MDA. For many years, much attention has been paid to the effects of NO in respiratory diseases 26 but recently, the focus has been shifted toward RNS in general, and to peroxynitrite (ONOO⁻) in particular (Sadeghi-Hashjin *et al.*, 1998). To probe the pathological contributions of ONOO⁻ to acute lung injury we have used the appearance of nitrotyrosine staining in the inflamed tissue. We have observed here that the immunoassaying of nitrotyrosine is reduced in the lung of carrageenan-treated mice and treated with 3,4-DHPEA-EA. Therefore, the inhibition of nitrotyrosine formation by oleuropein described in the present study is most likely attributed to the strong antioxidant activity of ole. During inflammation initiation, circulating leukocytes must first be able to adhere selectively and efficiently to vascular endothelium. This process is facilitated by induction of vascular cell adhesion molecules on the inflamed endothelium, such as vascular cell adhesion molecule VCAM-1, ICAM-1, E-selectin demonstrated that 3,4-DHPEA-EA was a more potent inhibitor of adhesion molecule expression on cultured human endothelial cells than was the glycoside(Carluccio *et al.*, 2003). Furthermore, the absence of an increased expression of the adhesion molecule in the lung from CAR mice treated with 3,4-DHPEA-EA was correlated with the reduction of leukocyte infiltration as assessed by the specific granulocyte enzyme MPO and with the attenuation of the lung tissue damage as evaluated by histological examination. Several studies also showed that in the auricular edema induced by either arachidonic acid (AA) or 12-O-tetradecanoylphorbol acetate (TPA), the topical application of the olive oil compounds such as ole also produced an inhibition of the

MPO in the inflamed tissue (de la Puerta *et al.*, 2000). Various studies have demonstrated that PARP activation after single DNA strand breakage induced by ROS plays an important role in the process of acute lung injury (Szabo *et al.*, 1998). In this study we confirm the increase in PAR formation in the lung tissues from carrageenan-treated mice as well as that 3,4-DHPEA-EA treatment attenuates PARP activation. In this regard, several studies demonstrated that hydroxytyrosol, a hydrolysis product of 3,4-DHPEA-EA, also exerts an inhibitory effect on peroxynitrite-dependent DNA base modifications and tyrosine nitration (Deiana *et al.*, 1999). Similarly, Salvini *et al.* (2006) showed a 30% reduction of oxidative DNA damage in peripheral blood lymphocytes during intervention on postmenopausal women with virgin olive oil containing high amounts of phenols. In conclusion, concomitant with inflammation is the generation of free radicals, which increase oxidation of proteins and lipids, resulting in signals that trigger more inflammation. Taken together, the results of the present study enhance our understanding of the role of ROS generation in the pathophysiology of carrageenan-induced pleurisy implying that olive oil compounds such as 3,4-DHPEA-EA may be useful in the therapy of acute inflammation.

3. 3,4-DHPEA-EA and chronic inflammation

Premise - Reactive oxygen species (ROS) are produced in cells by several physiological and environmental stimulations, such as infections, ultraviolet radiation and pollutants, known collectively as oxidants. Interestingly, ROS have also been considered as risk and enhancer factors for autoimmune diseases (Filippin *et al.*, 2008) as there is a significant relation between the oxidative stress and such diseases (Filippin *et al.*, 2008; Avalos *et al.*, 2007). Rheumatoid arthritis (RA) is an autoimmune disease characterized by the sequestration of various leukocyte subpopulations within both the developing pannus and synovial space. The chronic nature of this disease results in multiple joint inflammation with subsequent destruction of joint cartilage and erosion of bone. While this disease has a worldwide distribution, its pathogenesis is not clearly understood (Harris, 1990). Type II collagen-induced arthritis (CIA) in the mouse has proven to be a useful model of RA, as it possesses many of the cell and humoral immunity characteristics found in human RA (Holmdahl *et al.*, 1990). The pathogenesis of CIA is dependent upon the host's response to type II collagen challenge and the subsequent generation of antibodies that recognizes collagen rich joint tissue (Holmdahl *et al.*, 1990). The chronic activities initiated by immune complexes trigger a variety of cell-mediated and humoral events. Moreover, the recruitment and activation of neutrophils, macrophages, and lymphocytes into joint tissues and the formation of the pannus are hallmarks of the pathogenesis of both CIA and human RA. Recently, it has been demonstrated that interleukin (IL)-8, macrophage inflammatory protein (MIP)-1α, MIP-1β, and RANTES are differentially chemotactic for lymphocyte subsets (Taub *et al.*, 1993). Chemokines may play prominent roles in RA, as neutrophil and mononuclear cell stimulation and activation are prevalent in this disease. Concomitant with inflammation is the generation of ROS (Trichopoulou *et al.*, 2003)which increase oxidation of proteins and lipids, resulting in signals that trigger more inflammation. Thus, the study model was designed to evaluate the effects of 3,4-DHPEA-EA in a mice model of chronic inflammation (development of CIA in the mice). We have evaluated the following endpoints cf the inflammatory process: (1) clinical score; (2) body weight; (3) inducible oxide nitric synthase (iNOS) and cyclooxygenase expression (COX-2); (4) nitrotyrosine formation and activation of the nuclear enzyme poly (ADP-ribose) polymerase (PARP); (5) cytokine and chemokines production; (6) neutrophil infiltration; (7) joint histopathology.

Animals - DBA/1J mice (9 weeks, Harlan Nossan, Italy) were used for these studies. The animals were housed in a controlled environment and provided with standard rodent chow and water. Animal care was in compliance with Italian regulations on protection of animals used for experimental and other scientific purposes (Dlgs 116/92) as well as with the EEC regulations (O.J. of E.C. L 358/1 12/18/1986).

Experimental Design - Mice were divided into the following four experimental groups: (i) CIA-Control; mice were subjected to collagen-induced arthritis (as described below) and administered 200 μl of 10% ethanol solution (i.p., vehicle for 3,4-DHPEA-EA) every 24 h, starting from day 25 (n = 20); (ii) CIA-3,4-DHPEA-EA; mice subjected to collagen-induced arthritis (as described below) were administered 3,4-DHPEA-EA (40 μM/kg, i.p.) every 24 h, starting from day 25 (n = 20); (iii) Sham-Control; mice subjected to an intradermal injection at the base of the tail of 100 μl of 0.01 M acetic acid instead of the emulsion containing 100 μg of CII, were treated with 200 μl of 10% ethanol solution (i.p., vehicle for 3,4-DHPEA-EA), every 24 h starting from day 25 (n = 20); (iv) Sham-3,4-DHPEA-EA; mice subjected to an intradermal injection at the base of the tail of 100 μl of 0.01 M acetic acid instead of the emulsion containing 100 μg of CII, were administered 3,4-DHPEA-EA (40 μM/kg, i.p.), every 24 h starting from day 25 (n = 20). The dose of 3,4-DHPEA-EA used here to reduce joint injury was chosen based on a previous study (Procopio A, *et al.*, 2009). Collagen-induced arthritis (CIA) is induced in mice by two consecutive (interval 21 days) intradermal injection of 100 μl of the emulsion (containing 100 μg of bovine type II collagen) (CII) and complete Freund's adjuvant (CFA) at the base of the tail. Mice develop erosive hind paw arthritis with macroscopic clinical evidence of CIA as peri-articular erythema and edema in the hind paws. The incidence of CIA is 100% by day 27 in the CII challenged and the severity of CIA progressed over a 35-day period with a reabsorption of bone The histopathology of CIA include erosion of the cartilage at the joint.

Induction of CIA - Bovine CII was dissolved in 0.01 M acetic acid at a concentration of 2 mg/ml by stirring overnight at 4°C. Dissolved CII was frozen at -70°C until use. Complete Freund's adjuvant (CFA) was prepared by addition of Mycobacterium tuberculosis H37Ra at a concentration of 5 mg/ml. Before injection, CII was emulsified with an equal volume of CFA. CIA was induced as previously described (Szabo *et al.*, 1998). On day 1, mice were injected intradermally at the base of the tail with 100 μl of the emulsion containing 100 μg of CII. On day 21, a second injection of CII in CFA was administered.

Clinical assessment of CIA - The development of arthritis in mice in all experimental groups was evaluated daily starting from day 20 after the first intradermal injection by using a macroscopic scoring system: 0 = no signs of arthritis; 1 = swelling and/or redness of the paw or one digit; 2 = two joints involved; 3 = more than two joints involved; and 4 = severe arthritis of the entire paw and digits (Szabo *et al.*, 1998). Arthritic index for each mouse was calculated by adding the four scores of individual paws. Clinical severity was also determined by quantitating the change in the paw volume using plethysmometry (model 7140; Ugo Basile).

Histological examination - On day 35, animals were sacrificed while they were under anesthesia (sodium pentobarbital, 45 mg/kg, i.p), and paws and knees were removed and fixed in 10% formalin. The paws were then trimmed, placed in decalcifying solution for 24 h, embedded in paraffin, sectioned at 5 μm, stained with hematoxylin/eosin and Masson's trichrome stain and studied using light microscopy (Dialux 22 Leitz).

Immunohistochemical localization of nitrotyrosine, Poly ADP Ribose (PAR), iNOS, and COX-2 - On day 35, the joints were trimmed, placed in decalcifying solution for 24 h and 8 μm sections were prepared from paraffin embedded tissues. After deparaffinization, endogenous peroxidase was quenched with 0.3% H_2O_2 in 60% methanol for 30 min. The sections were permeabilized with 0.1% Triton X-100 in PBS for 20 min. Non-specific adsorption was minimized by incubating the section in 2% normal goat serum in phosphate buffered saline for 20 min. Endogenous biotin or avidin binding sites were blocked by sequential incubation for 15 min with avidin and biotin. Sections were incubated overnight with 1) anti-rabbit polyclonal antibody directed at iNOS (1:1000 in PBS, v/v) (DBA, Milan, Italy) or 2) anti-COX-2 goat polyclonal antibody (1:500 in PBS, v/v) or 3) anti- nitrotyrosine rabbit polyclonal antibody (1:1000 in PBS, v/v) or 4) with anti-PAR goat polyclonal antibody rat (1:500 in PBS, v/v) or 5). Controls included buffer alone or non-specific purified rabbit IgG. Specific labeling was detected with a biotin-conjugated goat anti-rabbit IgG (for nitrotyrosine and iNOS) or with a biotin-conjugated goat anti-rabbit IgG (for PAR and COX-2) and avidin-biotin peroxidase complex. In order to confirm that the immunoreaction for the nitrotyrosine was specific some sections were also incubated with the primary antibody (anti-nitrotyrosine) in the presence of excess nitrotyrosine (10 mM) to verify the binding specificity. To verify the binding specificity for PAR, COX-2 and iNOS, some sections were also incubated with only the primary antibody (no secondary) or with only the secondary antibody (no primary). In these situations, no positive staining was found in the sections indicating that the immunoreaction was positive in all the experiments carried out. Immunocytochemistry photographs (N=5) were assessed by densitometry by using Optilab Graftek software on a Macintosh personal computer.

Measurement of cytokines - Tumor Necrosis Factor-α (TNF-α) levels were evaluated in the plasma from CIA and sham mice as previously described (Cuzzocrea *et al.*, 2006). The assay was carried out using a colorimetric commercial ELISA kit (Calbiochem-Novabiochem Co., Milan, Italy) with a lower detection limit of 10 pg/ml.

Measurement of chemokines - Levels of chemokines MIP-1α and MIP-2 were measured in the aqueous joint extracts. Briefly, joint tissues were prepared by first removing the skin and separating the limb below the ankle joint. Joint tissues were homogenized on ice in 3 ml lysis buffer (PBS containing: 2 mM PMSF, and 0,1 mg/ml [final concentration], each of aprotinin, antipain, leupeptin, and pepstatin A) using Polytron (Brinkinarm Instr., Westbury, NY). The homogenized tissues were then centrifuged at 2,000 g for 10 min. Supernatant were sterilized with a millipore filter (0.2 μm) and stored at –80ºC until analyzed. The extracts usually contained 0.2-1.5 mg protein/ml, as measured by protein assay kit (Pierce Chemical Co., Rockford, IL). The levels of MIP-1α and MIP-2 were quantified using a modification of a double ligand method, as previously described (Kasama *et al.*, 1994). Briefly, flat-bottomed 96- well microtiter plates were coated with 50 µl/well of rabbit anti-cytokine antibodies (1 µg/ml in 0.6 mol/liter NaCl, 0.26 mol/liter H_3BO_4 and 0.08 N NaOH, pH 9.6) for 16 h at 4°C, and then washed with PBS, pH 7.5, 0.05% Tween 20 (wash buffer). Nonspecific binding sites on microtiter plates were blocked with 2% BSA in PBS and incubated for 90 min at 37°C. Plates were rinsed four times with wash buffer, and diluted aqueous joint samples (50 µl) were added, followed by incubation for 1 h at 37°C. After washing of plates, chromogen substrate was added. The plates were incubated at room temperature to the desired extinction, after which the reaction was terminated with 50 µl/well of 3 M H_2SO_4 solution. The plates were then read at 490 nm in an ELISA reader. This ELISA method consistently had a sensitivity limit of ~30 pg/ml.

Myeloperoxidase (MPO) assay - Neutrophil infiltration to the inflamed joints was indirectly quantitated using an MPO assay, as previously described for neutrophil elicitation (Mullane *et al.*, 1985). Tissue was prepared as described above and placed in a 50 mM phosphate buffer (pH = 6.0) with 5% hexadecyltrimethyl ammonium bromide (Sigma Chemical Co.). Joint tissues were homogenized, sonicated, and centrifuged at 12,000 g for 15 min at 4°C. Supernatants were assayed for MPO activity using a spectrophotometric reaction with O-dianisidine hydrochloride (Sigma Chemical Co.) at 460 nm.

Materials - 3,4-DHPEA-EA was obtained from Merck Biosciences (Calbiochem, Beecham, Nottingham, UK). Unless otherwise stated, other compounds were obtained from Sigma-Aldrich Company (Milan, Italy). All chemicals were of the highest commercial grade available. All stock solutions were prepared in nonpyrogenic saline (0.9% NaCl; Baxter Healthcare Ltd., Thetford, Norfolk, U.K.) or 10% ethanol (Sigma-Aldrich).

Data analysis - All values in the figures and text are expressed as mean ± standard error (s.e.m.) of the mean of n observations. For the in vivo studies *n* represents the number of animals studied. In the experiments involving histology or immunohistochemistry, the figures shown are representative of at least three experiments (histological or immunohistochemistry coloration) performed on different experimental days on the tissue sections collected from all the animals in each group. Data sets were examined by one- or two-way analysis of variance, and individual group means were then compared with *Student*'s unpaired t test. For the arthritis studies, Mann-Whitney U test (two-tailed, independent) was used to compare medians of the arthritic indices (Szabo *et al.*, 1998). A *p*-value of less than 0.05 was considered significant.

3.1. Results

Effect of 3,4-DHPEA-EA on joint injury during experimental arthritis - To imitate the clinical scenario of RA, mice were subjected to CIA. CIA developed rapidly in mice immunized with CII and clinical signs (periarticular erythema and edema) (Fig. 6B) of the disease first appeared in hind paws between 24 and 26 days post-challenge (Fig. 6D) leading to a 100% incidence of

CIA at day 28 (Fig. 6D). Hind paw erythema and swelling increased in frequency and severity in a time-dependent mode with maximum arthritis indices of approximately 10 observed between day 29 to 35 post immunization (Fig. 6D) in CIA-control mice. 3,4-DHPEA-EA treatment demonstrated a significant reduction of joint inflammation, as identified by a significant reduction in the incidence of arthritis (Figure 6C). CIA-3,4-DHPEA-EA mice showed a 40% reduction in the development of arthritis and a significantly lower arthritis index compared to CIA- control mice (Fig. 6D). There was no macroscopic evidence of either hind paw erythema or edema in the sham-control group mice (Fig. 6A and D). The data in Figure 6E demonstrate a time-dependent increase in hind paw volume (each value represents the mean of both hind paws). The CIA-3,4-DHPEA-EA mice showed a significant reduction of paw edema formation when compared to CIA-control mice (Fig. 6E). No increase in hind paw volume over time was observed in the sham-control mice (Fig. 6E). The rate and the absolute gain in body weight were comparable in sham-control and CIA-control mice in the first week (Fig. 6F). From day 25, the CII-challenged mice gained significantly less weight than the sham-control mice, and this trend continued through to day 35. 3,4-DHPEA-EA treatment determined a significant increase of the weight gain compared with the vehicle-treatment in CIA-control mice (Fig. 6F). The histological evaluation (on day 35) of the joint from CIA-control mice (Fig. 7B) revealed signs of severe arthritis, with inflammatory cell infiltration and bone erosion. The histological alterations of joint were significantly reduced in 3,4-DHPEA-EA-treated mice (Fig. 7C). Moreover Masson's trichrome stain reveals decreased collagen (blue stain) in bone and cartilage of arthritic joint due to bone erosion and cartilage degradation in CIA-control mice (Fig. 7E). The alterations of joint were significantly reduced in 3,4-DHPEA-EA-treated mice (Fig. 7F). There was no evidence of pathology in the sham-control mice (Fig. 7A and D). The histological score (Fig. 7G) was determined by an independent observer.

Effect of 3,4-DHPEA-EA on cytokines, chemokine expression and neutrophil infiltration - We initiated studies to assess the effect of 3,4-DHPEA-EA on the expression of chemokines in the aqueous joint extracts during the development of CIA. As shown in Fig. 8, A and B, the expression of MIP-1α and MIP-2, measured by ELISA, was significantly increased in the joint 35 days after CII immunization. MIP-1α and MIP-2 levels in CIA-3,4-DHPEA-EA mice on day 35 were significantly reduced in a dose-dependent manner in comparison with those in vehicle treated CIA-control mice. Assessment of neutrophil infiltration into the inflamed joint tissue was performed by measuring the activity of MPO. It was significantly elevated 35 days after CII immunization in vehicle-treated CIA-control mice (Fig. 8F), whereas in the CIA-3,4-DHPEA-EA group, MPO activity was markedly reduced in a dose-dependent manner (Fig. 8F). To test whether 3,4-DHPEA-EA modulates the inflammatory process through the regulation of cytokine secretion, we analyzed the plasma levels of the proinflammatory cytokines TNF-α, IL-1β, and IL-6. A substantial increase in TNF-α, (Fig. 8C), IL-1β (Fig. 8D), and IL-6 (Fig. 8E) production was found in CIA-control mice 35 days after CII immunization. Levels of TNF-α (Fig. 8C), IL-1β (Fig. 8D), and IL-6 (Fig. 8E) were significantly reduced in a dose-dependent manner in CIA-3,4-DHPEA-EA mice in comparison to CIA-control mice.

Effect of 3,4-DHPEA-EA treatment on iNOS, COX-2, PGE$_2$, nitrotyrosine, and PAR formation - Immunohistochemical analysis of the tibiotarsal joint sections obtained from CIA-control mice revealed positive staining for iNOS (Fig. 10, A and A1) and COX-2 (Fig. 9, A and A1), which were primarily localized in inflammatory cells. In contrast, staining for iNOS (Fig. 10B) and COX-2 (Fig. 9B) was markedly reduced in the tibiotarsal joints of CIA-3,4-DHPEA-EA (40 µg/kg) mice. No staining for either iNOS or COX-2 was detected in the tibiotarsal joints obtained from sham control mice (data not shown). Moreover, we also evaluated the

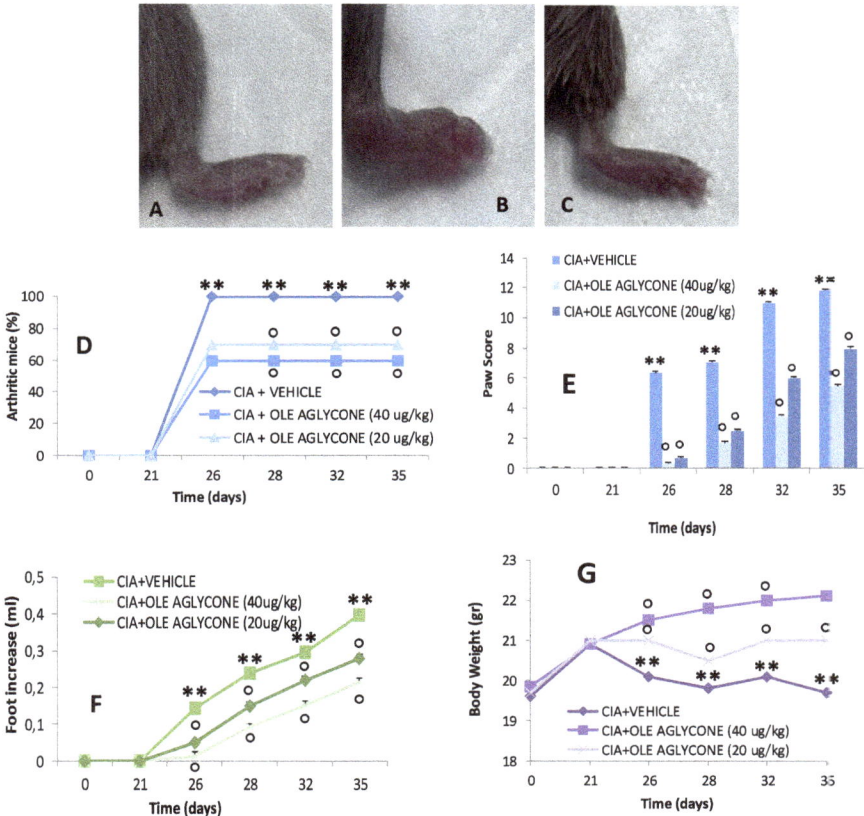

Figure 6. Effect of 3,4-DHPEA-EA (Ole aglycone) on the clinical expression of CIA and on body weight. A, no clinical signs were observed in sham mice. CIA developed rapidly in mice immunized with CII and clinical signs such as periarticular erythema and edema (B) were seen with a 100% incidence of CIA at day 28 (D). E, hind paw erythema and swelling increased in frequency and severity in a time dependent mode. CIA-3,4-DHPEA-EA mice demonstrated a significant reduction in the clinical signs of CIA (C), leading to a decrease in the incidence of arthritis in a dose-dependent manner (D). Swelling of hind paws (F) over time was measured at 2-day intervals. G, beginning on day 25, the CII-challenged mice gained significantly less weight and this trend continued through day 35. CIA-3,4-DHPEA-EA mice demonstrated a significant reduced incidence of weight loss (G) as well as less paw edema in a dose dependent manner (F). The figure is representative of all the animals in each group. Values are means ± S.E.M. of 20 animals for each group. **, P < 0.01 versus sham-control; °, P < 0.01 versus CIA.

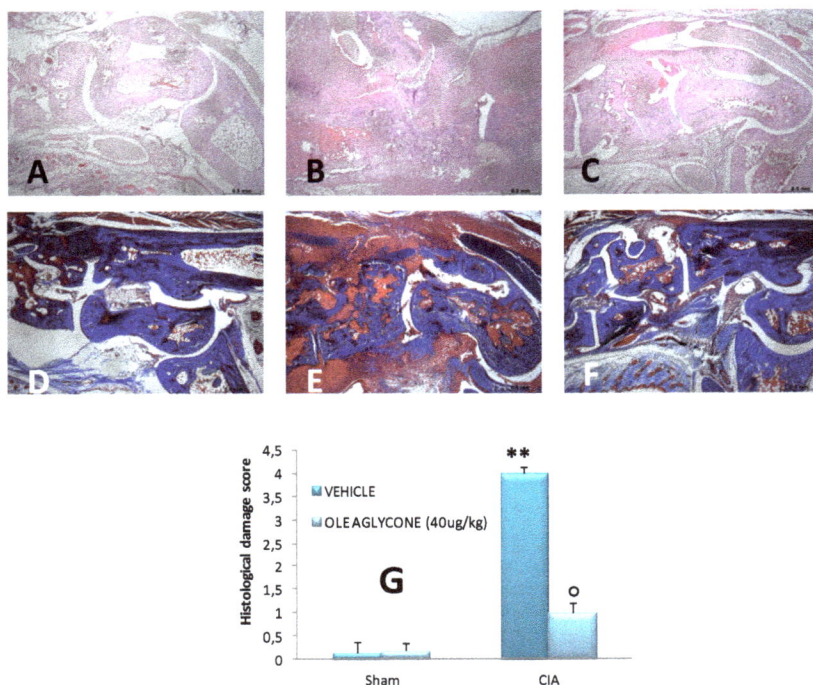

Figure 7. Morphological changes of CIA. Representative hematoxylin and eosin-stained section of joint was examined by light microscopy. The histological evaluation (on day 35) of joint from CIA-control mice (B and G) revealed signs of severe arthritis, with inflammatory cell infiltration and bone erosion. The histological alterations of the joint were significantly reduced in the tissues from CIA-3,4-DHPEA-EA (40 µg/kg)-treated mice (C and G). Masson's trichrome stain reveals decreased collagen in bone and cartilage of arthritic joint due to bone erosion and cartilage degradation in CIA-control mice (E and G). The alterations of joint were significantly reduced in 3,4-DHPEA-EA (40 µg/kg)-treated mice (F and G). There was no evidence of pathology in the sham-control mice (A, D, and G). The histological score (G) was made by an independent observer. The figure is representative of at least three experiments performed on different experimental days. Values are means ± S.E.M. of 20 animals for each group. **, P < 0.01 versus sham-control; °, P < 0.01 versus CIA.

levels of PGE$_2$, the metabolite of COX-2, in the serum during the development of CIA. A substantial increase in PGE$_2$ production was found in CIA-control mice 35 days after CII immunization (Fig. 9E). Levels of PGE-2 were significantly reduced in CIA-3,4-DHPEA-EA mice in a dose dependent manner compared with those in CIA-control mice (Fig. 9E). The release of free radicals and oxidant molecules during chronic inflammation has been suggested to contribute significantly to the tissue injury (Cuzzocrea et al., 2001). On day 35, positive staining for nitrotyrosine, a marker of nitrosative injury, was found in the tibiotarsal joints of vehicle-treated CIA-control mice (Fig. 11, A and A1). 3,4-DHPEA-EA (40 µg/kg) treatment significantly reduced the formation of nitrotyrosine (Fig. 11B). Immunohistochemical analysis of joint sections obtained from CII-challenged mice revealed positive staining for PAR (Fig. 12, A and A1). In contrast, no positive PAR was found in the tibiotarsal joints of CII-challenged mice treated with 3,4-DHPEA-EA (40 µg/kg) (Fig. 12B). There was no staining for either nitrotyrosine or PAR in the tibiotarsal joints obtained from sham-control mice (data not shown).

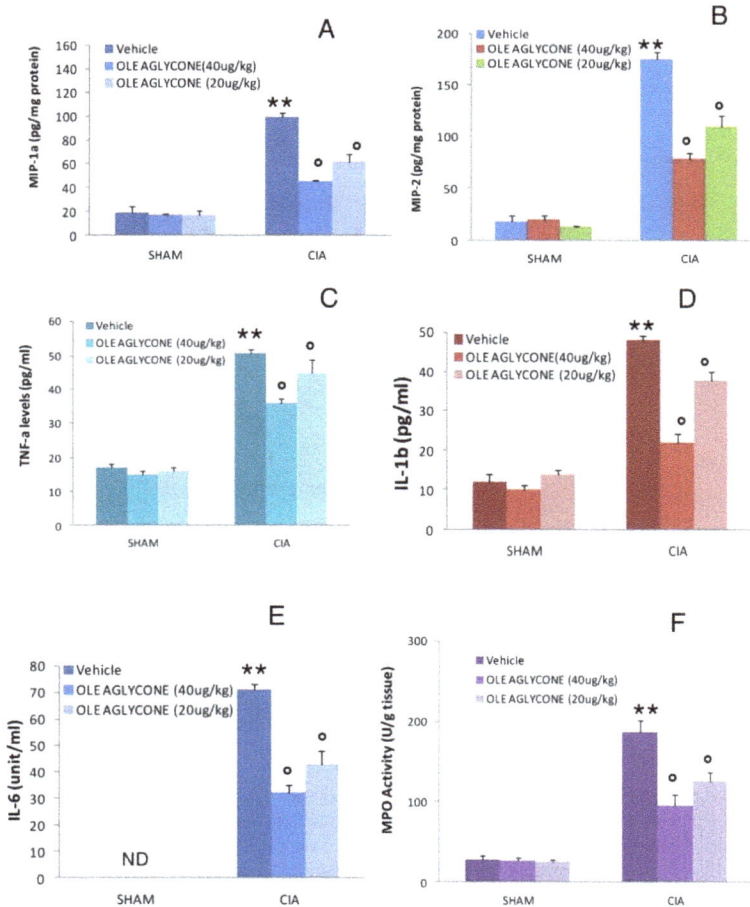

Figure 8. Effect of 3,4-DHPEA-EA (Ole aglycone) treatment on cytokine and chemokine expression and neutrophil infiltration. A substantial increase in the expression of MIP-1 (A), MIP-2 (B), MPO activity (F), plasma TNF-α (C), IL-1β (D), and IL-6 levels (E) was found in CIA-control mice 35 days after CII immunization. CIA-3,4-DHPEA-EA mice demonstrated a significant reduction in the expression of MIP-1 (A), MIP-2 (B), MPO activity (F), plasma TNF-α (C), IL-1β (D), and IL-6 levels in a dose dependent manner (E). Values are means ± S.E.M. of 20 animals for each group. **, $P < 0.01$ versus sham control; °, $P < 0.01$ versus CIA-control. ND, not detectable.

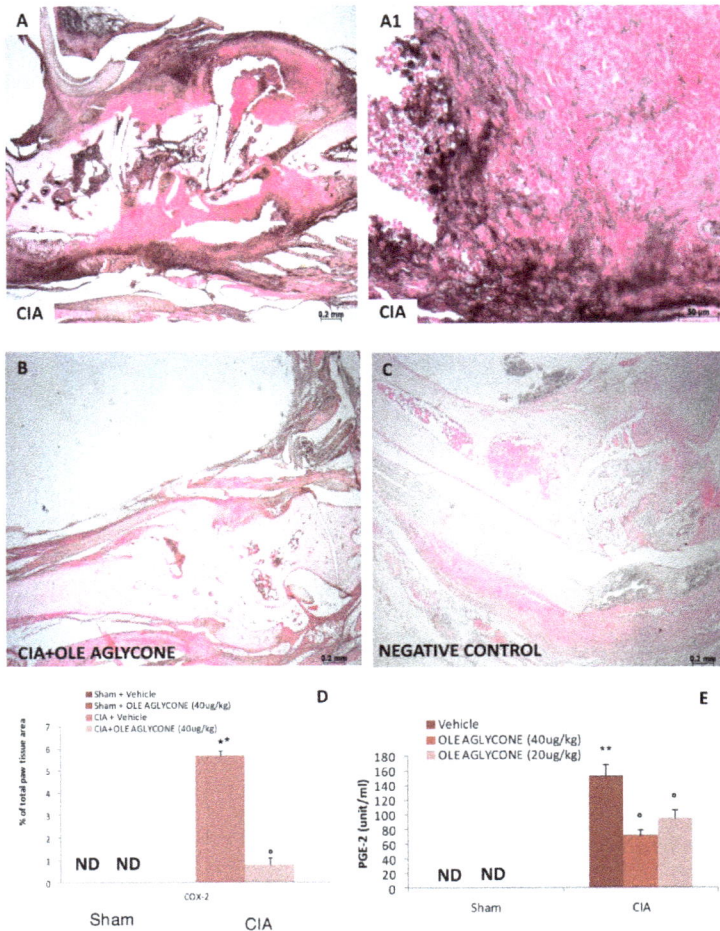

Figure 9. Effect of 3,4-DHPEA-EA (Ole aglycone) treatment on COX-2 immunostaining and on serum PGE$_2$ levels. A marked increase in COX-2 (A and in particular A1) staining was evident in the paw 35 days after initiation of CIA. There was a marked reduction in the immunostaining for COX-2 (B) in the paw of CIA-3,4-DHPEA-EA (40 µg/kg) mice. To verify the binding specificity for COX-2, some sections were also incubated with only the secondary antibody (no primary antibody). No positive staining for COX-2 was found in the sections indicating that the immunoreaction was positive (see negative control C). In addition, a marked increase of PGE$_2$ levels was found in the serum of CIA control mice 35 days after CII immunization (E). The treatment with 3,4-DHPEA-EA also caused a significant reduction in a dose-dependent manner of the serum levels of the metabolite of COX-2 (E). The figure is representative of at least three experiments performed on different experimental days. Densitometry analysis of immunocytochemistry photographs (n = 5) for COX-2 from paw section was assessed (D). The assay was performed by using Optilab Graftek software on a Macintosh personal computer (CPU G3-266). Data are expressed as a percentage of total tissue area. **, P < 0.01 versus sham control; °, P < 0.01 versus CIA. ND, not detectable.

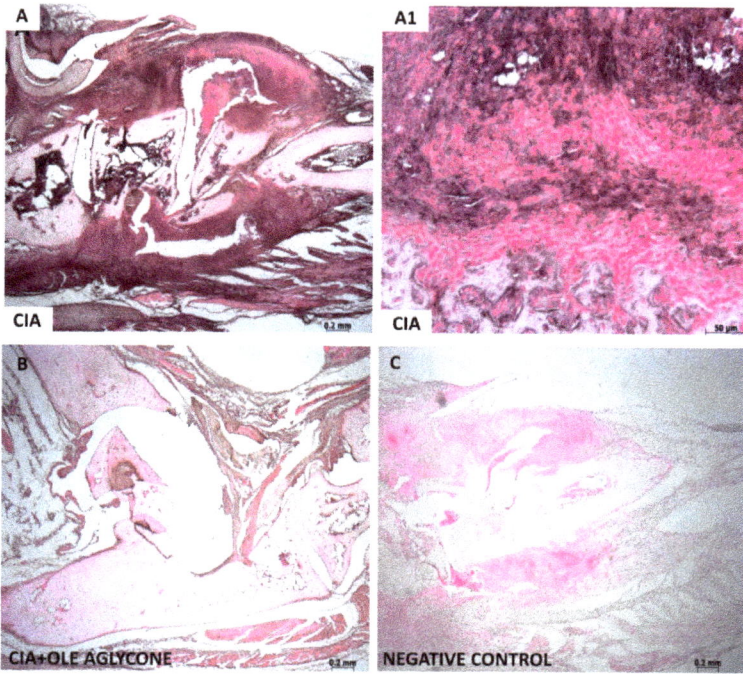

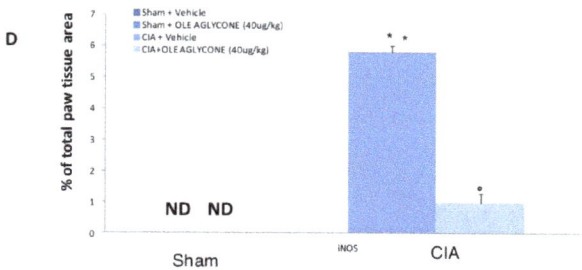

Figure 10. Effect of 3,4-DHPEA-EA (Ole aglycone) treatment on iNOS immunostaining. A marked increase in iNOS (A and in particular A1) staining was evident in the paw 35 days after initiation of CIA. There was a marked reduction in the immunostaining for iNOS (B) in the paw of CIA-3,4-DHPEA-EA (40 μg/kg) mice. To verify the binding specificity for iNOS, some sections were also incubated with only the secondary antibody (no primary antibody). No positive staining for iNOS was found in the sections, indicating that the immunoreaction was positive (see negative control C). The figure is representative of at least three experiments performed on different experimental days. Densitometry analysis of immunocytochemistry photographs (n = 5) for iNOS from paw section was assessed (D). The assay was performed by using Optilab Graftek software on a Macintosh personal computer (CPU G3-266). Data are expressed as a percentage of total tissue area. **, P < 0.01 versus sham-control; °, P < 0.01 versus CIA. ND, not detectable.

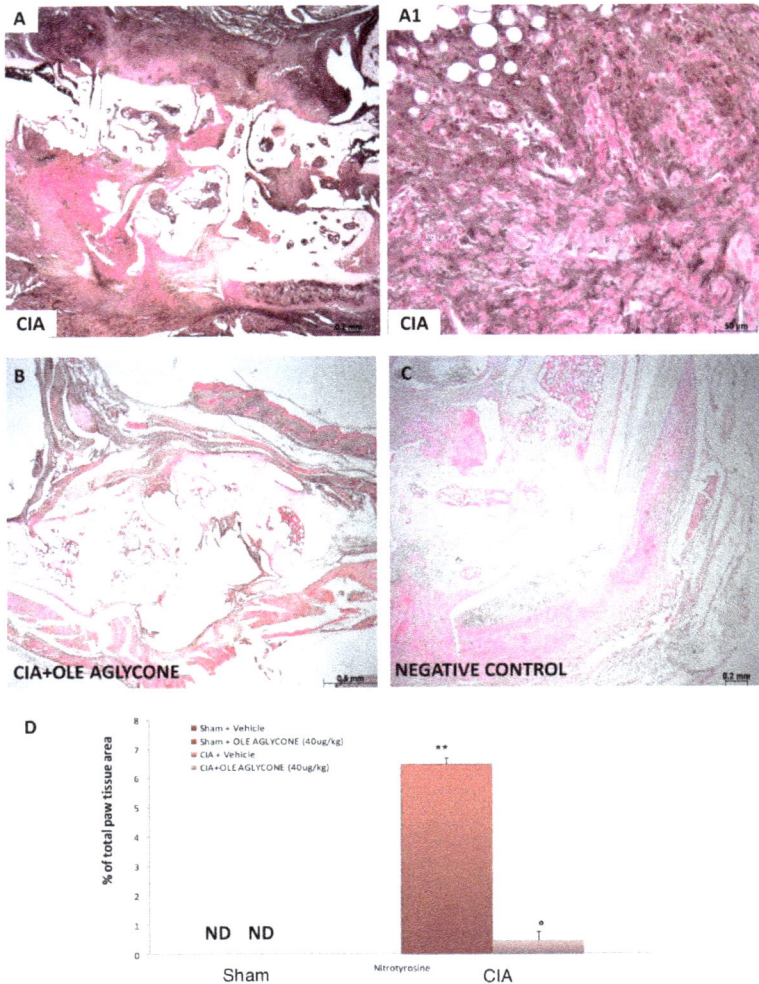

Figure 11. Effect of 3,4-DHPEA-EA (Ole aglycone) treatment on nitrotyrosine immunostaining. A marked increase in nitrotyrosine (A and in particular A1) staining was evident in the paw 35 days after initiation of CIA. There was a marked reduction in the immunostaining for nitrotyrosine (B) in the paw of CIA-3,4-DHPEA-EA (40 µg/kg)-treated mice. To verify the binding specificity for nitrotyrosine, some sections were also incubated with only the secondary antibody (no primary antibody). No positive staining for nitrotyrosine was found in the sections, indicating that the immunoreaction was positive (see negative control C). The figure is representative of at least three experiments performed on different experimental days. Densitometry analysis of immunocytochemistry photographs (n = 5) for nitrotyrosine from paw section was assessed (D). The assay was performed by using Optilab Graftek software on a Macintosh personal computer (CPU G3-266). Data are expressed as a percentage of total tissue area. **, $P < 0.01$ versus sham control; °, $P < 0.01$ versus CIA. ND, not detectable.

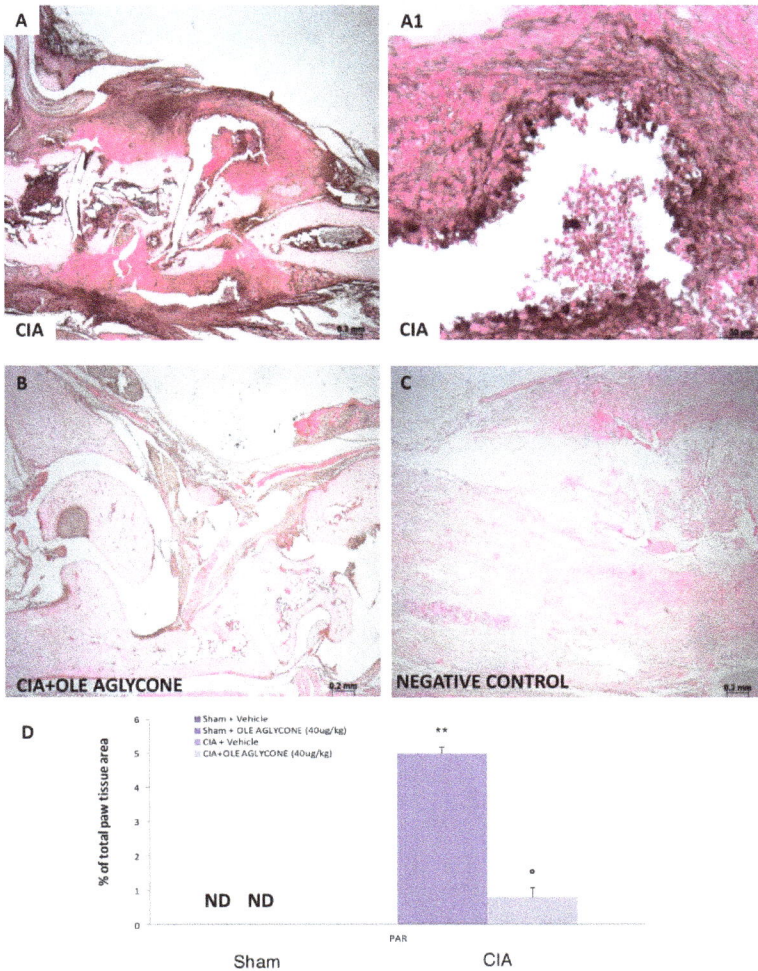

Figure 12. Effect of 3,4-DHPEA-EA (Ole aglycone) treatment on PARP immunostaining. A marked increase in PARP (A and in particular A1), staining was evident in the paw 35 days after initiation of CIA. There was a marked reduction in the immunostaining for PARP (B) in the paw of CIA-3,4-DHPEA-EA (40 µg/kg)-treated mice. To verify the binding specificity for PARP, some sections were also incubated with only the secondary antibody (no primary antibody). No positive staining for PARP was found in the sections, indicating that the immunoreaction was positive (see negative control C). The figure is representative of at least three experiments performed on different experimental days. Densitometry analysis of immunocytochemistry photographs (n = 5) for PARP from paw section was assessed (D). The assay was performed by using Optilab Graftek software on a Macintosh personal computer (CPU G3-266). Data are expressed as a percentage of total tissue area. **, P <0.01 versus sham-control; °, P<0.01 versus CIA. ND, not detectable.

3,4-DHPEA-EA Inhibits the Progression of Established Arthritis - To confirm that 3,4-DHPEA-EA exerts beneficial effects in the experimental model of collagen-induced arthritis, we have also evaluated its effect in a therapeutic regimen of post-treatment (40 μg/kg), starting the treatment at day 28. CIA-3,4-DHPEA-EA post-treatment mice also showed a reduction in the development of arthritis and a significantly lower arthritis score compared with those in CIA-control mice as shown in Fig. 13A. 3,4-DHPEA-EA post-treatment also significantly reduced paw edema formation (Fig. 13B). In addition, 3,4-DHPEA-EA post-treated mice showed significantly reduced histological alterations of the tibiotarsal joint as shown in the histological score (Fig. 13C) and increased body weight (Fig. 13D).

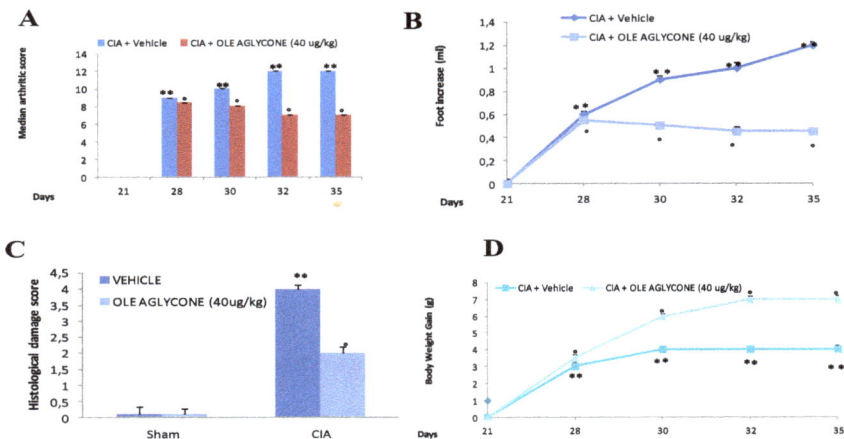

Figure 13. Effect of 3,4-DHPEA-EA (Ole aglycone) post-treatment on joint inflammation. Starting the treatment at day 28, we have also demonstrated that 3,4-DHPEA-EA post-treatment (40 μg/kg) caused a significantly lower arthritis score (A) and a reduction of foot increase (B) compared with the CIA-control. In addition, we have also shown a reduction in the histological damage (C) and increased body weight (D) in 3,4-DHPEA-EA -treated mice. Data are expressed as a percentage of total tissue area. **, $P< 0.01$ versus sham-control; °, $P< 0.01$ versus CIA.

3.2. Discussion

Rheumatoid arthritis is an inflammatory disease characterized by chronic inflammation of the synovial joints associated with proliferation of synovial cells and infiltration of activated immunoinflammatory cells, including memory T cells, macrophages, and plasma cells, leading to progressive destruction of cartilage and bone (Hitchon *et al.*, 2004). Another central feature of RA synovitis is the transformation of fibroblast-like synovial cells into autonomously proliferating cells with a tissue-infiltrating nature, forming hyperplastic tissue with the potential for bone erosion and cartilage degradation known as pannus (Filippin *et al.*, 2008). Therefore, it is necessary to establish and characterize experimental animal models to assess cellular and molecular events that contribute to the pathogenesis of joint inflammation. Of interest, CII-induced arthritis in the mouse has proven to be a useful model, because it possesses many of the cellular and humoral immune events found in human rheumatoid arthritis. Oxidative stress describes an imbalance between ROS synthesis and antioxidants. Many studies have demonstrated a role of ROS in the pathogenesis of inflammatory chronic arthropathies, such as rheumatoid arthritis (Filippin *et al.*, 2008). In this regard, we investigate here the effects of 3,4-DHPEA-EA, a hydrolysis product of oleuropein, in a mouse model of CII-induced arthritis. Although T-cell and antibody responses against CII are a crucial event for the initiation of CIA (Holmdahl *et al.*, 1989), it has been demonstrated that several cytokines also appear to direct cell-to-cell communication in a cascade fashion during the progression of CIA such as IL-1 (Hom *et al.*, 1992), TNF-α (Dong *et al.*, 2010), and IL-6 (Ferraccioli *et al.*, 2010). In addition, it has been demonstrated that monocyte chemotactic protein-1, MIP-α, MIP-1β, and regulated on activation normal T cell expressed and secreted are differentially chemotactic for lymphocyte subsets and are expressed in tissue from the inflamed joints of patients with rheumatoid arthritis (Koch *et al.*, 1994). In this study, we have confirmed that the cytokines (TNF-α, IL-1β, and IL-6) as well as the chemokines (MIP-1α and MIP-2) are expressed at sites of inflamed joints and probably contribute in different capacities to the progression of chronic joint inflammation. Several cytokines, including TNF-α and IL-1β, are known initiators of the nuclear factor (NF-κB) activation cascade (Filippin *et al.*, 2008) and are under its transcriptional control, constituting a positive feedback loop. Recent studies have observed that the acute consumption of olive oil decreased the activation of the NF-κB system on mononuclear cells from healthy men (Perez-Martinez *et al.*, 2007) and that 3,4-DHPEA-EA, *trans*-resveratrol, and hydroxytyrosol incubated with human umbilical vein endothelial cells inhibit lipopolysaccharide-triggered NF-κB and activator protein-1 activation (Carluccio *et al.*, 2003). Of interest, using 3,4-DHPEA-EA, we have demonstrated an inhibition of the release of proinflammatory cytokines and chemokines and a reduction of leukocyte infiltration measured by MPO activity. Several studies also showed that the potential cardioprotective activity of oleuropein in acute cardiotoxicity induced by doxorubicin treatment was determined in vivo in rats (Andreadou *et al.*, 2007) by inhibiting lipid peroxidation products, decreasing oxidative stress, and reducing iNOS in cardiomyocytes and that the olive oil polyphenols are capable of down-regulating COX-2 expression in colonic cancer cells by a mechanism involving the early inhibition of p38 mitogen-activated protein kinase and downstream inhibition of the transcription factor cAMP response element-binding protein (Corona *et al.*, 2007). We show here that 3,4-DHPEA-EA decreased iNOS and COX-2 expression by immunohistochemical staining and also reduced the levels of the metabolite of COX-2, PGE$_2$, in the serum of 3,4-DHPEA-EA treated mice. Reactive nitrogen species, such as the peroxynitrite radical (ONOO-) generated by the reaction between O2 . and nitric oxide, can also cause oxidative damage (Soneja *et al.*, 2005). The addition of ONOO- to body cells, tissues, and fluids leads to fast protonation, which may result in the depletion of -SH groups and other antioxidants, oxidation and nitration of lipids, DNA disruption, and nitration and deamination of DNA bases (Filippin *et al.*, 2008). In this report, an intense immunostaining of nitrotyrosine formation also suggested that a structural alteration of joint had occurred, most probably due to the formation of highly reactive nitrogen derivatives ROS produce strand breaks in DNA, which triggers energy-consuming DNA repair mechanisms and activates the nuclear enzyme poly(ADP-ribosyl) polymerase (PARP). There is various evidence that the activation of PARP may also play an important role in inflammation (Genovese *et al.*, 2005). Continuous or excessive activation of PARP produces extended chains of ADP-ribose (PAR) on nuclear proteins and results in a substantial depletion of intracellular NAD and subsequently, ATP, leading to cellular dysfunction and, ultimately, cell death (Chiarugi, 2002). We demonstrate here that 3,4-DHPEA-EA treatment reduced the activation of PARP with a decrease in PAR expression in the joint during CIA. In this regard, several studies demonstrated that hydroxytyrosol, a hydrolysis product of 3,4-DHPEA-EA, also exerts an inhibitory effect on peroxynitrite-dependent DNA base modifications and tyrosine nitration (Deiana *et al.*, 1999). Likewise, Salvini *et al.* (2006) showed a 30% reduction of oxidative DNA damage in peripheral blood lymphocytes during intervention in postmenopausal women with virgin olive oil containing high amounts of phenols. Thus, 3,4-DHPEA-EA, given at the onset of the disease, reduced paw swelling, clinical score, and the histological severity of the disease when injected after the onset of clinical arthritis. Amelioration of joint disease was associated with near to full inhibition of cytokines as well as inhibition of neutrophil infiltration, which is a key player in RA. Therefore, 3,4-DHPEA-EA was also administered from day 28 after collagen immunization, targeting this early initiation phase of CIA. Then, with treatment starting at day 28, 3,4-DHPEA-EA post-treatment caused a significant reduction of inflamed joints collected at day 35. In conclusion, RA is a complex chronic inflammatory disease dependent on multiple interacting environmental and genetic factors, making it difficult to understand its pathogenesis and thereby to find effective therapies. Taken together, the results of the present study enhance our understanding of the role of ROS generation in the pathophysiology of CII-induced arthritis, implying that olive oil compounds such as 3,4-DHPEA-EA may be useful in the therapy of inflammation.

4. Overall conclusion

Oxidative stress is described as an imbalance between ROS synthesis and antioxidant system in the mammal body where the normal production of oxidants is counteracted by several antioxidative mechanisms. Food constituents are the normal substrate for energy generation but a hypercaloric diet may result in higher ROS, thus inducing oxidative stress. Epidemiological studies have shown that populations consuming a predominantly olive oil-based Mediterranean-style diet exhibit lower incidences of breast cancer and other chronic diseases than those eating a northern European or North American diet. Habitual high intakes of olive oil, especially virgin olive oil, provide a continuous supply of antioxidants, which may reduce oxidative stress via inhibition of lipid peroxidation, a factor that is currently linked to a host of diseases such as cancer, heart disease, and ageing.

Author details

Domenico Britti and Antonio Procopio
Dept. of Health Sciences, University of Catanzaro, Catanzaro, Italy

Daniela Impellizzeri and Salvatore Cuzzocrea
*Dept. of Clinical and Experimental Medicine and Pharmacology, University of Messina,
Torre Biologica – Policlinico Universitario, Messina, Italy*

5. References

Andreadou I, Sigala F, Iliodromitis EK, Papaefthimiou M, Sigalas C, Aligiannis N, Savvari P, Gorgoulis V, Papalabros E, Kremastinos DT. Acute doxorubicin cardiotoxicity is successfully treated with the phytochemical oleuropein through suppression of oxidative and nitrosative stress. J Mol Cell Cardiol 2007;42:549-558.

Angerosa F, d'Alessandro N, Corana F, Mellero G. Characterization of phenolic and secoiridoids aglycones present in virgin olive oil. J Chromatogr 1996; 736: 195-203.

Artajo LS, Romero MP, Morello JR, Motilva MJ. Enrichment of refined olive oil with phenolic compounds: evaluation of their antioxidant activity and their effect on the bitter index. J Agric Food Chem. 2006; 54: 6079-6088.

Avalos I, Chung CP, Oeser A, Milne GL, Morrow JD, Gebretsadik T, Shintani A, Yu C, Stein CM: Oxidative stress in systemic lupus erythematosus: relationship to disease activity and symptoms. Lupus 2007, 16:195-200.

Bendini A, Cerratani L, Carasco-Pancorbo A, Gomez-Caravaca AM, Segura-Carretero A, Fernandez-Gutierrez A, Lerker G. Phenolic molecules in virgin olive oils: a survey of their sensory properties, health effects, antioxidant activity and analytical methods. An overview of the last decade. Molecules 2007, 12: 1679-1719.

Bowie A, O'Neill LA. Oxidative stress and nuclear factor-kappaB activation: a reassessment of the evidence in the light of recent discoveries. Biochem Pharmacol 2000;59:13-23.

Carasco-Pancorbo A, Cerratani L, Bendini A, Segura-Carretero A, Del Carlo M, Gallina-Toschi T, Lerker G, Compagnoni D, Fernandez-Gutierrez A. Evaluation of the antioxidant capacity of individual phenolic compounds in virgin olive oil. J Agric Food Chem. 2005; 53: 8918-8925.

Carluccio M A, Siculella L, Ancora MA, Massaro M, Scoditti E, Storelli C, Visioli F, Distante A, De Caterina R. Olive oil and red wine antioxidant polyphenols inhibit endothelial activation: antiatherogenic properties of Mediterranean diet phytochemicals. Arterioscler Thromb Vasc Biol 2003; 23: 622-29.

Chen CC, Chow MP, Huang WC, Lin YC, Chang YJ. Flavonoids inhibit tumor necrosis factor-alpha-induced up-regulation of intercellular adhesion molecule-1 (ICAM-1) in respiratory epithelial cells through activator protein-1 and nuclear factor-kappaB: structure-activity relationships. Mol Pharmacol 2004; 66: 683-693.

Chiarugi A. Poly(ADP-ribose) polymerase: killer or conspirator? The 'suicide hypothesis' revisited. Trends Pharmacol Sci 2002; 23:122–129.

Corona G, Deiana M, Incani A, Vauzour D, Dessi MA, Spencer JP. Inhibition of p38/CREB phosphorylation and COX-2 expression by olive oil polyphenols underlies their anti-proliferative effects. Biochem Biophys Res Commun 2007, 362:606-611.

Covas MI. Bioactive effects of olive oil phenolic compounds in humans: reduction of heart disease factors and oxidative damage. Inflammopharmacology 2008; 16: 216–8

Cuzzocrea S, Crisafulli C, Mazzon E, Esposito E, Muià C, Abdelrahman M, Di Paola R, Thiemermann C. Inhibition of glycogen synthase kinase-3beta attenuates the development of carrageenan-induced lung injury in mice. Br J Pharmacol 2006; 149: 687-702.

Cuzzocrea S, Riley D P, Caputi A P, Salvemini D. Antioxidant therapy: a new pharmacological approach in shock, inflammation, and ischemia/reperfusion injury. Pharmacol Rev 2001;53:135-159.

Cuzzocrea S, Sautebin L, De Sarro G, Costantino G, Rombolà L, Mazzon E, Ialenti A, De Sarro A, Ciliberto G, Di Rosa M, Caputi A P, Thiemermann C. Role of IL-6 in the pleurisy and lung injury caused by carrageenan. J Immunol 1999;163: 5094-5104.

de la Puerta R, Martinez-Dominguez E, Ruiz-Gutierrez V. Effect of minor components of virgin olive oil on topical antiinflammatory assays. Z Naturforsch C 2000; 55: 814-819.

Deiana M, Aruoma O I, Bianchi ML, Spencer JP, Kaur H, Halliwell B, Aeschbach R, Banni S, Dessi MA, Corongiu FP. Inhibition of peroxynitrite dependent DNA base modification and tyrosine nitration by the extra virgin olive oil-derived antioxidant hydroxytyrosol. Free Radic Biol Med 1999; 26: 762-9.

Della Ragione F, Cucciolla V, Borriello A, Della Pietra V, Pontoni G, Racioppi L, Manna C, Galletti P, Zappia V. Hydroxytyrosol, a natural molecule occurring in olive oil, induces cytochrome c-dependent apoptosis. Biophys Res Commun 2000; 278: 733-739.

Dong J, Gao Y, Liu Y, Shi J, Feng J, Li Z, Pan H, Xue Y, Liu C, Shen B et al: The protective antibodies induced by a novel epitope of human TNF-alpha could suppress the development of collagen-induced arthritis. PloS one 2010, 5:8920.

Esti M, Cinquanta L, La Notte E. Phenolic compounds in different olive varieties. J Agric Food Chem. 1998; 46: 32-35.

Fabiani R, De Bartolomeo A, Rosignoli P, Servili M, Montedoro GF, Morozzi G. Cancer chemoprevention by hydroxytyrosol isolated from virgin olive oil through G1 cell cycle arrest and apoptosis. Eur J Cancer Prev 2002; 11: 351-358.

Fabiani R, De Bartolomeo A, Rosignoli P, Servili M, Selvaggini R, Montedoro GF, Di Saverio C, Morozzi G. Virgin olive oil phenols inhibit proliferation of human promyelocytic leukemia cells (HL60) by inducing apoptosis and differentiation. J Nutr 2006; 136: 614-619.

Fabiani R, Rosignoli P, De Bartolomeo A, Fuccelli R, Servili M, Montedoro GF, Morozzi G. Oxidative DNA damage is prevented by extracts of olive oil, hydroxytyrosol, and other olive phenolic compounds in human blood mononuclear cells and HL60 cells. J Nutr 2008; 138: 1411-1426.

Ferraccioli G, Bracci-Laudiero L, Alivernini S, Gremese E, Tolusso B, De Benedetti F. Interleukin-1beta and Interleukin-6 in arthritis animal models. Roles in the early phase of transition from the acute to chronic inflammation and relevance for human rheumatoid arthritis. Mol Med 2010, 16: 552-7.

Fialkow L, Wang Y, Downey G P. Reactive oxygen and nitrogen species as signaling molecules regulating neutrophil function. Free RadicBiol Med 2007;42:153-164.

Filippin LI, Vercelino R, Marroni NP, Xavier RM: Redox signalling and the inflammatory response in rheumatoid arthritis. Clin Exp Immunol 2008, 152: 415-422.

Genovese T, Mazzon E, Muia` C, Patel NS, Threadgill MD, Bramanti P, De Sarro A, Thiemermann C, Cuzzocrea S. Inhibitors of poly(ADP-ribose) polymerase modulate signal transduction pathways and secondary damage in experimental spinal cord trauma. J Pharmacol ExpTher 2005; 312: 449–457.

Gutierrez-Rosales F, Rios JJ, Gomez-Rey ML. Main polyphenols in the bitter taste of virgin olive oil. Structural confirmation by on-line High-Performance Liquid Chromatography Electrospray Ionization Mass Spectrometry. J Agric Food Chem 2003; 51: 6021-6025.

Harris ED, Jr.: Rheumatoid arthritis. Pathophysiology and implications for therapy. N Engl J Med 1990, 322:1277-1289.

Hitchon CA, El-Gabalawy HS: Oxidation in rheumatoid arthritis. Arthritis Res Ther 2004, 6:265-278.

Holmdahl R, Andersson M, Goldschmidt TJ, Gustafsson K, Jansson L, Mo JA. Type II collagen autoimmunity in animals and provocations leading to arthritis. Immunol Rev 1990;118:193-232.

Holmdahl R, Mo J, Nordling C, Larsson P, Jansson L, Goldschmidt T, Andersson M, Klareskog L. Collagen induced arthritis: an experimental model for rheumatoid arthritis with involvement of both DTH and immune complex mediated mechanisms. Clin Exp Rheumatol 1989; 7: S51-55.

Hom JT, Cole H, Estridge T, Gliszczynski VL. Interleukin-1 enhances the development of type II collagen-induced arthritis only in susceptible and not in resistant mice Clin Immunol Immunopathol 1992; 62: 56-65.

Impellizzeri D, Esposito E, Mazzon E, Paterniti I, Di Paola R, Bramanti P, Morittu VM, Procopio A, Britti D, Cuzzocrea S. The effects of oleuropein aglycone, an olive oil compound, in a mouse model of carrageenan-induced pleurisy. Clin Nutr 2011; 30: 533–40.

Impellizzeri D, Esposito E, Mazzon E, Paterniti I, Di Paola R, Morittu VM, Procopio A, Britti D, Cuzzocrea S. Oleuropein aglycone, an olive oil compound, ameliorates development of arthritis caused by injection of collagen type II in mice. J Pharmacol Exp Ther. 2011; 339: 859-69.

Kasama T, Strieter RM, Lukacs NW, Burdick MD, Kunkel SL: Regulation of neutrophil-derived chemokine expression by IL-10. J Immunol 1994, 152:3559-69.

Koch AE, Kunkel SL, Harlow LA, Mazarakis DD, Haines GK, Burdick MD, Pope RM, Strieter RM: Macrophage inflammatory protein-1 alpha. A novel chemotactic cytokine for macrophages in rheumatoid arthritis. J Clin Invest 1994; 93:21-928.

Li JM, Gall NP, Grieve DJ, Chen M, Shah AM. Activation of NADPH oxidase during progression of cardiac hypertrophy to failure. Hypertension 2002; 40: 477-484.

Menendez JA, Vazquez-Martin A, Colomer R, Brunet J, Carrasco-Pancorbo A, Garcia-Villalba R, Fernandez-Gutierrez A, Segura-Carretero A. Olive oil's bitter principle reverses acquired autoresistance to trastuzumab (Herceptin) in HER2-overexpressing breast cancer cells. BMC Cancer 2007;(7) 80.

Miles EA, Zoubouli P, Calder PC. Differential anti-inflammatory effects of phenolic compounds from extra virgin olive oil identified in human whole blood cultures. Nutrition 2005;21:389-394.

Montedoro GF, Servili M, Baldioli M, Miniati E. Simple and hydrolyzable phenolic compounds in virgin olive oil. 1. Initial characterization of the hydrolyzable fraction. J Agric Food Chem 1992; 40: 1571-1577; *ibidem* 1578-1580.

Mullane K M, Kraemer R, Smith B. Myeloperoxidase activity as a quantitative assessment of neutrophil infiltration into ischemic myocardium. J Pharmacol Methods 1985;14:157-167.

Nantel F, Denis D, Gordon R, Northey A, Cirino M, Metters KM, Chan CC. Distribution and regulation of cyclooxygenase-2 in carrageenan-induced inflammation. Br J Pharmacol 1999;128: 853-9.

Ohkawa H, Ohishi N, Yagi K. Assay for lipid peroxides in animal tissues by thiobarbituric acid reaction. Anal Biochem 1979; 95: 351-358.

Omar SH. Oleuropein in olive and its pharmacological effects. Sci Pharm 2010; 78: 133-54.

Owen RW, Mier W, Giacosa A, Hull WE, Spiegelhalder B, Bartsch H. Phenolic compounds and squalene in olive oils: the concentration and antioxidant potential of total phenols, simple phenols, secoiridoids, lignansand squalene. Food Chem Toxicol 2000; 38: 647-659.

Paiva-Martins F, Fernandes J, Rocha S, Nascimento H, Vitorino R, Amado F, Borges F, Belo L, Santos-Silva A. Effects of olive oil polyphenols on erythrocyte oxidative damage. Mol Nutr Food Res. 2009; 53: 609-616.

Paiva-Martins F, Gordon MH. Isolation and characterization of the antioxidant component 3,4-dihydroxyphenylethyl 4-formyl-3-formylmethyl-4-hexenoate from olive (*Olea europaea*) leaves. J Agric Food Chem. 2001; 49: 4214-4219.

Perez-Martinez P, Lopez-Miranda J, Blanco-Colio L, Bellido C, Jimenez Y, Moreno JA, Delgado-Lista J, Egido J, Perez-Jimenez F. The chronic intake of a Mediterranean diet enriched in virgin olive oil, decreases nuclear transcription factor kappa B activation in peripheral blood mononuclear cells from healthy men. Atherosclerosis 2007;194: 141-146.

Petroni A, Blasevich M, Salami M, Papini N, Montedoro GF, Galli C. Inhibition of platelet aggregation and eicosanoid production by phenolic components of olive oil. Thromb Res 1995; 78:151-160.

Procopio A, Alcaro S, Nardi M, Oliverio M, Ortuso F, Sacchetta P, Pieragostino D, Sindona G. Synthesis, biological evaluation, and molecular modeling of oleuropein and its semisynthetic derivatives as cyclooxygenase inhibitors. J Agric Food Chem 2009; 57: 11161-67.

Sadeghi-Hashjin G, Folkerts G, Henricks P A, Muijsers R B, Nijkamp FP. Peroxynitrite in airway diseases. Clin Exp Allergy 1998;28:1464-1473.

Saija A, Tomaino A, Lo Cascio R, Rapisarda P, Dederen JC. In vitro antioxidant activity and in vivo photoprotective effect of a red orange extract. Int J Cosmet Sci 1998; 20: 331-42.

Salvini S, Sera F, Caruso D, Giovannelli L, Visioli F, Saieva C, Masala G, Ceroi M, Giovacchini V, Pitozzi V, Galli C, Romani A, Mulinacci N, Bortolomeazzi R, Dolara P, Palli D. Daily consumption of a high-phenol extra-virgin olive oil reduces oxidative DNA damage in postmenopausal women. Br J Nutr 2006; 95: 742-751.

Servili M, Selvaggini R, Esposto S, Taticchi A, Montedoro GF, Morozzi G. Health and sensory properties of virgin olive oil hydrophilic phenols: agronomic and technological aspects of production that affect their occurrence in the oil. J Chromatogr A. 2004; 1054: 113–27.

Soneja A, Drews M, Malinski T: Role of nitric oxide, nitroxidative and oxidative stress in wound healing. Pharmacol Rep 2005, 57:108-119.

Suárez M, Romero M-P, Macià A, Valls RM, Fernández S, Solà R, Motilva M-J. Improved method for identifying and quantifying olive oil phenolic compounds and their metabolites in human plasma by microelution solid-phase extraction plate and liquid chromatography–tandem mass spectrometry. J Chromatogr B 2009: 877: 4097-4106.

Szabo C, Dawson VL. Role of poly (ADP-ribose) synthetase in inflammation and ischaemia-reperfusion. Trends Pharmacol Sci 1998;19: 287-298.

Taub DD, Conlon K, Lloyd AR, Oppenheim JJ, Kelvin DJ: Preferential migration of activated CD4+ and CD8+ T cells in response to MIP-1 alpha and MIP-1 beta. Science 1993, 260:355-358.

Trichopoulou A, Costacou T, Bamia C, Trichopoulos D: Adherence to a Mediterranean diet and survival in a Greek population. N Engl J Med 2003;348: 2599-2608.

Tripoli E, Giammanco M, Tabacchi G, Di Majo D, Giammanco S, La Guardia M. The phenolic compounds of olive oil: structure, biological activity and beneficial effects on human health. Nutr Res Rev 2005; 18: 98–112.

Visioli F, Poli A, Galli C. Antioxidant and other biological activities of phenols from olives and olive oil. Med Res Rev 2002; 22: 65-75.

Table Olive

Chapter 15

Nutritional and Sensory Quality of Table Olives

Barbara Lanza

Additional information is available at the end of the chapter

1. Introduction

Table olives are greatly consumed mainly by Mediterranean Sea area populations. The italian cuisine, for example, offers many dishes, aperitifs and appetizer in which olives are an essential ingredient: fish and meat cooked with olives, table olive-based condimens for *pasta* and *pizza*, bread-dough mixed with green/black olives, *bruschetta* with olive paste, and so on. The gastronomic uses of olives are widely and very well-known, but the same cannot be said about its nutritional, nutraceutical and sensory properties. The various Italian olive cultivars with remarkable aptitude for processing as table olives have allowed the development of specific and diversified process technologies. Both chemical and biological treatments perform the dual function of hydrolyzing the compounds responsible for the bitter taste of olive fruit and stabilizing the end product to overcome the constraint of seasonal production. All technological interventions have negative collateral effects that must be minimized or at least kept under control to avoid compromising the quality of the processed product to satisfy the expectation, of the consumer, directed towards natural or minimally processed products, whose nutritional properties and health benefits remain unaltered.

The most important production zones of table olives are located in the Mediterranean area and their consumption is expanding, due to the increasing popularity of the Mediterranean diet (Table 1). In Italy, during the last three years, the average consumptior of table olives was approximately 124.000 tonnes/year with a *pro-capite* assumption of 2.1 kg/year. The Italian production covers only 48.1% of the consumers demand, the remaining part is imported from Spain and Greece.

The Italian table olive-growing boasts millenary traditions and its history is an integral part of Italian culture, since its first inhabitants, in prehistoric epoch, used the alimentary resources offered by the primitive Mediterranean "Macchia" and began to domesticate the wild olive tree. Among the over 500 varieties of *Olea europaea* on the national territory only a fraction is specifically suitable for processing as table olives. Italy is rich in typical table olive products, obtained by traditional methods, many of which have obtained or aspired to the recognition of trademarks (PDO, PGI, TAP). The reference to the origin through the Protected Designation of Origin and the Protected Geographical Indication gives to the product added value and it is greatly appreciated by the consumer. The Italian PDO recognized are only three, "Nocellara del Belice" (Sicilia), "La Bella della Daunia" (Puglia) and "Oliva Ascolana del Piceno" (Marche and Abruzzo). The trademark "Traditional Agrifood Products-TAP" (Decree of Italian Ministry of Agriculture 18 July 2000; XII Revision of 7 June 2012) indicate those food products whose methods of processing are consolidated and have been used for a period not less than 25 years. Italy recognize many traditional table olive products, thanks to the country tradition that has passed to the present unique and singular processing technologies for table olives, reflecting the different realities that characterize the different Italian regions (Table 2).

Country/Year	1990/1	1995/6	2000/1	2005/6	2008/9	2009/10	2010/11*	2011/12**
Spain	100.0	116.0	163.6	214.1	147.7	107.9	150.0	200.0
Greece	33.0	30.0	25.0	28.0	20.0	20.0	25.0	23.0
Portugal	19.0	11.0	12.9	9.4	12.7	7.0	7.0	7.0
Italy	**138.0**	**125.0**	**145.0**	**139.0**	**138.5**	**122.4**	**125.0**	**125.0**
France	31.7	28.0	39.0	53.9	53.0	56.8	58.0	59.5
Germany	15.8	21.5	35.0	40.0	47.2	57.8	67.2	72.0
Romania	5.0	10.0	9.5	10.5	16.5	22.9	23.0	23.0
UK	3.4	6.4	13.8	26.8	28.7	33.7	35.6	35.6
Other UE countries	0.5	4.2	17.0	42.9	85.0	81.6	83.9	75.6
Total UE	**346.4**	**352.1**	**460.8**	**564.6**	**549.3**	**510.1**	**574.7**	**620.7**
Egypt	11.0	48.0	57.0	170.0	360.0	340.0	200.0	300.0
Turkey	110.0	129.0	125.0	221.0	240.0	260.0	260.0	290.0
Syria	60.0	71.0	110.0	102.0	94.0	116.0	119.0	125.0
Algeria	14.0	14.5	33.0	80.0	97.5	134.0	129.0	137.0
USA	169.5	149.0	185.0	220.0	210.0	203.0	240.0	240.0
Russia	4.0	6.0	21.0	45.0	90.0	67.0	72.5	75.0
Brazil	41.0	46.8	45.0	55.5	69.0	79.0	87.0	87.0
Canada	12.0	13.5	20.5	25.0	26.0	27.5	27.5	28.0
Chile	6.0	8.0	9.0	11.5	21.0	29.0	30.0	32.0
Peru	14.0	15.0	9.0	19.0	25.5	50.0	50.0	50.0
Australia	7.5	9.5	14.0	18.5	19.0	20.5	22.5	22.5
Other countries	161.6	180.4	214.2	314.9	308.7	362.9	392.8	380.3
Total World	**957.0**	**1042.8**	**1303.5**	**1829.0**	**2110.0**	**2199.0**	**2205.0**	**2387.5**

Table 1. Table olive consumption. Data are expressed as 1000 tonnes. Source: International Olive Council (November 2011). * provisional; **previsional.

Region	Traditional Agrifood Product (TAP)	Cultivar
Sicilia	oliva nebba, oliva nera passuluni	Nocellara del Belice
Calabria	olive alla calce, olive in salamoia, olive nella giara, olive nere infornate, olive schiacciate, olive sotto sale	Carolea Tonda di Strongoli Cassanese Borgese
Basilicata	olive nere secche, oliva da forno di Ferrandina	Majatica
Puglia	oliva da mensa (mele di bitetto, ualie dolc), olive cazzate o schiacciate, olive celline di Nardò in concia tradizionale - olive in concia (ciline alla capàsa – volie alla capàsa), olive in salamoia, olive verdi, Peranzana da mensa di Torremaggiore	Termite di Bitetto Ogliarola leccese Bella di Cerignola Cellina di Nardò Peranzana
Campania	olive pisciottane schiacciate sott'olio, oliva caiazzara, oliva vernacciola di Melizzano	Pisciottana Caiazzana Vernacciola
Lazio	olive da mensa bianche e nere (olive calce e cenere, olive di Gaeta, oliva bianca di Itri, olive al fumo, olive sott'olio, olive spaccate e condite, olive in salamoia, olive essiccate) , pestato di olive di Gaeta	Itrana Carboncella
Abruzzo	olive Intosso (olive n'dosse, olive in salamoia)	Intosso
Molise	olive al naturale (live curvate, olie all'acqua e sale)	Sperone di gallo Olivone Olivoncello Leccino
Marche	olive nere marinate (olive nere strinate), salsa di olive	Raggiola Raggia Piantone di Falerone Leccino
Toscana	olive in salamoia	Leccino Frantoio
Liguria	olivo taggiasca	Taggiasca
Sardegna	olive a scabecciu, olive verdi in salamoia	Tonda di Cagliari Pizz'e Carroga

Table 2. TAP table olives (XII Revision – 2012).

2. Preparations and styles

The numerous Italian olive cultivars with remarkable aptitude for processing as table olives have allowed the development of specific and diversified process technologies. The Italian table olive production is mainly concentrated in the Southern and Central regions of Italy (Figure 1).

Figure 1. Regional distribution of Italian table olive production.

The main Italian table olive preparations are the following:

2.1. Treated olives

The "treated olives", according to "Trade Standard Applying to Table Olives" (IOC, 2004) are "green olives, olives turning color or black olives that have undergone alkaline treatment, then packed in brine in which they undergo complete or partial fermentation, and preserved or not by the addition of acidifying agents".

2.1.1. Treated green olives in brine by Sevillan-style

This method is also known as Sevillan or Spanish-style and it is one of the most common and oldest methods for Italian green table olives (Tavanti, 1819). To obtain the "treated green olives in brine", the fruits are debittered with NaOH aqueous solution ranging from 2.0% to 3.5%, mainly depending on the variety and ripeness of the olives. The alkaline treatment performs the function of hydrolyzing the compound principally responsible for the bitter taste (oleuropein). The lye solution completely covers the fruits and the olives remain in this solution until the lye has penetrated between 2/3 and 3/4 of the distance between the skin and the pit. To verify a correct lye treatment, olives are cut with a particularly sharp blade such as razor blade or scalpel near the pit, checking the surface after air exposition. After alkaline treatment, the olives are washed with potable water. The sequence of washings is the following:

- the first one is a strong washing with potable water made using a shower lasting 15-20 minutes in order to eliminate the lye attached to the fruit surface; the olives washed are then left in the final washing water
- after 2-3 hours a second faster and simpler washing takes place using a filling-emptying procedure
- in the following 24-48 hours 3-4 other washings are carried out, using always a filling-emptying procedure.

After water-washings to eliminate the residual lye, olives are covered with a sodium chloride solution (brine) and left to develop a spontaneous lactic fermentation. Initial brine concentrations are 8-10% NaCl but rapidly drop to 5% due to the high content of interchangeable water in the olives. A spontaneous fermentation starts as soon as the olives are placed in brine. After alkaline treatment, the pH of olive flesh reaches the value of 11.0-13.0 down to the value of 8.0-9.0 after the repeated washings. In this broth culture, a complex and variable microbiota grows. Reducing sugars and glucosides, the basic sources of carbon needed in the development of lactobacilli and other microorganisms, pass from olive flesh to the brine, where they are used by heterofermentative or homofermentative microorganisms to produce lactic acid. In the first phase of fermentation, when Gram-negative bacteria prevail, the pH ranges from 8.0-9.0 to about 6.0. This low pH promotes the growth of lactic acid bacteria that are aciduric with optimal growth between pH 5.5 and 5.8. At the end of lactic fermentation, the pH reaches values <4.0 and acidity increases, ensuring thus the preservation of the product. The lactic fermentation ends with the exhaustion of available carbohydrates (glucose from glucosides and reducing sugars). Several researches have been carried out to evaluate the technological functionality of selected lactic acid bacteria or enterococci in Spanish-style green olive processing. *Enterococcus casseliflavus* and *Lactobacillus pentosus* (de Castro *et al.*, 2002; Sánchez *et al.*, 2001) have been proposed as starter cultures to accelerate lactic acid formation at pH 9 (immediately after washings). In this case, obviously, the strain used as starter is not necessarily oleuropeinolytic because lye has just demolished the bitter glucoside. The importance of this type of starter is to reduce the lag phase and the risk of spoilage (Bevilacqua *et al.*, 2010; Ruiz-Barba *et al.*, 1994; Leal-Sanchez *et al.*, 2003). As a result of ill-treatment alkali set or poorly conducted, the tissues of the fruit may also undergo profound changes in structural and nutritional constituents (Marsilio & Lanza, 1995).

2.1.2. Sweet green olives by Castelvetrano-style

This is a production method used in Sicilia, almost exclusively in the Castelvetrano district using the *Nocellara del Belice* cultivar. The product is mainly consumed in central and southern Italy (Cappello & Poiana, 2005; Lanza & Poiana, 2011). The olives are hand-harvested at the green stage of ripening and, once they arrive at the processing plant, they are graded, since only fruits of more than 19 mm in diameter are used. The selected olives are put into plastic vessels and covered with 2.5-3.5 °Bè NaOH solution (1.7-2.4%), depending on the fruit ripeness and size. These vessels have 220 L total capacity, and are filled with around 140 kg of fruits. Eight hours after the lye treatment begins, 6-7 kg of grinded salt is added to each container, and the olives are kept in this "alkaline brine" for 8-10 days. Salting is performed by placing the grinded salt over the "press-fruit" of each barrel (Figure 2).

Figure 2. Addition of grinded salt directly on the "press-fruit" at the surface of the barrel (photograph by A. Cappello from Lanza & Poiana, 2011).

The sodium chloride in contact with the liquid is slowly dissolved and is distributed by gravity in the whole volume of the container. The salt traditionally used in this type of procedure is sea salt coming from the saltworks of Trapani. It is particularly rich in trace elements such as K, Mg, Ca, Fe and I, unlike the rock salt composed almost exclusively of NaCl. In this environment, the olives are softened quickly highlighting a deep green color. When the barrel is opened, the lye/salt brine is drained and the olives are washed to remove excess lye. A mild washing step, carried out before marketing, does not totally eliminate the lye, whose "soapy" taste is appreciated by the consumers of these olives. Under ambient storage conditions the Castelvetrano olives have a shelf-life of only a few months, especially under hot conditions. Deteriorated olives lose colour and develop off-odours. As this olive style is a seasonal product, long-term storage is not required. Future research, however, may refine the process and include steps for long-term storage under refrigeration, with the temperature maintained between 4 °C and 7 °C, or in packing solutions.

2.1.3. Green olives by lime-and-ash treatment

In some Italian regions (Puglia, Campania, Lazio and Calabria), olives are prepared with an ancient and traditional method that includes a debittering phase obtained using a lime-ash mixture, preparing these treated green olives by mixing CaO (lime) with olive wood ash and adding water to a paste, leaving submerged the green olives for some hours at room temperature. The ratio of lime and ash is different (1:4, 1:8, 1:10, etc) and depends on the cultivar and shape of the fruit. The action of lime-ash is similar to that of NaOH. After alkaline treatment, the olives are washed with potable water. The sequence of washings is similar to the sequence previously described for the treated green olives by Sevillan-style. It is possible to eat these olives right after the debittering treatment or after brining. In this case, after the washing step, olives were placed in an initial brine solution of NaCl (8% w/v), where a spontaneous fermentation takes place. NaCl concentration is carefully monitored during this phase. The brine concentration decreases quickly as a consequence of the osmotic phenomena between the brine and the fruits from 8% to about 5% in 48-72 h. For this reason after 4-5 days from brining, grinded salt is added to the brine to restore the initial concentration. After 1-2 months of fermentation olives are ready to eat.

2.2. Natural olives

The "natural olives", according to the "Trade Standard Applying to Table Olives" (IOC, 2004) are "green olives, olives turning color or black olives placed directly in brine in which they undergo complete or partial fermentation, preserved or not by the addition of acidifying agents". The most important industrial preparation for natural black olives takes the name "Greek-style" because it is traditionally practiced in Greece utilizing *Conservolea* cv. (Balatsouras, 1990).

2.2.1. Natural turning colour olives

Italy has a long tradition in producing "natural turning colour olives". In this process the olives are directly brined in 8-10% of sodium chloride. The brine stimulates the microbial activity for fermentation and reduces the bitterness of the oleuropein. Fermentation of these olives takes a long time because diffusion of soluble components through the epidermis, in fruits not treated with alkali, is slow. A diverse microbiota grows in these brines, although yeasts are the microorganisms always present throughout the process. *Enterobacteriacee* can be found during the first 7-15 days, but they disappear as the brine characteristics do not support their growth. The presence of lactic acid bacteria depends on the salt concentration and the polyphenol content of the variety used. The traditional brining is carried out under anaerobic conditions. However, an aerobic method can be applied, using a central column in the fermenter through which air is bubbled. This system changes the ratio between fermentative and oxidative yeasts, and a final product with better quality is attained (Garcia Garcia *et al.*, 1985 ; Garrido Fernández *et al.*, 1997). The colour fades during the process, but it is corrected by aerating the olives for two or three days; sometimes they are treated with 0.1% ferrous gluconate or lactate to make them deep black. Lastly, the olives are selected and packed in barrels or in varnished cans, which are filled with 8% fresh brine. They are popular because of their slightly bitter taste and aroma. When the bitterness has been sufficiently weakened the fruit can be marketed. To reduce the debittering phase, several researches have evaluated the use of selected oleuropeinolytic lactic acid bacteria (LAB) as starter cultures in Greek-style olive processing. *Lactobacillus plantarum* (Ciafardini *et al.*, 1994; Marsilio *et al.*, 1996; Marsilio & Lanza, 1998; Marsilio *et al.*, 2005; Panagou *et al.*, 2008) and *Lactobacillus pentosus* (Panagou *et al.*, 2003; Panagou *et al.*, 2008; Servili *et al.*, 2006), in addition to developing a fast growth and a good acidifying capacity even in the presence of high concentrations of salt (brine), were able to grow in the presence of phenolic compounds (known for their antimicrobial action) and, thanks to the marked oleuropeinolytic activity, degraded the oleuropein in no-bitter compounds, reducing thus considerably the time of debittering.

2.2.2. Natural green and black olives by Itrana-style

The *Itrana* variety is grown mainly in Lazio region and affects the hilly area of Ausoni, Lepini and Aurunci mountains. The production is concentrated in Latina district (Itri, Cori, Rocca Massima and Sonnino). The olives destined to produce the famous *"Oliva nera di Gaeta"* are harvested at the stage of full maturity in the months of February-March (very late compared to most table olive cultivars). This method provides an initial step of immersion in water for about 1 month to stimulate the growth of specific microflora that contributes to the debittering of the fruits (Lanza, 2010). After 10-30 days salt is added to the liquid, in quantities not exceeding 7 kg per 100 kg of fresh olives. After 4-6 months of storage in brine the olive flesh shows a typical red-wine colour and acidic taste probably due to the contribution of heterofermentative bacteria and yeasts. In recent years another type of product named *"Oliva bianca di Itri"* has evolved. The processing system is basically the same but the Itrana fruits are collected at the beginning of ripening in the months of November-December, when they reach their final size and look green or turning-color, and immersed immediately in water. After 6-8 months, the product results more bitter and acid respect to Gaeta olives and kinesthetic characteristics (hardness, fibrousness and crunchiness) are more pronounced.

2.2.3. Cracked/crushed olives

This type of processing is typical of the Calabria, Puglia and Campania regions. In Puglia (Salento) the olives of *Ogliarola salentina* cv., harvested at the green stage of ripening, are crushed with a wooden hammer (or a crushing machine) taking care not to break the stone, then they are put in an earthenware pot with water for at least 2 weeks changing the water at least twice a day. The olives suitable for this preparation must have a crunchy flesh with a strong detachment of the flesh from the stone. The finished product is seasoned with garlic, pepper, oregano and other spices (local name: *volia cazzata*). In Calabria (Crotonese) olives of *Carolea* cv. are similarly prepared (local name: *aliva maccata*). In Campania (Cilento) the olives of *Pisciottana* cv. are still prepared according to an ancient recipe that has very special ingredients and an elaborate preparation process. The green olives are picked, washed with a lot of water and left in a bowl, in which powdered lime and ashes are added; everything is then covered with water. The bowl is left to rest for 48 hours, then everything is drained and rinsed; the next step is to let the olives steep for 24 hours in a container with salt and bay leaves. After that the olives are drained and laid out on wood or stone board and squashed with a wood pin, until the seeds come away. Once squashed, the olives are gathered and placed in a glass container, they are pressed with the hands and dressed with chili pepper and other spices and condiments and covered with extra virgin olive oil. During the fermentation of cracked green olives, the predominant microorganisms are yeasts while lactic acid bacteria are not detected. The main role of yeasts in the processing of fermented olives, is associated with the production of alcohols, ethyl acetate, acetaldehyde and organic acids, compounds that are relevant for the development of taste and aroma and the preservation characteristics of this fermented food (Alves *et al.*, 2012; Arroyo-Lopez *et al.*, 2008).

2.2.4. "Scabecciu" olives

The Scabecciu olives are a traditional preparation typical of Sardegna region. The ripe olives of *Tonda di Cagliari* and *Pizz'e Carroga* varieties are engraved in three points, put in brine for about 3 days, washed with water and blanched in vinegar-water and then dried in the sun. Finally, they are fried with garlic and parsley and placed in oil. After about 1 month of preparation they can be consumed as appetizer.

2.3. Oven-dried black olives by Ferrandina-style

In Basilicata region, olives are prepared with an ancient and traditional method, Ferrandina method, which name derives from the small city Ferrandina, near Matera, where olive trees of *Olea europaea* L. *Majatica* cv. are cultivated (Savastano, 1937; Di Gioia, 1959; Cucurachi *et al.*, 1971). The olives are hand-harvested at the black stage of ripening at the end of November-early December. Initially the olives are blanched in water at 90°C for 3 min to make the skin more permeable and facilitate the osmotic processes. After blanching, the olives are salted with grinded NaCl (10:1 w/w) for 3 days and finally they are placed in single layer and oven-dried in an air-oven at 50 °C for 36 hours until the humidity is around 8 % and A_w around 0.7. This product is ready to eat in 1 week. The first two steps of the process (blanching and salting) contribute to fruit debittering; the second and the final steps (salting and oven-drying) result in water loss for better preservation of the product. This trade preparation is in accordance with the "Trade Standard Applying to Table Olives" (IOC, 2004) as "dehydrated and/or shrivelled black olives", black olives that have undergone or not mild alkaline treatment, preserved in brine or partially dehydrated in dry salt and/or by heating or by any other technological process. Today, olives from cv. Majatica are almost entirely destined to the production of oil. The production of oven-dried black olives by Ferrandina-style is in drastic decline and manufacturers are few. The product is prevalently exported in the USA and not all olives sold in the Italian market as oven-dried olives originate from cv. *Majatica*.

Textural changes occurring in oven-dried olive -tissues during each processing step were visualized by scanning electron microscope, texture analyzer, analysis of cell wall polysaccharides, pectins and biophenols (Borzillo *et al.*, 2000; Marsilio *et al.*, 2000; Cardoso *et al.*, 2009; Piga *et al.*, 2005). Recently, some authors (Lanza *et al.*, 2012) evaluated chemical and nutrient characteristics of Ferrandina oven-dried table olives with the aim to enhance the value of this end product from a nutritional point of view.

2.4. Dry-salted black olives

Salt-dried olives are prepared by packing naturally black-ripe olives in alternating layers with dry coarse salt (equivalent to 10–20% w/w of the weight of olives) and spices (oregano, orange peel, bay leaves, fennel, garlic, etc) in slatted containers that allow drainage of the vegetable water drawn out by the salt. The resulting olives, or 'date olives', are shrivelled in appearance and have a salty bitter-sweet taste (Panagou, 2006). Salt is also taken up by the olive, which acts as a preservative. Processing time is around four to six weeks and the olives are best eaten within three months of processing. Addition of olive oil enhances the flavour of the olive; however, oxidation of the oil can give the olives a rancid taste. In all Italian regions the most common cultivar that is prepared by salt-drying is *Leccino* cv. but there are also some local variations:

2.4.1. "Strinate" olives

In Marche region (areas near Ascoli Piceno and Macerata) black olives from *Leccino, Raggiola, Raggia, Piantone di Falerone* and *Lea* cvs. are stored in a jute bag after being combined with coarse salt. The bag, tied with a string, is hang outdoors sheltered from the rain and in a very cold place, taking care to mix the olives twice daily to prevent formation of mold and to promote a better distribution of salt. Under the bag a basin is placed to collect the vegetation water produced by the olives. The low winter temperatures favor the loss of bitter taste and cause dehydration of the olives, which assume a shriveled appearance. After 20-40 days, they are placed in glass containers for storage and with the addition of all the ingredients already described above for the preparation of dry-salted olives. The product thus obtained is called *strinate* olives from dialectal word that means "dead of cold".

2.4.2 "Passuluna" olives

In Sicilia region (Palermo), the *passuluna olives* are natural over-ripe olives, left on the tree, harvested in December-January, washed with hot water and leave to air-dry. Really, those olives are shriveled and debittered as the result of attack of the fungus *Camarosporium dalmaticum*, introduced in the drupe from dipteran *Prolasioptera berlesiana*, parasite of *Bactrocera oleae* eggs.

2.5. Stuffed olives by Ascolana-style

The stuffed olives by Ascolana-style are prepared from treated green olives in brine from *Ascolana tenera* cv. (PDO "Oliva Ascolana del Piceno). The meat (beef 40-70%; pork 30-50%;

chicken/turkey max 10%) is triturated and browned with onion, carrot and celery in olive oil and cooked on low heat with the addition of dry white wine and salt. When cooked, meat and add-ingredients are shredded and combined with beaten egg, grated cheese and ground nutmeg. The pre-pitted olives are filled with the mixture, are dipped in flour, then beaten egg and finally in breadcrumbs. The final product is intended for frying.

3. Nutritional and nutraceutical characterization

Table olives are a complete food from a nutritional point of view (Cannata, 1939). It is a drupe consisting primarily of water, fat, carbohydrates, protein, fiber, pectin, biophenols, vitamins, organic acids and mineral elements. The quality of this product is linked to the combined effect of various factors, such as the suitability of raw materials, the processing technologies, the nutritional composition and, in no small measure, the sensory properties.

Olive fruit is a drupe, constituted by three distinct anatomical zones: epicarp (skin), mesocarp (pulp) and endocarp (stone) containing the seed. All three influence the quality of the end product (Garrido Fernández et al., 1997; Montaño et al., 2010). The epicarp and mesocarp constitute the edible part of the olive fruit that is around 70-85 %. Information on the nutritional composition is related to 100g of edible portion (e.p.) (Table 2).

The energy value of 100 g of e.p. of Italian olives is around 200-250 kilocalories with some exceptions (455 kcal for *Majatica* olives and 164 kcal for *Bella di Cerignola* olives; Table 3). This value, referred to a diet of 2000 kcal for an adult of average build with limited physical activity, accounts for 10-12.5 % of total calories.

Nutrients/ 100g e.p.	Sevillan green olives *Intosso*	Sevillan green olives *Bella di Cerignola*	Castelvetrano green olives *Nocellara B.*	Ferrandina black olives *Majatica*	Natural black olives *Taggiasca*	Natural black olives *Peranzana*	Natural black olives *Itrana*	Natural green olives *Itrana*	Natural black olives *Celina N.*
Energy *kcal*	190	164	204	455	226	247	235	193	223
Proteins *g*	1.0	1.2	1.0	2.2	1.5	1.7	1.4	1.5	1.3
Carbohydrates *g* Sugars *g*	2.8 tr	2.5 <0.6	3.6 0.4	n.d. 4.4	8.9 tr	5.8 0.6	6.5 0.3	5.0 0.6	7.2 1.7
Fats *g*	17.5	15.5	19.8	46.9	19.9	23.2	21.7	17.7	19.9
SFA *g*	2.7	2.1	3.9	6.3	3.7	4.1	2.7	2.8	4.4
MUFA *g*	13.6	12.5	13.9	36.7	15.2	17.0	17.7	14.0	14.5
PUFA *g*	1.2	0.9	2.0	4.0	0.9	2.1	1.3	0.9	1.0
Fiber *g*	2.6	4.8	3.8	3.4	2.6	4.1	4.0	3.6	4.8
Sodium *g*	1.3	1.1	0.9	0.9	1.8	1.4	1.5	1.2	1.5
Calcium *mg*	33.6	34.9	n.d.	168.1	92.7	83.1	28.9	21.9	58.7
Polyphenols *mg*	168	104	24	263	206	334	211	109	299

Table 3. Nutritional characteristics of some Italian table olives. n.d. = not detected; tr = traces.

The protein content is low (1.0-2.2 g; Table 3), but nutritional quality is high for the presence of essential amino acids for adults, threonine, valine, leucine, isoleucine, phenylalanine and lysine (Young, 1994), and for childrens, arginine, histidine and tyrosine (Imura & Okada,

1998). Aspartic and glutamic acids are the most representative amino acids, however in some preparations threonine, valine, leucine and arginine contents are >100mg (Table 4). These results are confirmed by other authors for other cultivars and treatments (Lanza *et al.*, 2010; Lazovic *et al.*, 1999; López *et al.*, 2007; López-López *et al.*, 2010b; Montaño *et al.*, 2005).

Amino acid (mg/100g e.p.)	Treated green olives *Intosso*	Ferrandina black olives *Majatica*	Natural black olives *Peranzana*
Aspartic acid	150	214	131
Threonine	70	129	80
Serine	80	124	74
Glutamic acid	150	226	128
Proline	50	tr	tr
Glycine	70	115	65
Alanine	80	115	66
Valine	60	104	63
Isoleucine	100	86	50
Leucine	140	173	98
Phenylalanine + Tyrosine	100	168	117
Lysine	10	18	tr
Histidine	30	26	28
Arginine	70	120	68
Other amino acids	tr	tr	tr

Table 4. Amino acid pattern of some Italian table olives. tr = traces.

The carbohydrate content in the olive fruit is, by itself, lower than any other edible fruit (Marsilio *et al.*, 2001). However, table olives have even lower proportions of these compounds since during the fermentation process or brine storage the microorganisms present in brines consume sugars. Then table olives can be considered as practically free of sugar products. Therefore, the calculation of total carbohydrates as difference, as must be made for nutritional labeling in the US, may lead to an overestimation of the amounts of these compounds in table olives, resulting in an error for consumers. However, unfermented products such as Ferrandina olives, appreciable quantities (4.4g) of simple sugars are present (Table 3).

Table olives are a good source of dietary fibre, which in addition, has a high digestibility rate (Jimenez *et al.*, 2000; López-López *et al.*, 2007). In European Union countries (Reg. CE 1924/2006 and Reg. UE 116/2010) it is possible to write on the label the claim "source of fibre" if the product contains at least 3g of fiber/100g of e.p. If the product contains at least

6g of fiber/ 100 g of e.p. (as it can be the case for some varieties) the claim "with high content of fiber" could be used. Most preparations have a content of fibre $\geq$ 3g/100g of e.p., so they can be considered as a source of fibre (Table 3).

It is worth highlighting on table olives mineral content. Discrete calcium content was found in some samples (168.1 mg for Majatica, 92.7 mg for *Taggiasca* and 83.1 mg for *Peranzana*; Table 3). The contribution of this element, albeit insignificant, contributes along with other foods to reach the daily requirement of calcium which is 800 mg. The high Na content of some traditional preparations ($\geq$ 1.5 g; Table 3), resulting by fermenting and packing brines, is not in contrast with sodium RDV by Dietary Guidelines for Italians (LARN, 1996), that consists of 2.27 g/day. The consumption of these table olives may be unadvisable only in hypertensive pathologies. There are some studies (Bautista Gallego *et al.*, 2011; Marsilio *et al.*, 2000; Panagou *et al.*, 2011) on use of alternative chloride salt mixtures in substitution of common salt (NaCl) to reduce the intake of sodium, with an exception of the dry-salted olives (4-10g/100g of e.p.) (Figure 3).

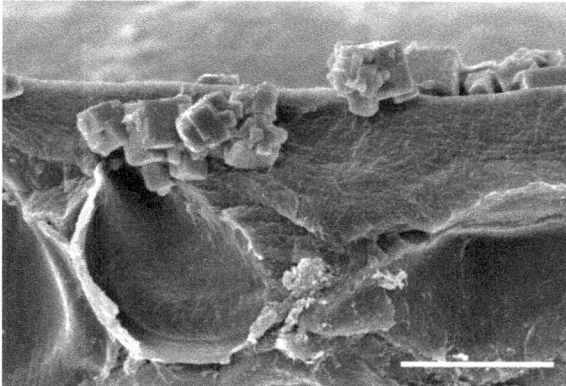

Figure 3. NaCl crystals on olive skin. Bar = 15 µm.

Table olives are also rich in natural antioxidants such as vitamins. They provide small amounts of B group vitamins as well as liposoluble vitamins such as pro-vitamin A and vitamin E, considered to have great antioxidant effects. The vitamin C content is low (<1 mg/kg of e.p.). Many green olive commercial presentations add, as antioxidant, ascorbic acid, which becomes a part of their final ingredients and increases the vitamin C content of the product. This compound may be progressively lost during shelf life, but depending on the time elapsed from packing, such table olives may eventually represent an interesting source of vitamin C.

All table olives analyzed (except *Nocellara del Belice* olives) are, despite treatments, still rich in natural antioxidants such as polyphenols (Table 3) and their antioxidant capacity and functional effects on human wellbeing is well ascertained (Baiano *et al.*, 2009; Ben Othman *et al.*, 2008; Boskou *et al.*, 2006). Table olives contain simple and complex phenolic compounds (at least 30 different phenolic compounds) in amounts ranging between 100 and 350 mg/100g of e.p. (the same quantity of 1kg of extra virgin olive oil!). It was demonstrated that the variability of those main phenolic compounds can be related to a combination of agronomic and/or technological processes. The polyphenol content and composition depends on several factors such as cultivar, stage of ripening, location and processing (Blekas et al., 2002; Bianchi, 2003; Romero et al., 2004). Recent research has focused on a phenolic compound of nutraceutical value, oleocanthal (dialdehydic form of deacetoxy-ligstroside aglycon) as NSAI-like drugs (non-steroidal anti-inflammatory drugs), with Ibuprofene-like activity (Beauchamp et al., 2005).

Organic acids (oxalic, succinic, malic, citric and lactic) are present in rather low percentage, such as to give the olive pulp a total acidity between 4-10g/kg (expressed as citric acid) and a pH between 3.8 and 5.0. The content of oxalic and malic acids decreases in the course of maturation and the content of citric acid increases, while succinic acid seems to remain constant. In addition, the ratio of citric and malic acids decreases in the course of maturation to reach, at the moment of maximum oil accumulation (inolition), values close to 1 (Garrido Fernández et al., 1997).

Crude fat content was determined by extraction in a Soxhlet apparatus, according to the procedures described previously for olives (Lanza et al., 2010a). The olive product with the lowest fat content was that of Bella di Cerignola processed with the Sevillan method (15.5 g), while the highest was that of Majatica oven-drying by Ferrandina-style (46.9 g). It is noticeable the difference between the fat content of Itrana processed as "Oliva Bianca di Itri" and "Oliva Nera di Gaeta" (17.7 and 21.7 g, respectively) is certainly due to the different periods of harvesting and maturation.

To analyze the fat composition, the olive fruits were pitted and triturated with a grinder. The olive paste was warmed up in a water bath at 28±2 °C for 30 min and the oil was extracted by centrifugation at 5000 rpm for 30 min. The resulting surnatant oil, preleased with a Pasteur pipette, was filtered in the presence of Na-sulphate anhydrous and stored at 4 °C in aluminum foil wrapped falcon tubes until analyses. This procedure simulates the extraction of olive oil in olive mill (crushing, mixing and centrifugation) and was used to prevent changes in the oil quality as much as possible (Lanza et al., 2012).

Table 5 shows the detailed fatty acid composition (relative percentage within the lipid fraction) of oil extracted from for different table olives. Oleic acid is the predominant one (63.4-80.7 %), palmitic acid was the second most abundant fatty acid (9.8-19.6 %), followed by linoleic acid (4.7-13.6 %) and stearic acid (1.6-3.0 %), a pattern common to most reported data (Lanza et al., 2010a; Sousa et al., 2011; Sakouhi et al., 2008; Borzillo et al., 2000; López et al., 2006; López-López et al., 2010a; Ünal & Nergiz, 2003; Issaoui et al., 2011). Recent research showed that intestinal mucosal cells utilize dietary oleic acid as a substrate to produce the lipid messenger oleoylethanolamide (OEA) that plays an important role in the regulation of animal food intake and body weight in human physiological and pathophysiological conditions (Schwartz et al., 2008).

Fatty acids	Sevillan green olives *Intosso*	Sevillan green olives *Bella di Cerignola*	Ferrandina black olives *Majatica*	Natural black olives *Moresca*	Natural black olives *Peranzana*	Natural black olives *Itrana*	Natural green olives *Itrana*	Natural black olives *Cellina N.*
Myristic acid C14:0	n.d.	n.d.	n.d.	n.d.	n.d.	n.d.	n.d.	n.d.
Palmitic acid C16:0	11.8	11.4	9.8	15.1	14.7	10.1	13.6	19.6
Palmitoleic acid C16:1	0.6	0.8	1.1	2.9	1.3	1.1	1.4	1.9
Heptadecanoic acid C17:0	0.1	0.1	n.d.	0.1	n.d.	n.d.	0.1	n.d.
Heptadecenoic acid C17:1	0.3	0.1	0.1	0.3	0.1	0.1	0.1	n.d.
Stearic acid C18:0	2.9	2.2	3.0	1.6	1.7	1.7	1.8	2.2
Oleic acid C18:1	76.4	79.6	76.8	63.4	70.2	80.7	76.5	70.8
Linoleic acid C18:2 ω6	6.3	4.7	8.0	13.6	10.7	5.2	5.5	4.7
Arachic acid C20:0	0.8	0,4	0.4	0.3	0.3	0.2	0.3	0.4
Linolenic acid C18:3 ω3	0.5	0.7	0.5	0.6	0.7	0.6	0.5	0.2
Eicosenoic acid C20:1	0.3	0.2	0.3	0.2	0.2	0.2	0.3	0.2
Behenic acid C22:0	0.1	n.d.	0.1	0.1	0.1	0.1	0.1	0.1
Lignoceric acid C24:0	n.d.	0.1	n.d.	n.d.	n.d.	n.d.	0.1	0.1
Trans fatty acids	0.02	tr	0.02	0.01	tr	tr	0.01	tr

Table 5. Fatty acid composition percentage of some Italian table olives. n.d. = not detected; tr = traces.

Monounsaturated fatty acids (MUFA) were the major group (66.8-82.1 %), saturated fatty acids (SFA) represented less than 22.4 % and polyunsaturated fatty acids (PUFA) range 4.9-14.2 %. The trans fatty acids had a very limited occurrence (less than 0.02%). The intake of α-linolenic acid (C18:3 ω3), precursor for the synthesis of long chain ω3 fatty acids, is appreciable but the ratio ω6:ω3 is still too far towards ω6 and depends on the cultivar (6.7-23.5; Table 5). Several sources of information suggest that human beings evolved on a diet with a ratio of ω6 to ω3 essential fatty acids of approximately 1 whereas in Western diets the ratio is 15-16.7. Western diets are deficient in omega-3 fatty acids, and have excessive amounts of omega-6 fatty acids compared with the diet on which human beings evolved and their genetic patterns were established (Simopoulos, 2008). The ratio of oleic to palmitic acid in dietary fats has a regulatory influence on certain thrombogenic and fibrinolytic markers during the postprandial state in healthy subjects (Pacheco *et al.*, 2006). It has a recommended ratio of at least 5. Also the polyunsaturated/saturated fatty acid (PUFA/SFA) ratio is used to assess the nutritional quality of the lipid fraction in foods. Consumption of saturated fatty acids has been associated with coronary heart disease (Serrano *et al.*, 2005). Consequently, nutritional guidelines have recommended that the PUFA/SFA ratio should be above 0.4–0.5 (Wood *et al.*, 2008). The mean values of the indexes (*cis* PUFA/SFA + TFA) and (*cis* MUFA + *cis* PUFA/SFA + TFA), which were most commonly used to express nutritional value of edible fats (Alonso *et al.* 2002; Anwar *et al.*, 2006; Garrido-Fernandez, 2008), presented very high values due to the absence of trans fatty acids. These results are in accordance, but even higher, with those found for "alcaparra" olives (0.3 and 5.0, respectively; Sousa *et al.*, 2011). Only for *Cellina di Nardò* olives these indexes are lower for "alcaparra" olives (Table 6).

Fatty acid index	Sevillan green olives *Intosso*	Sevillan green olives *Bella di Cerignola*	Ferrandin a black olives *Majatica*	Natural black olives *Moresca*	Natural black olives *Peranzana*	Natural black olives *Itrana*	Natural green olives *Itrana*	Natural black olives *Cellina N.*
SFA	15.7	14.2	13.3	17.2	16.8	12.1	16.0	22.4
MUFA	77.6	80.7	78.3	66.8	71.8	82.1	78.3	72.9
PUFA	6.8	5.4	8.5	14.2	11.4	5.8	6.0	4.9
MUFA/SFA	4.9	5.7	5.9	3.9	4.3	6.8	4.9	3.3
PUFA/SFA	0.4	0.4	0.6	0.8	0.7	0.5	0.4	0.2
(cis MUFA + cis PUFA/(SFA+TFA)	5.4	6.1	6.5	4.7	5.0	7.3	5.3	3.5
cis PUFA/(SFA+TFA)	0.4	0.4	0.6	0.8	0.7	0.5	0.4	0.2
Oleic/palmitic	6.5	7.0	7.8	4.2	4.8	8.0	5.6	3.6
ω6/ ω3	12.6	6.7	16.0	22.7	15.3	8.7	11.0	23.5

Table 6. Fatty acid indexes.

As regard to sterol, fatty and triterpenic alcohol composition, there are some studies relating to the changes during processing (López- López *et al.*, 2008, 2009). The main phytosterols and phytostanols found in Ferrandina table olives (Table 7) are β-sitosterol (59.1-89.6 %) and Δ5-avenasterol (1.5-34.3%), followed by campesterol (1.8-4.2%), Δ5,24-stigmastadienol (0.4-1.2%) and chlerosterol (1.0-1.5 %). The low content of β-sitosterol and high content of Δ5-avenasterol in *Majatica* olives is typical for this cultivar (Rotundo & Marone, 2002), but β-sitosterol (including Δ5,23-stigmastadienol, clerosterol, β-sitosterol, sitostanol, Δ5-avenasterol and Δ5,24-stigmastadienol) is higher than 93%, the limit fixed for extra virgin olive oil (EVOO) by the Commission Regulation (EEC) 2568/91 and its subsequent modifications. Epidemiologic and experimental studies suggest that dietary phytosterols, and in particular β-sitosterol, may offer protection from the most common cancers in Western societies, such as colon, breast and prostate cancer and contribute to lower the cardiovascular disease risk. (Awad & Fink, 2000; Woyengo *et al.*, 2009).

Finally, table olives could be utilized as a vehicle for incorporating probiotic bacteria and transporting bacterial cells into the human gastrointestinal tract. Food industries are now focusing on new foods which are part of a normal diet and can contribute to a regular assumption of probiotics (functional foods). The incorporation of health-promoting bacteria into table olives would add functional features to their current nutritional properties.

The consumption of table olives, in combination with the consumption of olive oil, which are basic components of the Mediterranean diet, provide a large amount of natural compounds of nutraceutical value (polyphenols, phytosterols and fatty acids) with antioxidant, anti-inflammatory or hormone-like properties.

Sterol	Sevillan green olives Bella di Cerignola	Castelvetrano green olives Nocellara B.	Ferrandina black olives Majatica	Natural black olives Moresca	Natural black olives Peranzana	Natural black olives Itrana	Natural green olives Itrana	Natural black olives Cellina N.
Cholesterol	0.8	1.4	0.4	0.8	0.5	0.4	0.5	0.6
Brassicasterol	0.1	0.3	0.1	0.1	n.d.	n.d.	0.1	n.d.
24-methylen cholesterol	n.d.	0.2	0.5	n.d.	0.2	0.1	0.1	0.2
Campesterol	3.1	4.2	1.8	2.1	2.0	2.6	2.4	3.1
Campestanol	n.d.	0.1	0.1	0.2	0.4	n.d.	n.d.	0.2
Stigmasterol	1.2	1.8	0.5	1.0	0.7	0.4	0.7	0.9
Δ7-campesterol	0.7	0.2	0.1	0.6	0.3	0.8	0.6	0.5
Δ5,23-stigmastadienol	n.d.	n.d.	0.1	n.d.	n.d.	n.d.	n.d.	0.1
Chlerosterol	1.5	1.4	1.0	1.5	1.2	1.0	1,0	1.0
β-sytosterol	89.6	84.5	59.1	80.2	77.4	87.8	88.7	85.4
Sitostanol	1.3	0.6	0.4	2.5	0.6	0.6	0.7	2.3
Δ5-avenasterol	1.5	4.7	34.3	8.5	15.9	5.6	4.3	5.3
Δ5,24-stigmastadienol	0.1	0.6	1.2	1.0	0.5	0.4	0.6	0.7
Δ7-stigmastenol	0.3	n.d.	0.3	0.7	0.2	0.2	0.3	0.1
Δ7-avenasterol	0.3	n.d.	0.1	0.8	0.2	0.2	0.2	0.1

Table 7. Sterolic composition (%) of some Italian table olives. n.d. = not detected.

3.1. The nutritional label

A separate discussion is made about nutritional labelling which, while optional until 13 December 2016, could add value to our product. What information do we expect to find on a nutrition label of a jar of olives? Currently the provision of a nutrition label is regulated by the Regulation (EU) 1169/2011 of the European Parliament and of the Council on nutrition labelling for foodstuffs as regards recommended daily allowances, energy conversion factors and definitions. The information written on the label should be referred to 100 g of product: in the case of whole olives reference should be made to 100 g of drained product (therefore considering the stone even if not edible) or 100 g of edible portion (in this case only the pulp of olives). For pitted olives and olive paste this problem does not arise. The nutritional information may also be referred to as a portion or "serving size", based on the amount of food consumed by a person. For table olives, a serving size could be formed by about 10 medium-sized olives and expressed in grams (taking into account the weight of the stone). It is also useful to relate the content of each nutrient with a daily reference value for a diet of 2000 kcal of an average weight adult performing limited physical activity. Percentages above 20% are considered significant, below 5% modest.

4. The IOC Method for the sensory analysis of table olives

Since 21 November 2008 the procedure for the classification of table olives based on parameters of quality has become official: *COI/OT/MO/Doc. No 1. Method for the sensory analysis of table olives.* On 25 November 2011 (Decision No DEC-18/99-V/2011) the International Olive Council adopted the revised version of the method (COI/OT/MO No 1/Rev. 2). The method establishes the necessary criteria and procedures for the sensory analysis of the odour, taste and texture of table olives and sets out the systematics for their commercial classification. It is applicable solely to the fruit of the cultivated olive tree (*Olea europaea* L.) which has been suitably treated or processed and which has been prepared for trade or for final consumption as table olives in accordance with the trade standard applying to table olives referenced COI/OT/NC No 1 (2004).

4.1. The panel

The sensory evaluation of table olives is done by a group of 8-10 expert tasters selected on the basis of aptitude and led by a panel leader. This group constitutes the taste panel. The tasters are chosen by means of a selection process implemented in accordance with an international standard according to his/her sensitivity and discriminatory power with regard to the organoleptic characteristics of table olives, who becomes skilled after suitable training and whose performance is objectively evaluated on the basis of rules established beforehand by the leader of the panel to which the taster belongs. The panel leader is the person whose chief duties are to lead panel activities, including taster recruitment, selection, training, skill building and monitoring. He/she designs and leads the sensory tests and analyses and interprets the data and may be assisted by one or more panel technicians.

4.2. Main facilities and equipment of test room

Tasting booth, referred to standard COI/T.20/Doc. No 6/Rev. 1 Guide for the installation of a test room or to ISO 8589:2007 General guidance for the design of test rooms;

Glasses, according to standard COI/T.20/Doc. No 5 *Glass for oil tasting*, covered with watch-glasses;

Plastic or metal cocktail sticks, two-pronged forks, spoons or tongs;

Profile sheet on hard or soft copy. The line for each attribute must measure exactly 10 cm.

4.3. Sample presentation and tasting session

The sample of table olives for analysis shall be presented in standard tasting glasses (Figure 4). The glass shall contain as many olives as the bottom of the glass can hold when the olives are placed side by side in a single layer. When brined table olives are undergoing analysis, sufficient covering liquid shall be poured over the olives to cover them fully. When the olives are above the 91/100 size-grade, the volume of sample contained in the glass shall in no case be more than half the height of the glass (i.e. 30 mm). In the case of table olives belonging to a size-grade below 91/100, the sample for testing in the glass shall comprise no less than three olives. When brined table olives are undergoing analysis, the quantity of covering liquid in the glass shall come up to at least three-quarters of the height of the olives. The glass shall be covered with the attendant watch-glass.

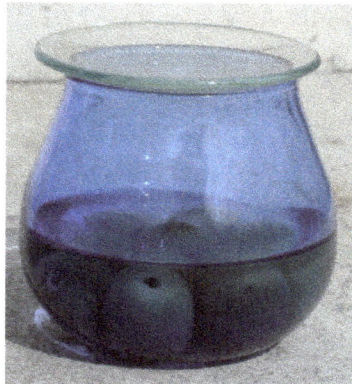
(a) olives above the 91/100 size-grade

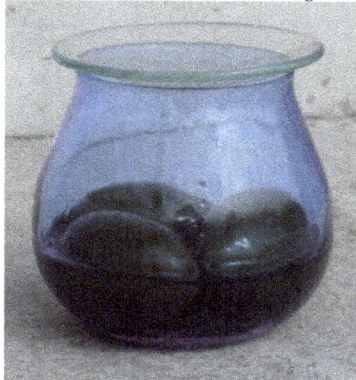
(b) olives below the 91/100 size-grade

Figure 4. Different distribution of olives and brine in the glass.

The samples of table olives intended for tasting shall be kept in the glasses at ambient temperature, between 20 and 25 °C, under white light (daylight). To avoid tasting fatigue and the appearance of bias or contrast effects, each tasting session should entail the sensory analysis of not more than three samples. Between each session the tasters should rinse out their mouth fully and take a break of at least fifteen minutes. No more than three tasting sessions should be conducted in any given day. It should be kept in mind that the morning, before lunch, is the period when olfactory-gustatory sharpness is optimal (between 10 a.m. and 12 noon).

4.4. Tasting session and use of the profile sheet by tasters

The tasters shall pick up the glass, keeping it covered with the watch-glass, and shall bend it gently to help the sample aromas to be released and blended. After doing so, they shall remove the watch-glass and smell the sample, taking slow deep breaths to evaluate the direct olfactory sensations cited in the profile sheet. Smelling shall not last more than 20 seconds. If no conclusion has been reached during this time, the tasters shall take a short rest before trying again. The tasters shall then assess the other sensations cited in the profile sheet. To do so, they shall place one of the olives contained in the glass in their mouth; they shall chew the olive after removing the stone, making sure to spread the chewed olive throughout the whole of the mouth cavity. They shall concentrate on the order of appearance of the salty, bitter and acid stimuli, the retronasal olfactory sensations and the kinaesthetic sensations of hardness, crunchiness and fibrousness and shall assess the intensity of each of these sensations by making the corresponding mark on the intensity scale of the tasting sheet. Next they shall spit out the chewed olives, rinse out their mouth with water and recommence the assessment of the sensations produced by each of the olives contained in the glass. They shall enter the intensity with which they perceive each of the attributes in the scales of the profile sheet provided. Tasters may refrain from placing the olives in their mouth when they observe an extremely intense negative attribute. They shall record this exceptional circumstance in the profile sheet. They must, however, mark the intensity of the odorous attribute on the corresponding scale. They shall enter in the profile sheet the intensity of each of the sensations perceived when smelling and chewing the olives.

4.5. Attributes to be perceived

4.5.1. Negative attributes

Unpleasant sensations are caused by the production of substances responsible for off-odours, which are not present in the fresh fruit or formed during well-performed processing treatments. The term "abnormal fermentation" includes all those olfactory sensations perceived directly or retronasally, reminiscent of the odour of decomposing organic matter, cheese, butter, rotten eggs, muddy sediment, sewer, rotten leather, caused by the development of contaminating microorganisms (butyric fermentation, putrid fermentation and zapateria). If the tasters perceive any negative attributes other than abnormal fermentations, they shall record them under the 'other defects' heading, using the following terms: musty, rancid, cooking effect, soapy, metallic, earthy and winey/vinegary. Musty sensation is an olfactory-gustatory sensation perceived directly or retronasally, characteristic of olives attacked by moulds. Rancid sensation is an olfactory sensation perceived directly or retronasally, characteristic of olives that have undergone a process of rancidity. Cooking effect is an olfactory sensation perceived directly or retronasally, characteristic of olives that have undergone excessive heating in terms of temperature and/or duration during pasteurisation or sterilisation. This taste may be typical of some cooked preparations as oven-dried black olives. Winey-vinegary is an olfactory–gustatory sensation reminiscent of wine or vinegar. Winey-vinegary taste should not be confused with the acid sensation. The defect winey-vinegary is due to alcoholic fermentation by yeasts, while the feeling of acid defines the taste associated with acids naturally present or produced during the lactic fermentation by homo and hetero fermentative lactic acid bacteria. Both sensations may be more or less marked by inappropriate use of acids used as acidity correctors (e.g., acetic or citric acid), but may also be typical of some preparations involving the use of vinegar (e.g., Kalamata olives). *Soapy* is an olfactory-gustatory sensation reminiscent of soap. This taste is found primarily in olives treated with lye (Spanish and Castelvetrano systems) and not sufficiently rinsed with water or consumed shortly after debittering. *Metallic* is an olfactory-gustatory sensation reminiscent of metals. It is found primarily in olives darkened by oxidation with the addition of iron salts such as gluconate or ferrous lactate, used as stabilizers of the color (eg. Californian black olives), but can also be perceived of olives preserved in metal packs. *Earthy* is an olfactory-gustatory sensation reminiscent of soil or dust. It is found in olives that have been in contact with soil and dust for a long time, usually harvested after they fall from the tree.

4.5.2. Descriptive gustatory attributes

The gustatory sensations involve distinct areas of tongue: the region affected by the perception of salty taste is the lateral-anterior, the region affected by the perception of acid taste is the posterior and the region affected by the perception of bitter taste is the basis of tongue. The *salty* sensation is associated to the basic taste produced by aqueous solutions of substances such as sodium chloride and depends on the concentration of fermentation or conditioning brines. The *bitter* sensation is associated to the basic taste produced by dilute aqueous solutions of substances such as quinine or caffeine and depends on the presence of bitter substances, mainly polyphenols. It may therefore be more intense in preparations in which the debittering, incomplete, is due to the action of microrganisms (natural olives). The *acid* sensation is associated to the basic taste produced by dilute aqueous solutions of most acid substances, such as tartaric or citric acids. It defines the taste associated with acids naturally present in the olive flesh (e.g., tartaric acid, malic acid or citric acid) or produced during the lactic fermentation by homo and heterofermentative lactic acid bacteria (e.g., lactic acid or acetic acid), but may also depend on the inappropriate use of acids as correctives of acidity (e.g., citric acid). High level of acid sensation is also found in the olives whose preparation involves the addition of vinegar (e.g., Kalamata olives).

4.5.3. Kinaesthetic sensations (texture)

With the term "kinaesthetic sensations" (from the greek κινηση = movement and αισθηση = sense) are indicated the sensations deriving from the contact of the fruit with the mouth. We could translate it as "musculoskeletal overall perception, in the oral cavity, of the mechanical characteristics of the fruit". Texture is defined as the set of rheological (related to the flow and deformation of matter) and structural (geometrical and surface) properties of a product perceptible to the mechanical receptors, tactile receptors and in some cases the visual and auditory receptors. The following attributes are assessed in table olives: hardness, crunchiness and fibrousness. *Hardness* is defined as the mechanical properties of consistency related to the force required to obtain the deformation or penetration of a product. It is perceived through the compression of the product between the teeth (solid products) or between the tongue and palate (semi-solid products); as regards the solid products such as table olives, the force required to compress the product is evaluated first between the incisors (Hardness 1), then between the molars (Hardness 2) and finally during mastication (Hardness 3) (Kim *et al.*, 2012). *Crunchiness* is the property related to the noise produced by friction or fracture between two surfaces. In our case, it is related to the force required to fracture a product with the teeth and is determined by compressing the fruit between the molars. *Fibrousness* is a geometric textural attribute relating to the perception of the shape and the orientation of particles in a product. Fibrousness refers to the elongated conformation of the particles, oriented in the same direction. It is evaluated by perceiving the fibres between the tongue and palate when chewing the olive. Hardness, crunchiness and fibrousness may also be assessed by instrumental puncture, compression, stretching and acoustic tests using Texture Analyzer. The texture of edible fruits and vegetables is strongly influenced by the chemical, physical and structural properties of the plant cell walls which surround every cell and, through adhesion to the adjacent ones, provide mechanical strength and protection (Brett & Waldron, 1996). Such properties will be affected by the stage of ripening (Mafra *et al.*, 2001) and processing technologies (Lanza & Marsilio, 2001; Tassou *et al.*, 2007). Tissues in which cell-to-cell adhesion is very strong may only be disrupted by breakage of the cell walls: such tissues are usually crunchy in texture. Tissues in which cell-to-cell adhesion is very weak may be disrupted through cleavage along the plane of the middle lamella: such tissues are usually soft in texture. The examination of the fracture surface can indicate whether tissue failure occurs as a result of cell rupture or cell-to-cell debonding. A simple olive fracturing method is described in Lanza *et al.* (2010a). In a recent research (Lanza & Di Serio, 2011), information provided by sensory and other instrumental measurements of tissue hardness were compared with that provided by scanning electron microscopy. The different levels of hardness applied to olives by sensory analysis are as follows: soft (low level), firm (moderate level) and hard (high level). Fractured epicarp and mesocarp olive fruits in longitudinal view, corresponding to the three levels of hardness, were examined by SEM (Figure 5). In *hard olives*, the thin-walled parenchyma cells were uniform and tightly packed. Tissue fracture involved cell walls breaking, both in epicarp and mesocarp and cell separation at the middle lamella level was not observed. The fracture surface of *firm olives* consisted of two regions: (a) one with 400–500 μm containing broken cell walls and some separated cells and (b) a second one containing intact cell walls. The region with broken cell walls extended from the olive epidermis as far as the eighth-ninth layer of mesocarp. In *soft olives*, the region of broken cell walls was smaller (100–200 μm) and involved only the epicarp and the first layers of the mesocarp (hypodermis and, sometimes, the following first layer). The other cells showed rounded outlines and were divided along the middle lamella, showing cell separation. Summarizing, in hard fruits, tissue fracture involves cell walls breaking and the thin-walled parenchyma cells olives are uniform and tightly packed (cell rupture). In soft olives the cells show rounded outlines and are divided along the middle lamella (cell separation). This is due to an increase in cell separation which consists mainly in the dissolution of middle lamella pectic polysaccharides.

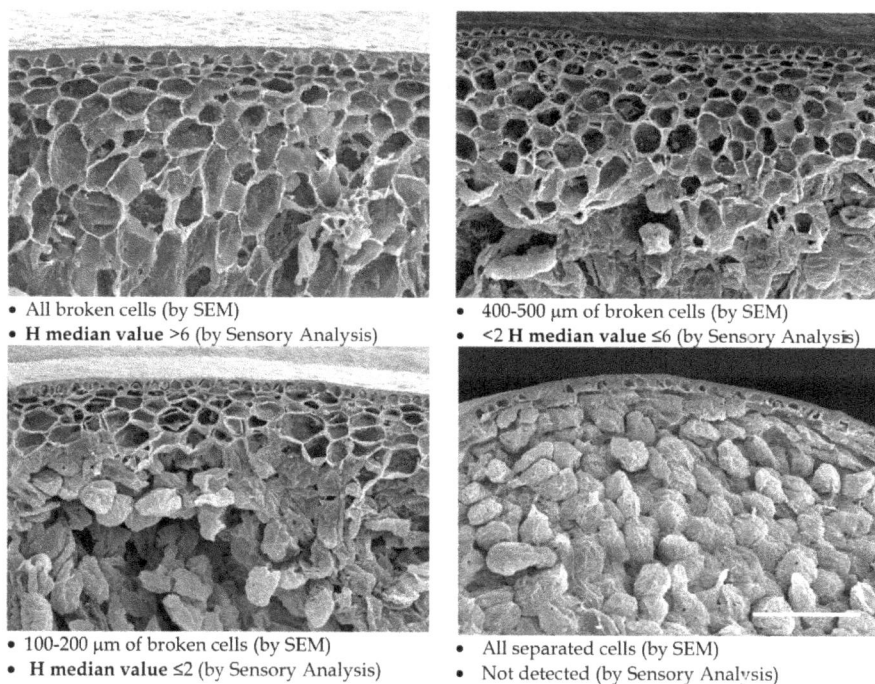

- All broken cells (by SEM)
- **H median value** >6 (by Sensory Analysis)

- 400-500 μm of broken cells (by SEM)
- <2 **H median value** ≤6 (by Sensory Analysis)

- 100-200 μm of broken cells (by SEM)
- **H median value** ≤2 (by Sensory Analysis)

- All separated cells (by SEM)
- Not detected (by Sensory Analysis)

Figure 5. The different levels of hardness (H). Bar = 300 μm.

4.6. Elaboration of sensory data

The panel leaders shall collect the profile sheets completed by each of the tasters and shall review the intensities recorded for each of the descriptors. If they find any anomaly, they shall invite the taster concerned to revise the profile sheet and, if necessary, to repeat the test. The panel leaders shall determine the intensities of the attributes listed in the profile sheet by using a ruler to measure the segment running from the origin of the scale to the mark made by the taster. When this mark lies between two notches on the ruler, they shall assign the value lying closest to one of the notches. The segment shall be expressed to one decimal place. The scale shall measure 10 cm long and the intensity shall range from 1 to 11. The panel leaders shall apply the method for calculating the median and the confidence intervals according to the method contained in Annex 1 (COI/OT/MO/n°1/Rev.2 Annex 1 *Method for calculating the median and the confidence intervals*) and shall only take into account those attributes with a robust coefficient of variation of 20% or less. The computer program for carrying out the calculations is presented in Annex 3 (COI/OT/MO/n°1/Rev.2 Annex 3 *Sensory analysis of table olives computer program*). When a defect is entered under the 'Other defects' heading by at least 50 percent of the panel tasters, the panel leaders shall carry out the statistical calculation of this defect and shall arrive at the corresponding classification if the coefficient of variation is 20% or less.

4.7. Classification according to the defect predominantly perceived (DPP)

For classification purposes, the panel leader shall solely take into account the median of the defect predominantly perceived (DPP) i.e. perceived with the greatest intensity, that satisfies the requirements specified in the preceding paragraph. According to the intensity of DPP, the samples shall be classified in four categories:

Extra or Fancy: DPP ≤ 3

First, 1st, Choice or Select: 3 < DPP ≤ 4.5

Second, 2nd or Standard: 4.5 < DPP ≤7.0

Olives that may not be sold as table olives: DPP > 7.0

The organoleptic analysis of table olives deriving by the same variety (*Itrana*) but processed at a different stage of ripening ("Oliva bianca di Itri" and "Oliva nera di Gaeta") shows a different sensory profile with regard to kinaesthetic properties (hardness, fibrousness and crunchiness) and bitter sensation (Figure 6 *a* and *b*) (Lanza *et al.*, 2010b). The organoleptic analysis of table olives deriving by the same variety (*Itrana*) and processing ("Oliva Bianca di Itri") show a different sensory profile between undefected and defected samples. Defected samples show a decrease in kinaesthetic properties (hardness, fibrousness and crunchiness) and an increase in acid sensation (Figure 7 *a* and *b*). The median value of DPP is less or equal to 3 and the olives remain of "Extra or Fancy" category.

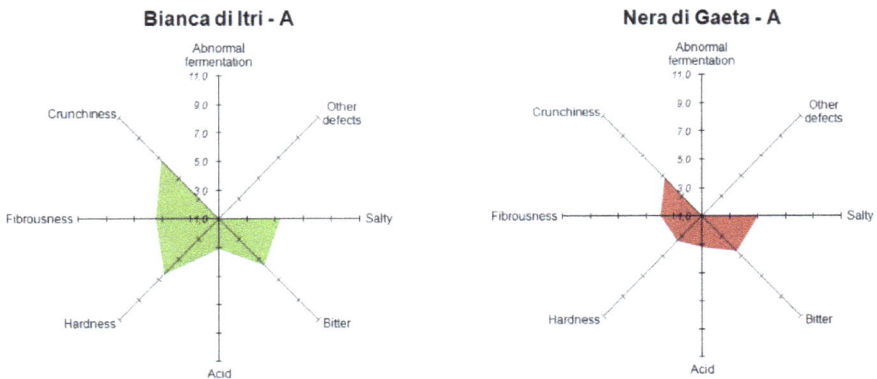

Figure 6. Sensory profiles of "Oliva bianca di Itri" and "Oliva nera di Gaeta".

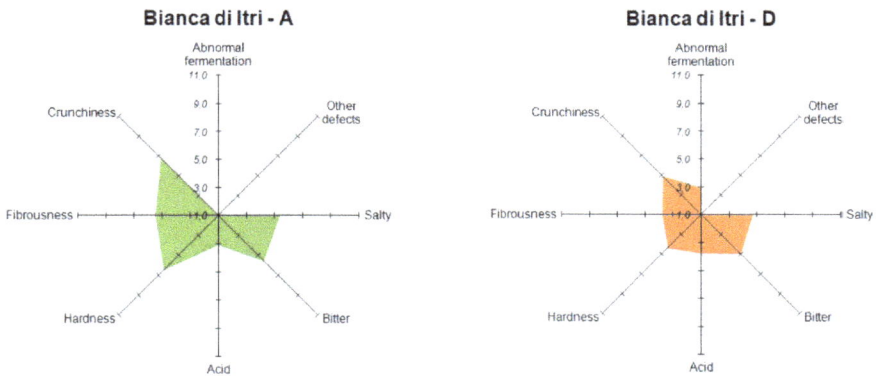

Figure 7. Sensory profiles of undefected (a) and defected (b) olives from "Oliva bianca di Itri".

Acknowledgement

Financial support for this study was provided by Italian Ministry of Agriculture, Food and Forestry through the project GERMOLI "Salvaguardia e valorizzazione del GERMoplasma OLIvicolo delle collezioni del CRA-OLI".

Author details

Barbara Lanza

Agricultural Research Council - Olive Growing and Oil Industry Research Centre, Città Sant'Angelo (PE), Italy

5. References

Alonso, L., Fraga, M.J., Juarez, M. & Carmona, P. (2002). Fatty acid composition of Spanish shortening with special emphasis on *trans* unsaturated content as determination by Fourier transform infrared spectroscopy and gas chromatography, *Journal of American Oil Chemistry Society* 79: 1-6.

Alves, M., Gonçalves, T. & Quintas, C. (2012). Microbial quality and yeast population dynamics in cracked green table olives fermentations, *Food Control* 23: 363-368.

Anwar, F., Bhanger, M.I., Iqbal, S. & Sultana, B. (2006). Fatty acid composition of different margarines and butters from Pakistan with special emphasis on trans unsaturated contents, *Journal of Food Quality,* 29: 87-96.

Arroyo López, F.N., Querol, A., Bautista Gallego, J. & Garrido Fernández, A. (2008). Role of yeasts in table olive production, *International Journal of Food Microbiology* 128: 189-96.

Awad, A.B. & Fink, C.S. (2000). Phytosterols as anticancer dietary components: evidence and mechanism of action, *Journal of Nutrition* 130: 2127-2130.

Baiano, A., Gambacorta, G., Terracone, C., Previtali, M.A. & La Notte, E. (2009). Characterization of drupes, phenolic content and antioxidant capacity of Italian olive fruits, *Journal of Food Lipids* 16: 209-226.

Balatsouras, G. (1990). Edible olive cultivars, chemical composition of fruit, harvesting, transportation, processing, sorting and packaging, styles of black olives, deteriorations, quality standards, chemical analyses, nutritional and biological value of the product. *Proceedings of International Seminar "Olio d'oliva e olive da tavola: tecnologia e qualità"*, Citta di S. Angelo (Pe), Italy.

Bautista Gallego, J., Arroyo López, F.N., Romero Gil, V., Rodríguez Gómez, F., García García, P. & Garrido Fernández, A. (2011). Chloride salt mixtures affect Gordal cv. green Spanish-style table olive fermentation, *Food Microbiology* 28: 1316-25.

Beauchamp, G.K., Keast, R.S.J., Morel, D., Lin, J., Pika, J., Han, Q., Lee, C-H., Smith, A.B. & Breslin, P.A.S. (2005). Ibuprofen-like activity in extra-virgin olive oil, *Nature* 437: 45-46.

Ben Othman, N., Roblain, D., Thonart, P. & Hamdi, M. (2008). Tunisian table olive phenolic compounds and their antioxidant capacity, *Journal of Food Science* 73: C235-40.

Bevilacqua, A., Altieri, C., Corbo, M.R., Sinigaglia, M. & Ouoba, L.I.I. (2010). Characterization of lactic acid bacteria isolated from Italian Bella di Cerignola table olives: selection of potential multifunctional starter cultures, *Journal of Food Science,* 75: M536–M544.

Bianchi, G. (2003). Lipids and phenols in table olives, *European Journal of Lipid Science and Technology* 105: 229-242.

Blekas, G., Vassilakis, C., Harizanis, C., Tsimidou, M. & Boskou D.G. (2002). Biophenols in table olives, *Journal of Agriculture and Food Chemistry* 50: 3688–3692.

Borzillo, A., Iannotta, N. & Uccella, N. (2000). Oinotria table olives: quality evaluation during ripening and processing by biomolecular components, *European Food Research and Technology* 212: 113-121.

Boskou, G., Salta, F.N., Chrysostomou, S., Mylona, A., Chiou, A. & Andrikopoulos, N.K. (2006). Antioxidant capacity and phenolic profile of table olives from the Greek market, *Food Chemistry* 94: 558-564.

Brett, C.T. & Waldron, K.W. (1996). *Physiology and biochemistry of plant cell walls,* Chapman & Hall, London, UK.

Cannata, U. (1939). Potenzialità nutritiva dell'oliva da mensa, *L'Italia Vinicola e Agraria* 12:1-4.

Cappello, A. & Poiana M. (2005). *Le olive da tavola in Sicilia.* Regione Siciliana Assessorato Agricoltura e Foreste – IX Servizio Regionale Servizi allo Sviluppo, Castelvetrano, Italy.

Cardoso, A.M., Mafra, I., Reis, A., Barros, A.S., Nunes, C., Georget, D.M.R., Smith, A., Saraiva, J., Waldron, K.W. & Coimbra, M.A. (2009). Traditional and industrial oven-dry processing of olive fruits: influence on textural properties, cell wall polysaccharide composition, and enzymatic activity, *European Food Research and Technology* 229: 415-425.

Ciafardini, G., Marsilio, V., Lanza, B. & Pozzi, N. (1994). Hydrolysis of oleuropein by *Lactobacillus plantarum* strains associated with olive fermentation, *Applied and Environmental Microbiology* 60: 4142-4147.

Cucurachi, A., Vitagliano, M. & Gervasi, P. (1971). Caratteristiche e utilizzazione delle olive "Majatica di Ferrandina", *Annali dell'Istituto Sperimentale per la Elaiotecnica* 1: 89-107.

De Castro, A., Montano, A., Casado F.J., Sanchez A.H. & Rejano L. (2002). Utilization of *Enterococcus casseliflavus* and *Lactobacillus pentosus* as starter cultures for Spanish-style green olive fermentation, *Food Microbiology* 19: 637-644.

Di Gioia, D. (1959). Le olive infornate di Ferrandina, *Olivicoltura* 8: 6-12.

Garcia Garcia, P., Duran Quintana, M.C. & Garrido Fernández, A. (1985). Fermentacion aerobica de aceitunas maduras en salmuera, *Grasas y Aceites* 36: 14-20.

Garrido Fernández, A. (2008). Revalorizacion nutricional de la aceituna de mesa. *Proceedings of II International Table Olive Conference*, Dos Hermanas, Sevilla, Spain.

Garrido Fernández, A., Fernandez Diez, M.J. & Adams, M.R. (1997). *Table Olives. Production and processing*, Chapman & Hall, London, UK.

Imura, K. & Okada, A. (1998). Amino acid metabolism in pediatric patients, Nutrition 14: 143-148.

Issaoui, M., Dabbou, S., Mechri, B., Nakbi, A., Chehab, H. & Hammami, M. (2011). Fatty acid profile, sugar composition, and antioxidant compounds of table olives as affected by different treatments, *European Food Research and Technology* 232: 867-876.

Jimenez, A., Rodriguez, R., Fernandez-Caro, I., Guillen, R., Fernandez-Bolanos, J. & Heredia, A. (2000). Dietary fibre content of table olives processed under different European styles: study of physico-chemical characteristics, *Journal of the Science of Food and Agriculture* 80: 1903-1908.

Kim, E.H-J., Corrigan, V.K., Wilson, A.J., Waters, J.R., Hedderley, D.I. & Morgenstern, M.P. (2012). Fundamental fracture properties associated with sensory hardness of brittle solid foods, *Journal of Texture Studies* 43: 49-62.

Lanza, B. & Di Serio, M.G. (2011). Table olive hardness: relationship between ultrastructural and sensory analysis. *Proceedings of 10th Multinational Congress on Microscopy*, Urbino, Italy.

Lanza, B. & Marsilio V. (2001). Microstructural changes in olive fruit as related to techno-processing, *Proceedings of 5th Multinational Congress on Electron Microscopy*, Lecce, Italy.

Lanza, B. & Poiana, M. (2011). *Olive da tavola: tecnologia*, Accademia Nazionale dell'Olivo e dell'Olio, Spoleto, Italy.

Lanza, B. (2010). L'Itrana non è solo nera. Le sei tipologie tradizionali, *Olivo e Olio* 9: 44-48.

Lanza, B. (2011). Enhancement of Italian traditional preparations of table olives and their nutraceutical properties. *Proceedings of Olivebioteq 2011*, Chania, Crete, Greece.

Lanza, B., Di Serio, M.G., Iannucci, E., Russi, F. & Marfisi, P. (2010a). Nutritional, textural and sensorial characterisation of Italian table olives (*Olea europaea* L. cv. 'Intosso d'Abruzzo'), *International Journal of Food Science and Technology* 45: 67-74.

Lanza, B., Di Serio, M.G., Russi, F., D'Achille, G. & Leonardi, G. (2010b). Organoleptic profile of Itrana table olives by IOC Method for the sensory analysis, *Proceedings of III International Table Olive Conference*, Sanlucar La Mayor, Seville, Spain.

Lanza, B., Di Serio, M.G., Russi, F., Iannucci, E., Giansante, L. & Di Giacinto, L. (2012). Evaluation of nutritional value of oven-dried table olives (cv. Maiatica) processed by Ferrandina style, *Journal of Food Composition and Analysis*, in press.

Lazovic, B., Miranovic, K., Gasic, O. & Popovic, M. (1999). Olive protein content and amino acid composition, *Acta Horticulturae* 474: 465-468.

Leal-Sanchez, M.V., Ruiz-Barba, J.L., Sanchez, A.H., Rejano, L., Jimenez-Diaz, R. & Garrido, A., (2003). Fermentation profile and optimization of green olive fermentation using *Lactobacillus plantarum* LPCO10 as a starter culture, *Food Microbiology* 20: 421-430.

López, A., Garrido, A. & Montaño, A. (2007). Proteins and aminoacids in table olives: relationship to processing and commercial presentation, *Italian Journal of Food Science* 19: 217-228.

López, A., Montaño, A., Garcia, P. & Garrido, A. (2006). Fatty acid profile of table olives and its multivarate characterization using unsupervised (PCA) and supervised (DA) chemometrics, *Journal of Agriculture and Food Chemistry* 54: 6747-6753.

López-López, A., Jiménez-Araujo, A., García-García, P. & Garrido-Fernández, A. (2007). Multivariate analysis for the evaluation of fiber, sugars, and organic acids in commercial presentations of table olives, *Journal of Agriculture and Food Chemistry* 55: 10803-10811.

López-López, A., Montaño, A. & Garrido-Fernández, A. (2010a). Nutrient profiles of commercial table olives: fatty acids, sterols, and fatty alcohols, *in* Preedy and Watson (eds.), *Olives and olive oil in health and disease prevention*. Elsevier, San Diego, CA. pp. 715-723.

López-López, A., Montaño, A. & Garrido-Fernández, A. (2010b). Nutrient profiles of commercial table olives: proteins and vitamins, *in* Preedy and Watson (eds.), *Olives and olive oil in health and disease prevention*. Elsevier, San Diego, CA. pp. 705-714.

López-López, A., Montaño, A., Ruíz-Méndez, M.V. & Garrido-Fernández, A. (2008). Sterols, fatty alcohols, and triterpenic alcohols in commercial table olives, *Journal of the American Oil Chemistry Society* 85: 253-262.

López-López, A., Rodriguez-Gomez, F., Ruíz-Méndez, M.V., Cortes-Delgado, A. & Garrido-Fernández, A. (2009). Sterols, fatty alcohol and triterpenic alcohol changes during ripe table olive processing, *Food Chemistry* 117: 127-134.

Mafra I., Lanza B., Reis A., Marsilio V., Campestre C., De Angelis M. & Coimbra M.A. (2001). Effect of ripening on texture, microstructure and cell wall polysaccharide composition of olive fruit (*Olea europaea*), *Physiologia Plantarum* 111: 439-447.

Marsilio, V. & Lanza, B. (1995). Effects of lye-treatment on the nutritional and microstructural characteristics of table olives (*Olea europaea* L.), *Revista Espanola de Ciencia y Tecnologia de Alimentos* 35: 178-190.

Marsilio, V. & Lanza, B. (1998). Characterization of an oleuropein degrading strain of *Lactobacillus plantarum*. Combined effect of compounds present in olive fermenting brines (phenols, glucose and NaCl) on bacterial activity, *Journal of the Science of Food and Agriculture* 76: 520-524.

Marsilio, V., Campestre, C., Lanza, B. & De Angelis, M. (2001). Sugar and polyol compositions of some European olive fruit varieties (*Olea europaea* L.) suitable for table olive purposes, *Food Chemistry* 72: 485-490.

Marsilio, V., Campestre, C., Lanza, B., De Angelis, M. & Russi, F. (2002). Sensory analysis of green table olives fermented in different saline solutions, *Acta Horticulturae* 586: 617-620.

Marsilio, V., Lanza, B. & Pozzi, N. (1996). Progress in table olives debittering: degradaticn in vitro of oleuropein and its derivatives by L. plantarum, Journal of American Oil Chemistry Society 73: 593-597.

Marsilio, V., Lanza, B., Campestre, C. & De Angelis, M. (2000). Oven-dried table olves: textural properties as related to pectic composition, Journal of the Science of Food and Agriculture 80: 1271-1276.

Marsilio, V., Seghetti, L., Iannucci, E., Russi, F., Lanza, B. & Felicioni, M. (2005). Use of a lactic acid bacteria starter culture during green olive (Olea europaea L., cv. Ascclana tenera) processing, Journal of the Science of Food and Agriculture 85: 1084-1090.

Montaño, A., Casado, F.J., de Castro, A., Sánchez, A.H. & Rejano L. (2005). Influence of processing, storage time, and pasteurization upon tocopherol and amino acid con-ents of treated green table olives, European Food Research and Technology 220: 255-260.

Montaño, A., Sánchez, A.H., López-López, A., de Castro, A. & Rejano, L. (2010). Chemical composition of fermented green olives: acidity, salt, moisture, fat, protein, ash, fiber, sugar and polyphenol, in Preedy and Watson (eds.), Olives and olive oil in health and disease prevention, Elsevier, San Diego, CA, pp. 291-297.

Pacheco, Y. M., Bermudez, B., Lopez, S., Abia, R., Villar, J. & Muriana, F. J. G. (2006). Ratio of oleic to palmitic acid is a dietary determinant of thrombogenic and fibrinolytic factors during the postprandial state in men, American Journal of Clinical Nutrition 84: 342-349.

Panagou, E.Z. (2006). Greek dry-salted olives: monitoring the dry-salting process and subsequent physico-chemical and microbiological profile during storage under different packing conditions at 4 and 20 °C, LWT-Food Science and Technology 39: 322-329.

Panagou, E.Z., Hondrodimou, O., Mallouchos, A. & Nychas, G.J. (2011). A study on the implications of NaCl reduction in the fermentation profile of Conservolea natural black olives, Food Microbiology 28: 1301-7.

Panagou, E.Z., Schillinger, U., Franz, C.M.A.P. & Nychas G-J. E. (2008). Microbiological and biochemical profile of cv. Conservolea naturally black olives during controlled fermentation with selected strains of lactic acid bacteria, Food Microbiology 25: 348-358.

Panagou, E.Z., Tassou, C.C. & Katsaboxakis C.Z. (2003). Induced lactic acid fermentation of untreated green olives of the Conservolea cultivar by Lactobacillus pentosus, Journal of the Science of Food and Agriculture 83: 667-674.

Piga, A., Mincione, B., Runcio, A., Pinna, I., Agabbio, M. & Poiana, M. (2005). Response to hot air drying of some olive cultivars of the south of Italy, Acta Alimentaria 34: 427-440.

Romero, C., Brenes, M., Yousfi, K., Garcia, P., Garcia, A. & Garrido, A. (2004). Effect of cultivar and processing method on the contents of polyphenols in table olives, Journal of Agriculture and Food Chemistry 52: 479-484.

Rotundo, A. & Marone, E. (2002). Il germoplasma olivicolo lucano, Tipolitografia Olita, Potenza, Italy.

Ruiz-Barba, J.L., Cathcart, D.P., Warner, P.J. & Jimenez-Diaz, R. (1994). Use of Lactobacillus plantarum LPCO10, a bacteriocin producer, as a starter culture in Spanish-Style green olive fermentations, Applied and Environmental Microbiology 6: 2059-2064.

Sakouhi, F., Harrabi, S., Absalon, C., Sbei, K., Boukhchina, S. & Kallel, H. (2003). α-tocopherol and fatty acids contents of some Tunisian table olives (Olea europaea L.):

changes in their composition during ripening and processing, *Food Chemistry* 108: 833-839.

Sánchez, A.H., Rejano, L., Montaño, A., De Castro, A. (2001). Utilization at high pH of starter cultures of lactobacilli for Spanish-style green olive fermentation, *International Journal of Food Microbiology* 67: 115-122.

Savastano, G. (1937). Sull'essiccamento delle olive nere di Ferrandina, *L'olivicoltore* 9: 11-17.

Schwartz, G.J., Fu, J., Astarita, G., Li, X., Gaetani, S., Campolongo, P., Cuomo, V. & Piomelli, D. (2008). The lipid messenger OEA links dietary fat intake to satiety, *Cell Metabolism* 8: 281-288.

Serrano, A., Cofrades, S., Ruiz-Capillas, C., Olmedilla-Alonso, B., Herrero-Barbudo, C. & Jiménez-Colmenero, F. (2005). Nutritional profile of restructured beef steak with added walnuts, *Meat Science* 70: 647–654.

Servili, M., Settanni, L., Veneziani, G., Esposto, S., Massitti, O., Taticchi, A., Urbani, S., Montedoro, G.F. & Corsetti, A. (2006). The use of *Lactobacillus pentosus* 1MO to shorten the debittering process time of black table olives (cv. Itrana and Leccino): a pilot-scale application, *Journal of Agriculture and Food Chemistry* 54: 3869-3875.

Simopoulos, A.P. (2008). *The importance of the omega-6/omega-3 fatty acid ratio in cardiovascular disease and other chronic diseases, Experimental Biology and Medicine 233: 674-688.*

Sousa, A Casal, S., Bento, A., Malheiro, R., Oliveira, M.B.P.P. & Pereira, J.A. (2011). Chemical charactization of "Alcaparra" stoned table olives from Northeast Portugal, *Molecules* 16: 9025-9040.

Tassou, C., Katsaboxakis, C.Z., Georget, D., Parker, M.L, Waldron, K.W, Smith, A.C. & Panagou, E.Z. (2007). Effect of calcium chloride on mechanical properties and microbiological characteristics of cv.Conservolea naturally black olives fermented at different sodium chloride levels, *Journal of the Science of Food and Agriculture* 87: 1123-1131.

Tavanti, G. (1819). *Trattato teorico pratico completo sull'ulivo*, Stamperia Piatti, Firenze, Italy.

Ünal, K. & Nergiz, C. (2003). The effect of table olive preparing methods and storage on the composition and nutritive value of olives, *Grasas y Aceites* 54: 71-76.

Wood, J. D., Enser, M., Fisher, A. V., Nute, G. R., Sheard, P. R, Richardson, R. I., Huges, S. I. & Whittington, F. M. (2008). Fat deposition, fatty acid composition and meat quality: A review, *Meat Science* 78: 343- 358.

Woyengo, T.A., Ramprasath, V.R. & Jones, P.J. (2009). Anticancer effects of phytosterols, *European Journal of Clinical Nutrition* 63: 813-20.

Young, V.R. (1994). Adult amino acid requirements: the case for a major revision in current recommendations, *Journal of Nutrition* 124: 1517S–1523S.

Chapter 16

Microbiological Aspects of Table Olives

Flora Valeria Romeo

Additional information is available at the end of the chapter

1. Introduction

Table olives are the most important fermented vegetables because of their world-wide economic importance. The three main techniques for table olive production used in Italy concern 82% green olives, 16% black olives and 2% processed at the cherry ripened stage (UNAPROL, 2008). There are three main trade preparations of table olives: Spanish-style olives, Californian-style olives and naturally black or turning colour olives (Garrido-Fernández et al., 1997). The Spanish processing method includes treatment with sodium hydroxide solution, washing, brining, fermentation and packaging. The Greek-style method is milder and includes washing, natural fermentation in brine, air-oxidation for colour improvement, and packing. The Californian method includes lye treatment, washing, iron-salt treatment and air-oxidation, canning and heat treatment. This last method includes a final sterilization, so it is usually considered as a more safe production. Besides these most prominent preparations, there are many other traditional table olive elaboration recipes that are less known in the international market (Panagou et al., 2003) but very relevant at local market level, where they are frequently sold in in glass jars or plastic pouches.

In the South of Italy the main traditional process method for table olive production is the natural olive process, according which untreated, generally green, olives are washed, put into containers and then filled with freshly prepared brine. Both treated and natural (untreated) olives have to fermented, and in order to enhance their safety extent, the current practice requires the reduction of pH to a value of 4.5 or below. The fruits are maintained in the brine until they lose their natural bitterness, at least partially (Arroyo-López et al., 2008a), and where they undergo the fermentation process whose characteristics depend on the cultivar and on the applied conditions. At the end of the process the olives acquired typical characteristics of final products.

Olives contain a significant amount of oil, ranging from 12% to 30%, depending on the considered cultivar. The fermentable carbohydrates of flesh olives generally ranges from 2% to 6%, however, when the olives are washed or lye treated sugars are also lost along with other soluble compounds. Olive fermentation is considered to be over when the sugars are totally consumed by microorganisms. The months necessary for this process might change depending on several factors, such as the variety and olive size, the salt concentration and temperature (Cardoso et al., 2010). Nowadays there are neither physico-chemical nor microbiological controls to objectively determine the end of fermentation and producers decide, according to personal criteria, when olives are ready to eat (Hurtado et al., 2008).

Other important components of olives are polyphenols. The most important classes of phenolic compounds in table olives are phenolic acids, phenolic alcohols, flavonoids, and secoiridoids (Sousa et al., 2006). The olive phenols are nutritionally interesting due to their antioxidant activities, moreover, these compounds are determinant in the shelf-life of olive oil and sensory qualities of both oil and table olives. Some of these, such as oleuropein and its hydrolysis derivatives, have antimicrobial activities against a wide variety of microorganisms, including lactic acid bacteria (LAB). The inhibiting effect of many polyphenols on LAB growth has been widely studied (Fleming et al., 1973; Ruiz-Barba et al., 1993). Moreover, the increase of oleuropein content in the growth medium reduces the activity of bacteria to hydrolyse this glycoside (Romeo & Poiana, 2007). Recently several studies on antimicrobial activity of olive products have been carried out, namely with olive leaves, olive fruits and their pure compounds, such as oleuropein, hydroxytyrosol and aliphatic aldehydes (Sousa et al., 2006) and it has been found that ferulic acid exhibits toxicity effects toward several microorganisms (Sayadi et al., 2000).

Compared to the relatively few microbial species employed in other fermented foods, microorganisms evolved in vegetable fermentations are many and different. In olive fruits the epiphytic microbial population consists of yeast, fungi, and both Gram positive and Gram negative bacteria but throughout the fermentation process, *Enterobacteriaceae*, LAB and yeasts are the most relevant microorganisms (Garrido-Fernández et al., 1997).

It has been generally established that LAB are responsible for the fermentation of treated olives. While LAB and yeasts compete for the fermentation of untreated olives, and in some cases yeasts can be exclusively responsible for fermentation on untreated olives.

LAB usually isolated from fermentation brines of treated olives include both heterofermentative and homofermentative species (Hutkins, 2006), and *Lactobacillus plantarum* is considered essential to produce the lactic acid needed for preservation and typical flavour. *L. plantarum* generally coexists with a yeast population until the end of the fermentation process and during storage, although other microorganisms may be involved depending on the applied parameters of the process. Organic acids, such as lactic acid, and sodium chloride are primary preservatives for table olives. Olives show a water activity greater than 0.85 and a final pH close to 4.6 or below, which represent the most important hygienic limit to avoid microbiological risks for consumers.

The control of temperature during the fermentation steps often led to beneficial effects, especially in those region where the fermentation temperature follows environmental fluctuations. Unfortunately, in most companies the temperature control is not applicable because it is an expensive procedure.

The control of salt, temperature, anaerobiosis (or low oxygen percentage) and process hygiene is necessary for successful fermentation. Under appropriate conditions, most non-lactic acid bacteria will grow slower than LAB that are less affected and that will grow and rapidly produce acid compounds, mainly lactic acid. These acids, along with CO_2 that may also be produced, create an even more stringent environment for competitors. So the competitors disappear while LAB overcome the lactic acid fermentation.

In the past, fermentation was always considered to be an economical means for temporary preservation of different kind of foods. Nowadays, the consumption of fermented foods is also promoted because of their health benefits, nutritional value, sensorial properties and functionality. This last aspect has been studied for some foods, such as yogurt and fermented milk (Lavermicocca et al., 2005), but it must be improved with regards to vegetable fermentation. The challenge of the next years must be the enhancement of this research field and the design of new functional foods.

2. Microorganisms associated to table olives

2.1. Role of yeasts

The positive role of yeasts in table olive fermentation has recently been reconsidered. Yeasts are especially relevant in directly brined green and black natural olive fermentations, where fruits are not treated with NaOH solutions. In these conditions, in the first fermentation step the LAB growth is slow because of the presence of phenolic compounds in brine. Growth of oxidative yeasts and molds may occur in brine surfaces if the tanks are open. To prevent this growth, the air layer between the liquid and the top of tank must be reduced as much as possible.

The main roles of yeasts in the processing of fermented olives, are associated with the production of alcohols, ethyl acetate, acetaldehyde and organic acids, compounds that are relevant for the development of taste and aroma and for the preservation of the typical characteristics of this fermented food (Alves et al., 2012).

Some yeast species seem to improve the growth of LAB. Yeasts are able to synthesize substances such as vitamins, amino acids and purines, or breakdown complex carbohydrates, which are essential for the growth of *Lactobacillus* species that request a nutritionally rich environment for optimal growth (Viljoen, 2006). However, in table olive processing yeasts can also produce spoilage such as off-flavour production, clouding of brines and softening of fruits (Arroyo-López et al., 2008a).

Recently, molecular methods have been applied for the identification of yeast associated with table olives. These techniques confer a higher degree of accuracy in the final identification than classical biochemical methods. Deiana et al. (1992) observed that the species found depended on the degree of maturation of the olive fruits. The genera *Candida*, *Pichia*, *Rhodotorula*, *Saccharomyces*, *Debaryomyces*, *Kluyveromyces*, *Kloeckera*, *Torulopsis*, *Trichosporon* and *Cryptococcus* were found by several authors (Arroyo-López et al., 2008a; Rodríguez-Gómez et al., 2010). While the main frequently isolated species, in both naturally black and Spanish-style olive brines, are *Candida boidinii, Candida diddensiae, Pichia anomala, Pichia kluyveri, Pichia membranifaciens* and *Saccharomyces cerevisiae* (Oliveira et al., 2004; Coton et al., 2005; Arroyo-López et al., 2006). Recently Rodríguez-Gómez and co-workers (2012) drawn up a list of isolates representative of the yeasts of table olives, and the most suitable yeasts to be used as starters, alone or in combination with LAB.

The interrelationships between *Lactobacillus* species and yeasts in table olives may also play an essential role in product preservation. Several authors have recently focused their attention on yeast biodiversity associated with the different types of olive processes with particular regard to their enzymatic activities, in order to propose yeast as starters (Bautista-Gallego et al., 2011).

2.2. The microbiological hazards in table olives

Fermented vegetable technology is based on lactic acid and alcoholic fermentations, that convert sugars to different end-products, and the obtained food products take on new and different characteristics (Hutkins, 2006). To enhance the quality of final product is one of the main scientific and technological challenges for table olive production together with the reduction of cost of harvesting, of spoilage occurrence, and of environmental pollution (Brenes, 2004).

The olive-ecosystem is influenced by the indigenous microbial population, by the intrinsic factors related to the olives (pH, aw, phenols, sugar content, etc.) and extrinsic factors (temperature, oxygen and salt levels). The microbial population characterizing the first days of fermentation seem to be always the same: *Enterobacteriaceae*, lactic cocci, *Bacillaceae* and yeasts, whose evolution is strongly related to the pH value. In addition to these populations, Nychas and co-workers (2002) also reported the presence of *Pseudomonas* spp. at the start of fermentation, that decreased as other Gram-negative bacteria, within the first two weeks of fermentation.

Among the three main commercial preparations of table olives, there are some processing parameters affecting the fermentation process. The most important is the pH. The fermentation process in Spanish-style olives (treated olives) begins at an alkaline pH, higher than 9–10, because the fruits are previously treated with NaOH to hydrolyse the oleuropein (Medina et al., 2010). In this case, the microbial population is mainly composed of *Enterobacteriaceae*, lactic cocci and other epiphytic microorganisms which are able to drop the pH value below 7.0, creating the optimal conditions for LAB growth. Regarding the other two main preparations, Greek-style and California-style, the fermentation is influenced since the first steps by the processing conditions. In this case, the microbial population starts to grow at an acidic pH because organic acid (acetic, citric, lactic) are added to prevent the growth of Gram-negative bacteria.

No official microbiological criteria for table olives are available. However, the Standards of the Codex Alimentarius prescribes the minimum requirements related to hygiene for table olives. The final product shall be free from microorganisms and parasites in amounts which may represent a hazard to health and shall not contain any substance originating from microorganisms in amounts which may represent a hazard to health (Pereira et al., 2008). To reduce the risk of food-borne illness and spoilage phenomenon, good practices in agriculture (GAP), hygiene (GHP) and manufacturing (GMP) must be applied.

Although heat treatments have some negative effects such as alterations in consistency and colour (Romeo et al., 2009), the correct use of temperature during pasteurization and/or sterilization is essential to ensure microbiological safety and stability, inactivating enzymes, and lessens the oxidizing processes. If heat sterilization is applied to olives, the treatment must be sufficient both in time and temperature, to destroy spores of *Clostridium botulinum* (COI, 2004). While olives preserved by salt and acidification or natural fermentation are usually *C. botulinum* and its toxin free, only if the pH is constantly monitored and maintained below 4.6. Clostridial bacteria are relatively common in the environment because they are spore-forming. Spores of *C. botulinum* were detected both in pasteurized and sterilized olives (Pereira et al., 2008) indicating a poor attention to the application of sterilisation parameters. The occurrence of *C. botulinum* appears, however, to be rare. The sulphite reducing *Clostridium* spores are indicators of remote faecal contamination. Their presence in pasteurized olives is due to the occurrence of anaerobic fermentations or to the resistance of spores to pasteurization. However, the spores should be destroyed by sterilisation as its presence in a sterilised product indicates either inadequate heat treatment or post-sterilisation contamination.

The occurrence of *Listeria monocytogenes* in green table olives has been assessed, demonstrating that the product, despite its low pH and high salt concentration, can support *Listeria* survival for which an appropriate heat treatment must be applied (Caggia et al., 2004). Another hazard in table olives is *Escherichia coli* O157:H7, a pathogenic bacterium responsible for hemorrhagic colitis and hemolytic uremic syndrome. Its presence may be particularly associated to the Spanish-style method because the drop in the pH is slower than in natural fermented olive brine. The death rate of *E. coli* could be affected by using starter strain (Spyropoulou et al., 2001). More recently, the species *Enterobacter cloacae*, an opportunistic pathogen for humans, has been recovered in spontaneously fermented table olives (Bevilacqua et al., 2010a). The occurrence of *Listeria, Salmonella, Escherichia coli, Yersinia* pathogen strains and others are extensively reported in the scientific report of the European Food Safety Authority (EFSA) and European Centre for Disease Prevention and Control (ECDC), issued on 21 February 2012.

Other than the pH value, a parameter which strongly influences the storage and quality of table olives is NaCl concentration. Its level is important for achieving stability of the products because it prevents spoilage and growth of pathogens. During recent years, consumers have developed an attitude on low sodium intake principally because a diet rich in sodium leads to higher blood pressure. So, several scientific studies (Arroyo-López et al., 2008b; Romeo et al., 2009; Bautista-Gallego et al., 2010; Bautista-Gallego et al., 2011; Panagou et al., 2011) have focalized on the viability, application and consequences of replacement of sodium with calcium or potassium in table olive fermentation. Apparently, NaCl may be substituted in diverse proportions with KCl or CaCl$_2$ without substantially altering the usual fermentation profiles and producing good sensorial characteristics. In particular, a mixture of NaCl, CaCl₂ showed the ability to reduce both bacterial and yeast growth, while KCl showed similar effect of NaCl. Moreover, using different mixed salts, Tsapatsaris and Kotzekidou (2004) showed that the replacement of NaCl by KCl in Kalamon olives resulted in a strong synergy between calcium lactate and calcium acetate with higher growth rates of starter cultures of *Lactobacillus plantarum* and *Debaryomyces hansenii*.

The replacement of NaCl with other chlorides could be important in those productions traditionally processed in a high salt concentration, such as Greek-style olives, because this action could lower the NaCl concentration without reaching the lowest limits necessary to obtain a safe product. Therefore, besides the pH decrease and the NaCl concentration, several actions have been proposed in order to overcome all the fermentation problems: pasteurization, addition of sugars (glucose and sucrose), extra salt addition and use of starter cultures. Sugar supplements increase the pH drop rate reducing the dangerous early stage and ensuring the safety of the final product (Chorianopoulos et al., 2005).

2.3. LAB starter cultures

At industrial level LAB play a positive role in the production of wines and beers, but therefore they represent major spoilage organisms for such products (Bamforth, 2005). Table olive processing is still based on empirical methods despite its growing economical value. However, interest in developing starter cultures to be used in table olives is increasing. LAB have long been employed in fermentation as a food preservation technique owing to their progressive acidification of the fermenting brine with a consequent pH decrease (Marsilio et al., 2005). In addition, the use of LAB could standardize olive fermentation and reduce the use of highly polluting chemicals as NaOH (lye solution), contribute significantly to storage preventing microbial spoilage, and improve flavour.

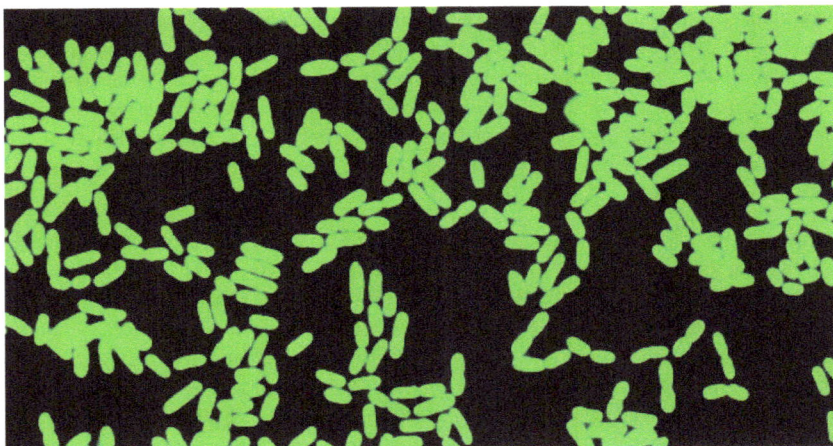

Figure 1. *L. plantarum* strain detected by fluorescence microscopy.

In table olive processing, starter cultures must have some properties such as good resistance to the inhibitory effect of polyphenols, good survival against wild strains of related species, rapid acid production, complete utilization of fermentable sugars, good tolerance against high levels of salt and low pH, and a possibly inhibitory effect against undesirable organisms. The latter effect is due to the production of bacteriocin, peptides that were found to be active against a number of natural competitors of *L. plantarum* in the fermentation brines and also against bacteria that can cause olive spoilage (Leal-Sánchez et al., 2003). This property is considered of importance in the development of new preservation technologies of foods (Devlieghere el al., 2004).

Moreover, lactobacilli are important members of the healthy human microbiota and exert several beneficial physiological effects, such as antimicrobial and antitumorigenic activities (Nguyen et al., 2007; Bevilacqua et al., 2010b). The reduction of cholesterol by LAB has been demonstrated in human, mouse, and pig studies (Nguyen et al., 2007). Nowadays, foods fortified with health-promoting probiotic bacteria are mainly produced with milk derivatives, so functional food industries are focusing on new non-dairy foods that can contribute to a regular assumption of probiotics (Lavermicocca et al., 2005).

The species of the genus *Lactobacillus* are widely occurring in many natural environments often playing important roles in fermentation processes and in the regulation of relationships among species of complex ecosystems. In particular, *L. plantarum* and *L. pentosus* are regarded as the main species leading this process (Table 1) often being used as a starter in guided olive fermentation (Sánchez et al., 2001; Leal-Sánchez et al., 2003; Hurtado et al., 2009). *L. pentosus* and *L. plantarum* are also the most frequently isolated species in table olives; the other species used as inocula, with little exception, have always been studied in conjunction with them (Hurtado et al., 2012).

However, a significant occurrence of *Leuconostoc* spp. on olive fruits and leaves was highlighted in the study of Ercolini and co-workers (2006), suggesting that *Lactobacillus* spp. may also originate from the environment or tools of production and not exclusively from the olives. Lavermicocca and others (2005) used table olives as a vehicle for delivering probiotic bacterial species, such as *Lactobacillus rhamnosus*, *L. paracasei*, *Bifidobacterium longum* and *B. bifidum*, but these strains are not involved in spontaneous fermentation and so they are not well adapted to the environmental conditions of table olives (Perricone et al., 2010).

Isolation from olive brines of *Enterococcus* strains has been reported by several authors, so mixed starters of *E. faecium* and *L. plantarum* (Lavermicocca et al., 1998) or *E. casseliflavus* and *L. pentosus* have been studied. The suggestion to inoculate *E. casseliflavus*, isolated from fermenting olives, is due to its good tolerance to the initial high pH (in case of lye treatment), without the drawback of transmissible antibiotic resistance shown by *E. faecium* (de Castro et al., 2002).

The selection of starters is based on diverse criteria including homo- and hetero-fermentative metabolism, acid production, salt tolerance, flavour development, temperature range growth, oleuropein-splitting capability and bacteriocin production (Panagou et al.,

2008). Furthermore, the ability to grow at low temperatures must be considered essential in cold regions, since heating brine is complex and expensive (Durán Quintana et al., 1999).

Genus	Species	Authors & year
Lactobacillus	plantarum	Randazzo et al., 2011
		Ruiz-Barba et al., 2010
		Perricone et al., 2010
		Hurtado et al., 2010
		Kumral et al., 2009
		Sabatini et al., 2008
		Panagou et al., 2008
		Saravanos et al., 2008
		Romeo & Poiana, 2007
		Marsilio et al., 2005
		Chorianopoulos et al., 2005
		Lamzira et al., 2005
		Caggia et al., 2004
		Leal-Sánchez et al., 2003
		Sánchez et al., 2001
	pentosus	Aponte et al., 2012
		Hurtado et al., 2010
		Medina et al., 2009; 2008
		Panagou et al., 2008
		Peres et al., 2008
		Romeo & Poiana, 2007
		Servili et al., 2006
		Caggia et al., 2004
		De Castro et al., 2002
		Sánchez et al., 2001
	casei	Randazzo et al., 2011
		Caggia et al., 2004
	paracasei	De Bellis et al., 2010
		Saravanos et al., 2008
	paraplantarum	Romeo & Poiana, 2007
	brevis	Kumral et al., 2009
		Romeo & Poiana, 2007
	coryniformis	Aponte et al., 2012
Leuconostoc	cremoris	Kumral et al., 2009
	paramesenteroides	Kumral et al., 2009
Pediococcus	pentosaceus	Ruiz-Barba et al., 2010
Enterococcus	faecium	Ruiz-Barba et al., 2010
	casseliflavus	De Castro et al., 2002

Table 1. LAB tested as starter cultures in table olives (references are shown for the last twelve years)

3. Enumeration of microorganisms in fermented olives: Methods of analysis

Each microorganism should grow and form a separate colony when the sample is plated in a solid medium during plate count procedures. Unfortunately, some organisms may not be capable to grow under the conditions used. Moreover, some chains of organisms could appear as a single colony (Swanson et al., 2001).

Up to now, methods available for detection and identification of microbial population involved in table olive fermentation have been very limited and generally culture-dependent, not providing reliable information on the composition of the entire microbial community (Randazzo et al., 2012). A culture-independent method, such as the denaturing gradient gel electrophoresis (DGGE), has the potential to study microbial population quickly and economically, avoiding the use of selective cultivation and isolation of bacteria (Rantsiou et al., 2005). So, plate count techniques are still the major but not the most representative method. Using aseptic techniques, brine samples may be taken and used directly to prepare serial dilutions. In order to analyse the whole olive, for example dry salted olives, the containers should be shaken and after mixing the olives, a sample should be taken aseptically with a sterile spoon, weighed and transferred to a sterile container such as plastic bag. After adding a sterile volume of quarter-strength Ringer's solution or 0.9% NaCl, the bag is then pummelled in a stomacher to prepare dilutions. Each mL collected by this bag represents 1 g of olive sample. The decimal dilutions are usually used for the calculation of results. Different ranges to obtain a readable number of colonies on plates may be necessary depending on the microorganism, procedures and initial contamination of olives. After pipetting the diluted sample into the petri plate, about 15 mL of each liquefied medium is added (the temperature must not exceed 45°C) into the plate. The medium is mixed with inoculum by carefully rotating. Several dilutions or replicate plates for each dilution should be prepared for each sample, making sure that the sample is tested, at least, in triplicates. After solidification, the petri plates should be inverted and placed in the incubator at specific temperature shown in the next sections. After incubation, the plates should contain between 25 and 250 colonies for the best count accuracy (Swanson et al., 2001). If microbial changes during fermentation are to be followed, the sampling must start at time zero, when the olives are salted or brined, and samples should be collected at regular intervals up to the end of fermentation. The sampling intervals may be of 1-3 days during the first month of fermentation, then these intervals may be extended up to 7 days. The brine samples, such as olives, should be examined as soon as possible, but if it is temporarily impossible, the samples must be refrigerated and analysed within 24 hours.

3.1. Lactic acid bacteria

The simply enumeration of LAB may be carried out on MRS agar. It is probably the most commonly used medium for the cultivation of lactobacilli and other LAB (Schillinger and Holzapfel, 2003). This medium must be supplemented with nystatin (50 mg/L) or cycloheximide (100 mg/L) as inhibitors of eukaryotic organisms (to prevent yeasts and molds growth) or with sodium azide (200 mg/L) as Gram-negative bacteria inhibitor. A number of differential and selective media were created for the isolation and characterization of certain groups of LAB. HHD medium is the most used for the differential enumeration of homofermentative and heterofermentative LAB. This medium contains fructose which is reduced to mannitol by heterofermentative but not by homofermentative LAB. Differences in the colour of the colonies are based on differences in the amounts of acids produced by these bacterial groups (Schillinger & Holzapfel, 2003). So in this medium, homofermentative LAB are blue to green, while heterofermentative LAB colonies are white (Fleming et al., 2001). M17 medium should be used for lactococci isolation, but isolation of these LAB is more frequent when analysing dairy products rather than fermented vegetables. All LAB isolation requires anaerobic conditions and incubation at 30-32°C for 48-72 hours, depending on the medium used.

3.2. Mesophilic aerobic bacteria

The mesophilic count, or standard plate count, is generally obtained on Plate Count agar (PCA) which is a generic medium for aerobic microorganisms that grow at mesophilic temperatures. Aerobic plate counts are poor indicators of safety in some products such as those fermented which commonly show a high aerobic count. However, this count gives information about the hygienic and sensorial quality, about the adherence to good manufacturing practice and shelf life of the product (Morton, 2001).

Alternatively to PCA, similar generic media are available as Nutrient agar. The growth conditions are 25-30°C for 24-48 hours, aerobically.

3.3. Yeasts and molds

Yeasts and molds are widely diffused eukaryotic microorganisms because of their adaptation to different environmental conditions. Yeasts and molds can cause various degrees of food decomposition. Invasion and growth may occur on virtually any type of food (Beuchat & Cousin, 2001). Their contamination of food can lead to product losses but, in particular, the highest risk is due to the mycotoxin production by the molds. Several yeasts genera may be important for the sensorial properties of fermented products and for their interrelation with *Lactobacillus* bacteria (see section 2.1) in fermented olives. However, film-forming yeasts as *Debaryomyces, Candida, Pichia* and *Endomycopsis*, are often associated with pickled products and vegetable brines (Fleming et al., 2001), representing the cause of olive defects and consequent product losses.

Available media for yeasts and molds count are several: Rose Bengal Chloramphenicol agar, YM agar, Oxytetracycline Glucose Yeast Extract agar, Sabouraud agar and others. When not included in the medium, the addition of 100 mg/L chloramphenicol is recommended to inhibit the growth of bacteria. For simple enumeration of yeasts and molds, the plates should be incubated aerobically at 25°C for 48 hours.

3.4. *Enterobacteriaceae* and coliform bacteria

The *Enterobacteriaceae* family consists of Gram-negative, facultative anaerobic rods widely distributed in the environment, but which are usually associated with intestinal infections.

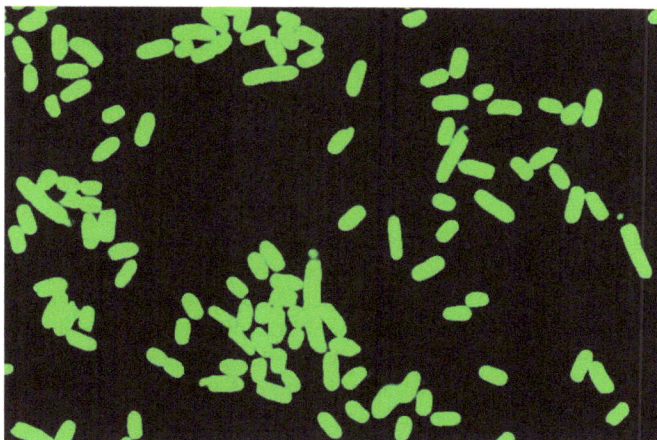

Figure 2. *E. coli* strain detected by fluorescence microscopy.

Numerous studies have determined that *Escherichia coli*, coliforms, faecal coliforms and *Enterobacteriaceae* are unreliable when used as index of pathogen contamination of food. So National and International advisory committees invalidated the prediction of food safety based on levels of *Enterobacteriaceae*, coliforms, faecal coliforms and *E. coli* (Kornacki & Johnson, 2001). The most important application of *Enterobacteriaceae* and coliforms is their enumeration to assess if pasteurization has been adequately performed for example in pasteurized milk, because a proper pasteurization under appropriate conditions inactivates *E. coli* cells present in raw material.

Total coliform bacteria belong to the *Enterobacteriaceae* family, and include *E. coli* as well as various members of the genera *Enterobacter*, *Klebsiella* and *Citrobacter*. All ferment lactose with gas and acid production in 48 hours at 35-37°C. While faecal coliforms are coliforms which can ferment lactose to acid and gas within 48 hours at 45°C and they are so called because are more closely associated with faecal pollution (Manafi, 2003). The term coliform is based only on lactose hydrolysis and has no taxonomy validity. In order to perform the best examination of olive samples, it is preferable enumerate both *Enterobacteriaceae* and coliforms, because some lactose-negative bacteria, such as *Salmonella,* are pathogens.

The most probable number (MPN) is a method which indicates the most likely number of microorganisms present in the analysed sample. This method, used for several years, has now been replaced with techniques on agar media because the MPN is based on a statistical approximation. So, in addition to Violet Red Bile agar or Violet Red Bile Glucose agar, other differential media based on a chromatic response were developed.

Mac Conkey's MUG agar and Eosin Methylene Blue agar (EMB) are differential and selective media, suitable to obtain at the same time the isolation of *Salmonella, Shigella* and coliform bacteria, in particular *E. coli*. In the first medium, lactose-negative colonies are colourless, lactose-positive colonies are red and often surrounded by a turbid zone due to the precipitation of bile acids. *E. coli* can be identified by fluorescence in UV due to its β-D-glucoronidase production. In EMB agar, lactose-fermenters form colonies with dark-blue centres (*E. coli* may also have a green metallic sheen) while the non-lactose fermenters form completely colourless colonies.

3.5. *Staphylococcus aureus*

The foods most associated with staphylococcal poisoning are meat products, dairy products and cream filled bakery products. In processed foods in which *S. aureus* is destroyed by processing, its presence usually indicates post-processing contamination from human skin, mouth, nose or food handlers (Lancette & Bennett, 2003). Due to the high salt tolerance of *S. aureus*, it can grow in table olives even though the low pH and the olive phenols may represent natural inhibitors (Tassou & Nychas, 1994). However, it may be isolated and enumerated in table olives for the same above mentioned reason (as contamination index).

A variety of coagulase-positive and coagulase-negative staphylococci are able to produce enterotoxins, but *S. aureus* still plays a predominant role in staphylococcal food poisoning.

Baird-Parker agar and Rabbit Plasma Fibrinogen agar are the media recommended by the International Organisation for Standardisation (ISO). Moreover, Baird-Parker agar is also used in the official AOAC method in the United States (Zangerl &Asperger, 2003). Tellurite reduction, egg yolk reaction and a high level of sodium chloride are the most applied selective chemicals added in media for *S. aureus* isolation and enumeration.

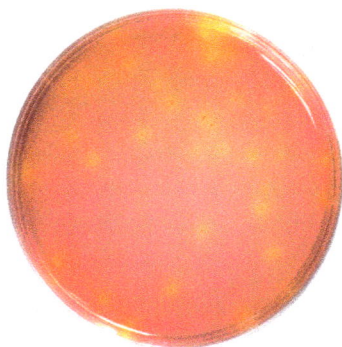

Figure 3. *S. aureus* colonies on Mannitol Salt Agar.

S. aureus colonies in Baird-Parker agar are black, with an opaque zone around the colony because of the egg yolk reaction. Another medium frequently used is the Mannitol Salt agar, which contains mannitol. Coagulase-positive staphylococci grow and produce acid from mannitol showing a yellow colony and halo in a red medium. For all these media, the chemical inhibitors usually used are not completely selective, so additional diagnostic tests may be necessary to identify *S. aureus* colonies. Microscopic examination, catalase test and coagulase test may be rapidly executed to identify *S. aureus* from isolates. Common MPN procedures may be used also for enumeration of *S. aureus*. In most cases, the methodologies need a liquid enrichment procedure to detect low numbers of staphylococci (<100 UFC/g).

Culture conditions are usually 37°C for 48 hours aerobically.

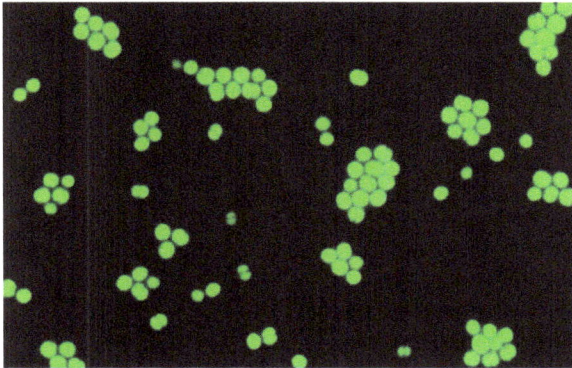

Figure 4. *S. aureus* strain detected by fluorescence microscopy.

3.6. Sulphite reducing clostridia

Clostridia are a widespread heterogeneous group of bacteria showing metabolic and nutritional differences. They easily contaminate foods because they produce resistant spores which can survive under mild processing conditions. The group of sulphite reducing clostridia may be used as marker of raw material quality and hygiene of manufacturing practices. Of particular concern for public health are *C. botulinum*, which forms a deadly toxin in foods, and *C. perfringens* which causes enteritis when present in high numbers. Other species or strains are also known to be toxinogenic or neurotoxinogenic (Bredius & Ree, 2003). *C. perfringens* is the agent of a food poisoning usually associated with consumption of cooked meats or poultry products. Sporulating cells of *C. perfringens* produce a heat-labile enterotoxin which appears to be released *in vivo* in the intestine (Labbe, 2001).

Most isolation media include sulphite and an appropriate iron salt. Sulphite is reduced to sulphide by the clostridial enzyme sulphite reductase; the sulphide will then precipitate as a black deposit in the presence of iron salt. This causes a blackening of the liquid media and clostridia will appear as black colonies. To obtain a higher selectivity for *C. perfringens* isolation, the Oleandomycin Polymixin Sulphadiazine Perfringens (OPSP) agar medium was performed. It contains selective antibiotics and utilises sodium metabisulphite and liver extract as sources of H₂S with ferric ammonium citrate as the indicator. In any case, subsequent confirmation tests such as motility, reduction of nitrate, lactose fermentation, gelatin liquefaction (Labbe, 2001) should be necessary. The plates are incubated anaerobically at 37°C for 18-48 hours.

3.7. Listeria monocytogenes

Listeria species are ubiquitous organisms widely distributed in the environment, especially in plant matter and soil. The microorganism is established as an important foodborne pathogen, which can grow at high salt concentration (up to 10% of NaCl) and at refrigerated temperatures. A study carried out by the U.S. Food and Drug Administration, U.S. Department of Agriculture (USDA, 2001), indicated that vegetables are able to support *L. monocytogenes* growth and, after a few years, Caggia and co-workers (2004) demonstrated that *L. monocytogenes* can survive and grow in green table olives. As indicated by the most recent scientific report of the European Food Safety Authority (2012), the number of listeriosis cases in humans in the EU slightly decreased, and 1,601 confirmed human cases were reported in 2010.

The minimum infective dose of *L. monocytogenes* has not yet been established and many authorities require that the organism must be absent in 25 g of product. This has led to the development of methods following the sequence of pre-enrichment, selective enrichment and diagnostic plating. The whole procedure may take about five days. Then, presumptive positive results need to be confirmed adding further time to complete the examination (Beumer & Curtis, 2003).

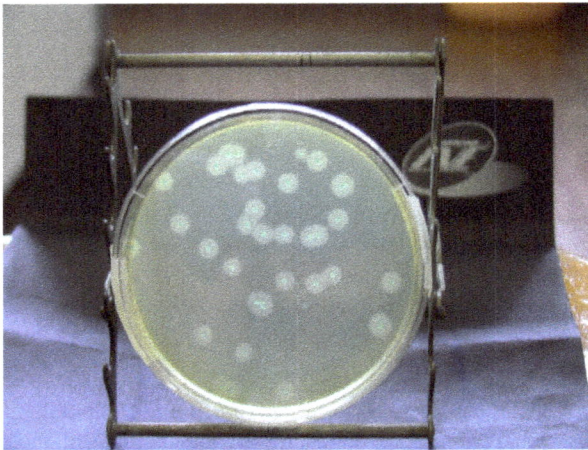

Figure 5. *L. monocytogenes* colonies on ALOA medium.

Most of the isolation media differentiate *Listeria* spp. by means of aesculin hydrolysis which, in the presence of iron, forms a black phenolic compound. According to the USDA method

(McClain & Lee, 1988) 25 mL of brine sample are added to 225 mL of Listeria Enrichment broth base for the resuscitation of stressed cells. After blending, the bag is incubated at 30°C and then, at intervals from 4 hours to 7 days, an aliquot is plated for enumeration in Listeria Selective agar base and incubated at 37°C for 24 h. Black colonies on selective agar base medium were considered presumptive *Listeria* colonies.

The presumptive *Listeria* isolates are to be tested for sugar fermentation, tumbling motility, hemolytic reaction and growth at different salt concentrations (Caggia et al., 2004). Alternatively to this medium, ALOA agar contains the chromogenic compound for the detection of β-glucosidase, common to all *Listeria*, which appear as blue coloured colonies. While *L. monocytogenes*, which possesses a specific phospholipase, hydrolyses the specific substrate added to the medium producing an opaque halo around the colonies (Beumer & Curtis, 2003). Another chromogenic medium used is the Rapid' L. mono agar.

Where counts < 100 CFU/g are expected, it is necessary to use the Most Probable Number (MPN) technique (USDA, 2001).

In order to rapidly perform additional confirmation tests, systems of strips are commercially available for *Listeria* (API Listeria) as for other pathogens.

3.8. Propionibacteria

Propionibacteria are often associated with food spoilage. The "zapateria" spoilage, which can occur in brined olives, is characterized by a malodorous fermentation due to propionic acid produced by certain species of *Propionibacterium* (Jay, 2000), alone or together with *Clostridium* spp.. The propionibacteria may be considered as marker of the end of shelf life of table olives, because the growth of these bacteria cause an increase in pH values creating conditions for the growth of spoilage or pathogen bacteria and the arising of off-odors (Plastourgos & Vaughn, 1957).

Unfortunately, propionibacteria are difficult to isolate because they grow very slowly on solid media. The complex medium usually used is not able to suppress competing organisms. This medium is Sodium Lactate agar (1% typticase, 1% yeast extract, 1% sodium lactate, 0.025% dipotassium phosphate, 1.5% agar) in which the propionibacteria growth appears in 5-7 days at 32°C under anaerobic or microaerophilic conditions (Richter & Vedamuthu, 2001).

4. Conclusion

A successful olive treatment depends on different factors. The olive cultivars show different fermentation behaviours when directly brined. In fact, besides the correct application of manufacturing practices, the knowledge of the chemical and physical characteristics of olive cultivar used and its attitude to treatments are decisive. A more complete knowledge of the olive cultivars is necessary to bring fermentation to a successful condition and attain a good final product. A correlation between each method and chemical-physical composition of each cultivar is needed. Moreover, the presence of potential pathogens detected in all olive treatments by several authors suggests the necessity of pH control, following good hygienic practices throughout the process, and the necessity of a heat treatment of traditional products which are often empirically performed. Control and verification systems should be employed in order to guarantee a safe and hygienic product.

Author details

Flora Valeria Romeo

Agricultural Research Council - Olive Growing and Oil Industry Research Centre, Rende (CS), Italy

Acknowledgement

The author greatly thanks Emanuela Fornasari and Barbara Bonvini of the Fodder and Dairy Productions Research Centre (CRA-FLC, Lodi) for helpfully providing the images of bacteria.

5. References

Alves, M.; Gonçalves, T. & Quintas, C. (2012). Microbial quality and yeast population dynamics in cracked green table olives' fermentations. *Food Control*, Vol.23, pp. 363-368.

Aponte, M.; Blaiotta, G.; Croce, F.L.; Mazzaglia, A.; Farina, V.; Settanni, L. & Moschetti, G. (2012). Use of selected autochthonous lactic acid bacteria for Spanish-style table olive fermentation. *Food Microbiology*, Vol.30, No.1, pp. 8-16.

Arroyo López, F.N.; Durán Quintana, M.C.; Ruiz Barba, J.L.; Querol, A. & Garrido Fernández, A. (2006). Use of molecular methods for the identification of yeast associated with table olives. *Food Microbiology*, Vol.23, pp. 791–796.

Arroyo-López, F.N.; Querol, A.; Bautista-Gallego, J. & Garrido-Fernández, A. (2008 a). Role of yeasts in table olive production. *International Journal of Food Microbiology*, Vol.128, pp. 189-196.

Arroyo-López, F.N.; Bautista-Gallego, J.; Durán-Quintana, M.C.; Rodríguez-Gómez, F.; Romero-Barranco, C. & Garrido-Fernández, A. (2008 b). Improvement of the storage process for cracked table olives. *Journal of Food Engineering*, Vol.89, pp. 479-487.

Bamforth, C.W. (2005). The Science Underpinning Food Fermentations, In: *Food, Fermentation and Micro-organisms*, Blackwell Publishing (Ed.), 1-2, ISBN-13: 978-0632-05987-4, Oxford, UK.

Bautista-Gallego, J.; Arroyo-López, F.N.; Durán-Quintana, M. C. & Garrido-Fernández, A. (2010). Fermentation profiles of Manzanilla-Aloreña cracked green table olives in different chloride salt mixtures. *Food microbiology*, Vol.27, No.3, pp. 403-12.

Bautista Gallego, J.; Arroyo López, F.N.; Romero Gil, V.; Rodríguez Gómez, F.; García García, P. & Garrido Fernández, A. (2011). Chloride salt mixtures affect Gordal cv. green Spanish-style table olive fermentation. *Food Microbiology*, Vol.28, pp. 1316-1325.

Beuchat, L.R. & Cousin, M.A. (2001). Yeasts and Molds, In: *Compendium of methods for the Microbiological examination of Foods*, Fourth edition, F.P. Downes & K. Ito (Eds.), American Public Health Association, 209-215, ISBN: 0-87553-175-x, Washington DC.

Beumer, R.R. & Curtis, G.D.W. (2003). Culture media and methods for the isolation of *Listeria monocytogenes*, In: *Handbook of Culture Media for Food Microbiology*, J.E.L. Corry, G.D.W. Curtis & R.M. Baird (Eds.), Elsevier Science, 79-90, ISBN: 0-444-51084-2, Amsterdam, Netherlands.

Bevilacqua, A.; Cannarsi, M.; Gallo, M.; Sinigaglia, M. & Corbo, M.R. (2010 a). Characterization and implications of *Enterobacter cloacae* strains, isolated from Italian table olives "Bella di Cerignola". *Journal of Food Science*, Vol.75, pp. 53-60.

Bevilacqua, A.; Altieri, C.; Corbo, M.R.; Sinigaglia, M. & Ouoba, L.I.I. (2010 b). Characterization of lactic acid bacteria isolated from Italian Bella di Cerignola table olives: selection of potential multifunctional starter cultures. *Journal of Food Science*, Vol.75, No.8, pp. 536-544.

Bredius, M.W.J. & de Ree, E.M. (2003). Media for the detection and enumeration of clostridia in foods, In: *Handbook of Culture Media for Food Microbiology*, J.E.L. Corry, G.D.W. Curtis & R.M. Baird (Eds.), Elsevier Science, 49-60, ISBN: 0-444-51084-2, Amsterdam, Netherlands.

Brenes, M. (2004). Olive fermentation and processing: scientific and technological challenges. *Journal of Food Science*, Vol.69, No.1, pp. 33-34.

Caggia, C.; Randazzo, C.L.; Di Salvo, M.; Romeo, F.V. & Giudici, P. (2004). Occurrence of *Listeria monocytogenes* in green table olives. *Journal of food protection*, Vol.67, No.10, pp. 2189-94.

Cardoso, S.M.; Mafra, I.; Reis, A.; Nunes, C.; Saraiva, J.A. & Coimbra M.A. (2010). Naturally fermented black olives: Effect on cell wall polysaccharides and on enzyme activities of Taggiasca and Conservolea varieties. *LWT - Food Science and Technology*, Vol.43, pp. 153-160.

Chorianopoulos, N.G.; Boziaris, I.S.; Stamatiou, A. & Nychas, G.-J.E. (2005). Microbial association and acidity development of unheated and pasteurized green-table olives fermented using glucose or sucrose supplements at various levels. *Food Microbiology*, Vol.22, No.1, pp. 117-124.

COI (2004). Trade Standard Applying to Table Olives. *International Olive Oil Council*, COI/OT/NC no. 1, Dezembro de 2004.

Coton, E.; Coton, M.; Levert, D.; Casaregola, S. & Sohier, D. (2005). Yeast ecology in French cider and black olive natural fermentations. *International Journal of Food Microbiology*, Vol.108, pp. 130–135.

De Bellis, P., Valerio, F., Sisto, A., Lonigro, S. L., & Lavermicocca, P. (2010). Probiotic table olives: microbial populations adhering on olive surface in fermentation sets inoculated with the probiotic strain *Lactobacillus paracasei* IMPC2.1 in an industrial plant. *International Journal of Food Microbiology*, Vol.140, No.1, pp. 6-13.

De Castro, A.; Montaño, A.; Casado, F.-J.; Sánchez, H. & Rejano, L. (2002). Utilization of *Enterococcus casseliflavus* and *Lactobacillus pentosus* as starter cultures for Spanish-style green olive fermentation. *Food Microbiology*, Vol.19, pp. 637-644.

Deiana, P.; Farris, G.A.; Catzeddu, P. & Madan, G. (1992). Impiego di fermenti lattici e lieviti nella preparazione delle olive da mensa. *Industrie Alimentari,* Vol.31, pp. 1011–1023.

Devlieghere, F.; Vermeiren, L. & Debevere, G. (2004). New preservation technologies: possibilities and limitations. *International Dairy Journal,* Vol.14, 273-285.

Durán Quintana, M.C.; García García, P. & Garrido Fernández, A. (1999). Establishment of conditions for green table olive fermentation at low temperature. *International Journal of Food Microbiology,* Vol.51, No.2-3, pp. 133-143.

EFSA & ECDC (2012). The European Union Summary Report on Trends and Sources of Zoonoses, Zoonotic Agents and Food-borne Outbreaks in 2010. *EFSA Journal,* Vol.10(3):2597, pp. 1-442, Available from: http://www.efsa.europa.eu/efsajournal.

Ercolini, D.; Villani, F.; Aponte, M. & Mauriello, G. (2006). Fluorescence in situ hybridisation detection of *Lactobacillus plantarum* group on olives to be used in natural fermentations. *International Journal of Food Microbiology,* vol.112, No.3, pp. 291-296.

Fleming, H.P.; Mcfeeters, R.F; Thompson, R.L. & Sanders, D.C. (1973). Storage stability of vegetables fermented with pH control. *Journal of Food Science,* Vol.48, pp. 975-981.

Fleming, H.P.; Mcfeeters, R.F. & Breidt, F. (2001). Fermented and acidified vegetables, In: *Compendium of methods for the Microbiological examination of Foods,* Fourth edition, F.P. Downes & K. Ito (Eds.), American Public Health Association, 521-532, ISBN: 0-87553-175-x, Washington DC.

Garrido Fernández, A.; Fernández Díaz, M.J. & Adams, R.M. (1997). *Table Olives. Production and Processing,* Chapman & Hall, London, UK.

Hurtado, A.; Reguant, C.; Esteve-Zarzoso, E.; Bordons, A. & Rozès, N. (2008). Microbial population dynamics during the processing of Arbequina table olives. *Food Research International,* Vol.41, pp. 738-744.

Hurtado, A.; Reguant, C.; Bordons, A. & Rozès, N. (2009). Influence of fruit ripeness and salt concentration on the microbial processing of Arbequina table olives. *Food Microbiology,* Vol.26, No.8, pp. 827-833.

Hurtado, A.; Reguant, C.; Bordons, A. & Rozès, N. (2010). Evaluation of a single and combined inoculation of a *Lactobacillus pentosus* starter for processing cv. Arbequina natural green olives. *Food Microbiology,* Vol.27, No.6, pp. 731-740.

Hurtado, A.; Reguant, C.; Bordons, A. & Rozès, N. (2012). Lactic acid bacteria from fermented table olives. *Food microbiology,* Vol.31, No.1, pp. 1-8.

Hutkins, R.W. (2006). Fermented vegetables, In: *Microbiology and Technology of fermented foods,* Blackwell Publishing (Ed.), 233-260, ISBN-13: 978-0-8138-0018-9, Oxford, UK.

Jay, J.M. (2000). Fruit and Vegetable Products: Whole, Fresh-Cut, and Fermented In: *Modern Food Microbiology,* Sixth Edition, Aspen Publishers, Inc., 131-161, ISBN 0-8342-1671-X, Gaithersburg, Maryland.

Kornacki, J.L. & Johnson, J.L. (2001). Enterobacteriaceae, coliforms and *Escherichia coli* as quality and safety indicators, In: *Compendium of methods for the Microbiological examination of Foods,* Fourth edition, F.P. Downes & K. Ito (Eds.), American Public Health Association, 69-82, ISBN: 0-87553-175-x, Washington DC.

Kumral, A.; Basoglu, F. & Sahin, I. (2009). Effect of the use of different lactic starters on the microbiological and physicochemical characteristics of naturally black table olives of Gemlike cultivar. *Journal of Food Processing and Preservation*, Vol.33, pp. 651-664.

Labbe, R.G. (2001). *Clostridium perfringens*, In: *Compendium of methods for the Microbiological examination of Foods*, Fourth edition, F.P. Downes & K. Ito (Eds.), American Public Health Association, 325-330, ISBN: 0-87553-175-x, Washington DC.

Lamzira, Z.; Asehraou, A.; Brito, D.; Oliveira, M.; Faid, M. & Peres, C. (2005). Bloater spoilage of green olives. *Food Technology and Biotechnology*, Vol.43, pp. 373-377.

Lancette, G.A. & Bennett, R.W. (2001). *Staphylococcus aureus* and Staphylococccal enterotoxins, In: *Compendium of methods for the Microbiological examination of Foods*, Fourth edition, F.P. Downes & K. Ito (Eds.), American Public Health Association, 387-403, ISBN: 0-87553-175-x, Washington DC.

Lavermicocca, P.; Gobbetti, M.; Corsetti, A. & Caputo, L. (1998). Characterization of lactic acid bacteria isolated from olive phylloplane and table olive brines. *Italian Journal of Food Science*, Vol.10, pp. 27–39.

Lavermicocca, P.; Valerio, F.; Lonigro, S.L.; De Angelis, M.; Morelli, L.; Callegari, M.L.; Rizzello, C.G. & Visconti, A. (2005). Study of Adhesion and Survival of Lactobacilli and Bifidobacteria on Table Olives with the Aim of Formulating a New Probiotic Food. *Applied and Environmental Microbiology*, Vol.71, No.8, pp. 4233–4240.

Leal-Sánchez, M.V.; Ruiz-Barba, J.L.; Sánchez, A.H.; Rejano, L.; Jiménez-Díaz, R. & Garrido, A. (2003). Fermentation profile and optimization of green olive fermentation using *Lactobacillus plantarum* LPCO10 as a starter culture, *Food Microbiology*, Vol.20, pp. 421-430.

Manafi, M. (2003). Media for detection and enumeration of 'total' Enterobacteriaceae, coliforms and Escherichia coli from water and foods, In: *Handbook of Culture Media for Food Microbiology*, J.E.L. Corry, G.D.W. Curtis & R.M. Baird (Eds.), Elsevier Science, 167-194, ISBN: 0-444-51084-2, Amsterdam, Netherlands.

Marsilio, V.; Seghetti, L.; Iannucci, E.; Russi, F.; Lanza, B. & Felicioni, M. (2005). Use of a lactic acid bacteria starter culture during green olive (*Olea europaea* L cv Ascolana tenera) processing. *Journal of the Science of Food and Agriculture*, Vol.85, No.7, pp. 1084-1090.

McClain, D. & Lee, W.H. (1988) Development of USDA-FSIS method for isolation of *Listeria monocytogenes* from raw meat and poutry. *Journal of the Association of Official Analytical Chemists*, Vol.71, pp. 660-664.

Medina, E.; Romero, C.; de Castro, A.; Brenes, M. & García, A. (2008). Inhibitors of lactic acid fermentation in Spanish-style green olive brines of the Manzanilla variety. *Food Chemistry*, Vol.110, No.4, pp. 932-937.

Medina, E.; García, A.; Romero, C.; de Castro, A. & Brenes, M. (2009). Study of the anti-lactic acid bacteria compounds in table olives. *International Journal of Food Science & Technology*, Vol.44, No.7, pp. 1286-1291.

Medina, E.; Gori, C.; Servili, M.; De Castro, A.; Romero, C. & Brenes, M. (2010). Main variables affecting the lactic acid fermentation of table olives. *International Journal of Food Science & Technology*, Vol.45, No.6, pp. 1291-1296.

Morton, R.D. (2001). Aerobic Plate Count, In: *Compendium of methods for the Microbiological examination of Foods*, Fourth edition, F.P. Downes & K. Ito (Eds.), American Public Health Association, 63-67, ISBN: 0-87553-175-x, Washington DC.

Nguyen, T.D.T.; Kang, J.H. & Lee, M. S. (2007). Characterization of *Lactobacillus plantarum* PH04, a potential probiotic bacterium with cholesterol-lowering effects. *International Journal of Food Microbiology*, Vol.113, No.3, pp. 358-61.

Nychas, G.-J.E.; Panagou, E.Z.; Parker, M.L.; Waldron, K.W. & Tassou, C.C. (2002). Microbial colonization of naturally black olives during fermentation and associated biochemical activities in the cover brine. *Letters in Applied Microbiology*, Vol.34, pp. 173-177.

Oliveira, M.; Brito, D.; Catulo, L.; Leitao, F.; Gomes, L.; Silva, S.; Vilas-Boas, L.; Peito, A.; Fernandes, I.; Gordo, F. & Peres, C. (2004). Biotechnology of olive fermentation of 'Galega' Portuguese variety. *Grasas Y Aceites*, Vol.55, pp. 219–226.

Panagou, E.Z.; Tassou C.C. & Katsaboxakis, C.Z. (2003). Induced lactic acid fermentation of untreated green olives of the Conservolea cultivar by *Lactobacillus pentosus*. *Journal of the Science of Food and Agriculture*, Vol.83, pp. 667-674.

Panagou, E.Z.; Schillinger, U.; Franz, C.M.A.P. & Nychas, G.-J.E. (2008). Microbiological and biochemical profile of cv. Conservolea naturally black olives during controlled fermentation with selected strains of lactic acid bacteria. *Food microbiology*, Vol.25, No.2, pp. 348-58.

Panagou, E.Z.; Hondrodimou, O.; Mallouchos, A. & Nychas, G.-J. E. (2011). A study on the implications of NaCl reduction in the fermentation profile of Conservolea natural black olives. *Food microbiology*, Vol.28, No.7, pp. 1301-1307.

Pereira, A.P.; Pereira, J.A.; Bento, A. & Estevinho, M.L. (2008). Microbiological characterization of table olives commercialized in Portugal in respect to safety aspects. *Food and Chemical Toxicology*, Vol.46, pp. 2895-2902.

Peres, C.; Catuloa, L.; Brito, D. & Pintadoa, C. (2008). *Lactobacillus pentosus* DSM 16366 starter added to brine as freeze-dried and as culture in the nutritive media for Spanish style green olive production. *Grasas y Aceites*, Vol.59, pp. 234-238.

Perricone, M.; Bevilacqua, A.; Corbo, M.R. & Sinigaglia, M. (2010). Use of *Lactobacillus plantarum* and glucose to control the fermentation of "Bella di Cerignola" table olives, a traditional variety of Apulian region (southern Italy). *Journal of Food Science*, Vol.75, pp. 430-436.

Plastourgos, S. & Vaughn, R.H. (1957). Species of *Propionibacterium* associated with zapatera spoilage of olives. *Applied Microbiology*, Vol.5, pp. 267-271.

Randazzo, C.L.; Fava, G.; Tomaselli, F.; Romeo, F.V.; Pennino, G.; Vitello, E. & Caggia, C. (2011). Effect of kaolin and copper based products and of starter cultures on table olive fermentation. *Food Microbiology*, Vol.28, No.5, pp. 910-919.

Randazzo, C.L.; Ribbera, A.; Pitino, I.; Romeo, F.V. & Caggia, C. (2012). Diversity of bacterial population of table olives assessed by PCR-DGGE analysis. *Food Microbiology*, In Press, 18.06.2012, Available from: http://dx.doi.org/10.1016/j.fm.2012.04.013.

Rantsiou, K.; Urso, R.; Iacumin, L.; Cantoni, C.; Cattaneo, P.; Comi, G. & Cocolin, L. (2005). Culture dependent and independent methods to investigate the microbial ecology of

Italian fermented sausages. *Applied and Environmental Microbiology*, Vol.71, pp. 1977-1986.

Richter, R.L. & Vedamuthu, E.R. (2001). Milk and milk products, In: *Compendium of methods for the Microbiological examination of Foods*, Fourth edition, F.P. Downes & K. Ito (Eds.), American Public Health Association, 483-495, ISBN: 0-87553-175-x, Washington DC.

Rodríguez-Gómez, F.; Arroyo-López, F.N.; López-López, A.; Bautista-Gallego, J. & Garrido-Fernández, A. (2010). Lipolytic activity of the yeast species associated with the fermentation/storage phase of ripe olive processing. *Food Microbiology*, Vol.27, pp. 604-612.

Rodríguez-Gómez, F.; Romero-Gil, V.; Bautista-Gallego, J.; Garrido-Fernández, A. & Arroyo-López, F.N. (2012). Multivariate analysis to discriminate yeast strains with technological applications in table olive processing. *World Journal of Microbiology and Biotechnology*, Vol.28, pp. 1761-1770.

Romeo, F.V. & Poiana, M. (2007). Ability of commercially available *Lactobacillus* strains as starters in brining and debittering of table olives. *Acta Alimentaria*, Vol.36, No.1, pp. 49-60.

Romeo, F.V.; De Luca, S.; Piscopo, A.; Perri, E. & Poiana, M. (2009). Effects of post-fermentation processing on the stabilisation of naturally fermented green table olives (cv Nocellara Etnea). *Food Chemistry*, Vol.116, pp. 873-878.

Ruiz-Barba, J.L.; Brenes-Balbuena, M.; Jiménez-Díaz, R.; García-García, P. & Garrido Fernández, A. (1993). Inhibition of *Lactobacillus plantarum* by polyphenols extracted from two different kind of olive brine. *Journal of Applied Bacteriology*, Vol.74, pp. 15-19.

Ruiz-Barba, J.L.; Caballero-Guerrero, B.; Maldonado-Barragán, A. & Jiménez-Díaz, R. (2010). Coculture with specific bacteria enhances survival of *Lactobacillus plantarum* NC8, an autoinducer-regulated bacteriocin producer, in olive fermentations. *Food Microbiology*, Vol.27, No.3, pp. 413-417.

Sabatini, N.; Mucciarella, M.R. & Marsilio, V. (2008). Volatile compounds in uninoculated and inoculated table olives with *Lactobacillus plantarum* (*Olea europaea* L., cv. Moresca and Kalamata). *LWT - Food Science and Technology*, Vol.41, No.10, pp. 2017-2022.

Sánchez, A.H.; Rejano, L.; Montaño, A. & de Castro, A. (2001). Utilization at high pH of starter cultures of lactobacilli for Spanish-style green olive fermentation. *International journal of food microbiology*, Vol.67, No.1-2, pp. 115-22.

Sayadi, S.; Allouche, N.; Jaoua, M. & Aloui, F. (2000). Determinal effects of high molecular-mass polyphenols on olive mill wastewater biotreatment. *Process Biochemistry*, Vol.35, pp. 725–735.

Saravanos, E.; Kagli, D.; Zoumpopoulou, G.; Panagou, E.Z. & Tassou, C.C. (2008). Use of probiotic lactic acid bacteria as starter cultures in Spanish-style green olive fermentation and determination of their survival using PFGE. *Food Microbiology*. 2008, 1-4 September, Aberdeen, UK.

Schillinger, U. & Holzapfel, W.H. (2003).Culture media for lactic acid bacteria, In: *Handbook of Culture Media for Food Microbiology*, J.E.L. Corry, G.D.W. Curtis & R.M. Baird (Eds.), Elsevier Science, 127-140, ISBN: 0-444-51084-2, Amsterdam, Netherlands.

Servili, M.; Settanni, L.; Veneziani, G.; Esposto, S.; Massitti, O.; Taticchi, A.; Urbani, S.; Montedoro, G.F. & Corsetti, A. (2006). The use of *Lactobacillus pentosus* 1MO to shorten the debittering process time of black table olives (Cv. Itrana and Leccino): a pilot-scale application. *Journal of Agricultural and Food Chemistry*, Vol.54, pp. 3869-3875.

Sousa, A.; Ferreira, I.C.F.R.; Calhelha, R.; Andrade, P.B.; Valentão, P.; Seabra, R.; Estevinho, L.; Bento, A. & Pereira, J.A. (2006). Phenolics and antimicrobial activity of traditional stoned table olives 'alcaparra'. *Bioorganic & Medicinal Chemistry*, Vol.14, pp. 8533–8538.

Spyropoulou, K. E.; Chorianopoulos, N. G.; Skandamis, P. N. & Nychas, G. J. (2001). Survival of *Escherichia coli* O157:H7 during the fermentation of Spanish-style green table olives (Conservolea variety) supplemented with different carbon sources. *International journal of food microbiology*, Vol.66, No.1-2, pp. 3-11.

Swanson, K.M.J.; Petran, R.L. & Hanlin, J.H. (2001). Culture methods for enumeration of microorganisms, In: *Compendium of methods for the Microbiological examination of Foods*, Fourth edition, F.P. Downes & K. Ito (Eds.), American Public Health Association, 53-62, ISBN: 0-87553-175-x, Washington DC.

Tassou, C.C. & Nychas, G.J.E. (1994). Inhibition of *Staphylococcus aureus* by Olive Phenolics in Broth and in a Model Food System. *Journal of Food Protection*, Vol.57, No.2, pp. 120-124.

Tsapatsaris, S. & Kotzekidou, P. (2004). Application of central composite design and response surface methodology to the fermentation of olive juice by *Lactobacillus plantarum* and *Debaryomyces hansenii*. *International Journal of Food Microbiology*, Vol.95, pp. 157-168.

UNAPROL (2008). Filiera olivicola - Analisi strutturale e monitoraggio di un campione di imprese. Campagna 2007-2008, Reg. CE 2080/2005, 05/12/2011, Available from http://www.unaprol.it/Pubblicazioni/FILIERA%20OLIVICOLA%20CAMP.% 202007-08.PDF

U.S. Food and Drug Administration, and Center for Food Safety and Applied Nutrition, Department of Health and Human Services (2001). Bacteriological analytical manual online, 18.06.2012, Available from: http://www.fda.gov/Food/ScienceResearch/LaboratoryMethods/BacteriologicalAnalytic alManualBAM/default.htm.

Viljoen, B.C. (2006). Yeast ecological interactions. Yeast–yeast, yeast bacteria, yeast–fungi interactions and yeasts as biocontrol agents, In: *Yeasts in Food and Beverages*, Querol, A. & Fleet, H. (Eds.), 83–110, Springer–Verlag, Berlin.

Zangerl, P. and Asperger, H. (2003). Media used in the detection and enumeration of *Staphylococcus aureus*, In: *Handbook of Culture Media for Food Microbiology*, J.E.L. Corry, G.D.W. Curtis & R.M. Baird (Eds.), Elsevier Science, 91-110, ISBN: 0-444-51084-2, Amsterdam, Netherlands.

Permissions

The contributors of this book come from diverse backgrounds, making this book a truly international effort. This book will bring forth new frontiers with its revolutionizing research information and detailed analysis of the nascent developments around the world.

We would like to thank Innocenzo Muzzalupo, for lending his expertise to make the book truly unique. He has played a crucial role in the development of this book. Without his invaluable contribution this book wouldn't have been possible. He has made vital efforts to compile up to date information on the varied aspects of this subject to make this book a valuable addition to the collection of many professionals and students.

This book was conceptualized with the vision of imparting up-to-date information and advanced data in this field. To ensure the same, a matchless editorial board was set up. Every individual on the board went through rigorous rounds of assessment to prove their worth. After which they invested a large part of their time researching and compiling the most relevant data for our readers. Conferences and sessions were held from time to time between the editorial board and the contributing authors to present the data in the most comprehensible form. The editorial team has worked tirelessly to provide valuable and valid information to help people across the globe.

Every chapter published in this book has been scrutinized by our experts. Their significance has been extensively debated. The topics covered herein carry significant findings which will fuel the growth of the discipline. They may even be implemented as practical applications or may be referred to as a beginning point for another development. Chapters in this book were first published by InTech; hereby published with permission under the Creative Commons Attribution License or equivalent.

The editorial board has been involved in producing this book since its inception. They have spent rigorous hours researching and exploring the diverse topics which have resulted in the successful publishing of this book. They have passed on their knowledge of decades through this book. To expedite this challenging task, the publisher supported the team at every step. A small team of assistant editors was also appointed to further simplify the editing procedure and attain best results for the readers.

Our editorial team has been hand-picked from every corner of the world. Their multi-ethnicity adds dynamic inputs to the discussions which result in innovative outcomes. These outcomes are then further discussed with the researchers and contributors who give their valuable feedback and opinion regarding the same. The feedback is then collaborated with the researches and they are edited in a comprehensive manner to aid the understanding of the subject.

Apart from the editorial board, the designing team has also invested a significant amount of their time in understanding the subject and creating the most relevant covers. They scrutinized every image to scout for the most suitable representation of the subject and create an appropriate cover for the book.

The publishing team has been involved in this book since its early stages. They were actively engaged in every process, be it collecting the data, connecting with the contributors or procuring relevant information. The team has been an ardent support to the editorial, designing and production team. Their endless efforts to recruit the best for this project, has resulted in the accomplishment of this book. They are a veteran in the field of academics and their pool of knowledge is as vast as their experience in printing. Their expertise and guidance has proved useful at every step. Their uncompromising quality standards have made this book an exceptional effort. Their encouragement from time to time has been an inspiration for everyone.

The publisher and the editorial board hope that this book will prove to be a valuable piece of knowledge for researchers, students, practitioners and scholars across the globe.

List of Contributors

Catherine Marie Breton
CNRS ISE-M UMR 5554, Montpellier, France
INR , TGU AGAP, Equipe DAVEM, Montpellier, France

Peter Warnock
Missouri Valley College, USA
André Jean Bervillé INRA, UMR DIAPC, Montpellier, France

Rosario Muleo
University of Tuscia, Dept. DAFNE, Viterbo, Italy

Michele Morgante
IGA-Institute of Applied Genomics, Udine, Italy

Riccardo Velasco
Edmund Mach Foundation, IASMA, San Michele all'Adige, Italy

Andrea Cavallini
Dept. Crop Species Biology, Pisa, Italy

Gaetano Perrotta
ENEA, Trisaia, Rotondella (MT), Italy

Luciana Baldoni
CNR- Institute of Plant Genetics, Perugia, Italy

Adriana Chiappetta
University of Calabria (UNICAL), Dept. of Ecology,Arcavacata di Rende (CS), Italy

Innocenzo Muzzalupo
Agricultural Research Council - Olive Growing and Oil Industry Research Centre, Rende (CS), Italy

Caterina Briccoli Bati, Elena Santilli and Pietro Toscano
Agricultural Research Council - Olive Growing and Oil Industry Research Centre, Rende (CS), Italy

Ilaria Guagliardi
National Research Council - Institute for Agricultural and Forest Systems in the Mediterranean, (ISAFOM), Rende (CS), Italy

Nino Iannotta and Stefano Scalercio
Agricultural Research Council - Olive Growing and Oil Industry Research Centre, Rende
(CS), Italy

Giuliana Albanese
Dipartimento di Gestione dei Sistemi Agrari e Forestali, Università degli Studi
Mediterranea di Reggio Calabria, Reggio Calabria, Italy

Maria Saponari
Istituto di Virologia Vegetale del CNR – Unita' Organizzativa di Supporto di Bari, Bari,
Italy

Francesco Faggioli
CRA-Centro di Ricerca per la Patologia Vegetale, Roma, Italy

Adolfo Rosati, Silvia Caporali and Andrea Paoletti
Agricultural Research Council - Olive Growing and Oil Industry Research Centre,
Spoleto (PG), Italy

Pietro Toscano
Agricultural Research Council, Olive Growing and Oil Industry Research Centre, Rende
(CS), Italy

Francesco Montemurro
Agricultural Research Council, Research Unit for the Study of Cropping Systems,
Metaponto (MT), Italy

Amalia Piscopo and Marco Poiana
Dipartimento di Biotecnologie per il Monitoraggio, Agroalimentare e Ambientale
(Bio.M.A.A.), Mediterranean University of Reggio Calabria, Italy

Giovanni Sindona and Domenico Taverna
University of Calabria, Dept. Chemistry, Arcavacata di Rende (CS), Italy

**Innocenzo Muzzalupo, Massimiliano Pellegrino, Enzo Perri, Cinzia Benincasa and
Enzo Perri**
Agricultural Research Council - Olive Growing and Oil Industry Research Centre, Rende
(CS), Italy

Domenico Britti and Antonio Procopio
Dept. of Health Sciences, University of Catanzaro, Catanzaro, Italy

Daniela Impellizzeri and Salvatore Cuzzocrea
Dept. of Clinical and Experimental Medicine and Pharmacology, University of Messina,
Torre Biologica – Policlinico Universitario, Messina, Italy

Barbara Lanza
Agricultural Research Council - Olive Growing and Oil Industry Research Centre, Città Sant'Angelo (PE), Italy

Flora Valeria Romeo
Agricultural Research Council - Olive Growing and Oil Industry Research Centre, Rende (CS), Italy